The
Plant Viruses

Volume 3
POLYHEDRAL VIRIONS WITH
MONOPARTITE RNA GENOMES

THE VIRUSES

Series Editors
HEINZ FRAENKEL-CONRAT, *University of California*
Berkeley, California

ROBERT R. WAGNER, *University of Virginia School of Medicine*
Charlottesville, Virginia

THE VIRUSES: Catalogue, Characterization, and Classification
Heinz Fraenkel-Conrat

THE ADENOVIRUSES
Edited by Harold S. Ginsberg

THE BACTERIOPHAGES
Volumes 1 and 2 • Edited by Richard Calendar

THE HERPESVIRUSES
Volumes 1–3 • Edited by Bernard Roizman
Volume 4 • Edited by Bernard Roizman and Carlos Lopez

THE PAPOVAVIRIDAE
Volume 1 • Edited by Norman P. Salzman
Volume 2 • Edited by Norman P. Salzman and Peter M. Howley

THE PARVOVIRUSES
Edited by Kenneth I. Berns

THE PLANT VIRUSES
Volume 1 • Edited by R. I. B. Francki
Volume 2 • Edited by M. H. V. Van Regenmortel and Heinz Fraenkel-Conrat
Volume 3 • Edited by Renate Koenig
Volume 4 • Edited by R. G. Milne

THE REOVIRIDAE
Edited by Wolfgang K. Joklik

THE RHABDOVIRUSES
Edited by Robert R. Wagner

THE TOGAVIRIDAE AND FLAVIVIRIDAE
Edited by Sondra Schlesinger and Milton J. Schlesinger

THE VIROIDS
Edited by T. O. Diener

The
Plant Viruses

Volume 3
POLYHEDRAL VIRIONS WITH MONOPARTITE RNA GENOMES

Edited by

RENATE KOENIG

Institute for Plant Viral Diseases
Braunschweig, Federal Republic of Germany

PLENUM PRESS • NEW YORK AND LONDON

Library of Congress Cataloging in Publication Data

Polyhedral virions with monopartite RNA genomes.

(The Plant viruses; v. 3) (The Viruses)
Bibliography: p.
Includes index.
1. Viruses, RNA. 2. Plant viruses. I. Koenig, Renate. II. Series. III. Series: Viruses.

| QR357.P58 | 1985 vol. 3 | 576′.6483 s | 87-32809 |
| [QR395] | | [576′.6483] | |

ISBN-13:978-1-4612-8246-4   e-ISBN-13:978-1-4613-0921-5
DOI: 10.1007/978-1-4613-0921-5

© 1988 Plenum Press, New York
Softcover reprint of the hardcover 1st edition 1988
A Division of Plenum Publishing Corporation
233 Spring Street, New York, N.Y. 10013

Contributors

J. C. Carrington, Department of Plant Pathology, University of California, Berkeley, California 94729

R. Casper, Plant Virus Institute, Federal Biological Research Center for Agriculture and Forestry, D-3300 Braunschweig, Federal Republic of Germany

H. Fraenkel-Conrat, Department of Molecular Biology and Virus Laboratory, University of California, Berkeley, California 94720

D. Gallitelli, Department of Plant Pathology, University of Bari, Bari, Italy

R. Gámez, Cellular and Molecular Biology Research Center, University of Costa Rica, Costa Rica

R. E. Gingery, United States Department of Agriculture, Agricultural Research Service, Department of Plant Pathology, The Ohio State University, Ohio Agricultural Research and Development Center, Wooster, Ohio 44691

L. Givord, Institute of Cellular and Molecular Biology of CNRS, 67084 Strasbourg Cédex, France

L. Hirth, Institute of Cellular and Molecular Biology of CNRS, 67084 Strasbourg Cédex, France

R. Hull, Department of Virus Research, John Innes Institute, Colney Lane, Norwich NR4 7 UH, United Kingdom

R. Koenig, Plant Virus Institute, Federal Biological Research Center for Agriculture and Forestry, D-3300 Braunschweig, Federal Republic of Germany

P. León, Cellular and Molecular Biology Research Center, University of Costa Rica, Costa Rica

G. P. Martelli, Department of Plant Pathology, University of Bari, Bari, Italy

T. J. Morris, Department of Plant Pathology, University of California, Berkeley, California 94720

A. F. Murant, Scottish Crop Research Institute, Invergowrie, Dundee DD2 5DA, Scotland, United Kingdom

M. Russo, Department of Plant Pathology, University of Bari, Bari, Italy

Contents

Chapter 3

Carnation Mottle Virus and Viruses with Similar Properties 73

T. J. Morris and J. C. Carrington

Chapter 4

The Sobemovirus Group 113

R. Hull

Chapter 5

Tobacco Necrosis, Satellite Tobacco Necrosis, and Related Viruses

H. Fraenkel-Conrat

Chapter 6

Tymoviruses 163

L. Hirth and L. Givord

Chapter 9

Maize Chlorotic Dwarf and Related Viruses

R. E. Gingery

Chapter 10

Parsnip Yellow Fleck Virus, Type Member of a Proposed New Plant Virus Group, and a Possible Second Member, Dandelion Yellow Mosaic Virus

A. F. Murant

CHAPTER 1

Polyhedral Plant Viruses with Monopartite RNA Genomes
Introduction and Summary of Important Properties

R. KOENIG

This volume describes eight established or anticipated groups of polyhedral plant viruses with single-stranded monopartite RNA genomes. It differs from other volumes in the series *The Plant Viruses* in that all aspects of an individual virus group have been reviewed by the same author or group of authors. In order to make it easy for the reader to obtain comparative information on specific properties for different groups of viruses, all chapters follow the same basic outline. An introductory Section I describes the history of the recognition of the group and summarizes its main properties. Section II describes host range and geographical distribution for individual group members and the diseases caused by them. Section III deals with transmission, epidemiology, and control. Purification methods are described in Section IV. Section V summarizes the information on the morphological, physical, and chemical properties of the virions and their components (except for the genome). Section VI deals with serological observations, and Section VII with genome properties and replication. Relations with tissues and cells are described in Section VIII. Further optional sections deal with specific aspects, e.g., the occurrence of satellites, evolutionary considerations, etc., or present general conclusions.

In order to provide the reader with some quick information, the most

R. KOENIG • Plant Virus Institute, Federal Biological Research Center for Agriculture and Forestry, D-3300 Braunschweig, Federal Republic of Germany.

important properties of the individual virus groups have been summarized in Tables I to VII of this chapter. Most of the information given in these tables was obtained from Chapters 2 to 10 of this volume. A few missing data were obtained from the reviews by Kurstak (1981), Matthews (1982), Harrison (1984), and Francki *et al.* (1985).

Tombusviruses (Chapter 2), viruses of the proposed carnation mottle virus group (Chapter 3), and even Sobemoviruses (Chapter 4) and tobacco necrosis virus and its satellite virus (Chapter 5) show a number of similarities in structure and other properties (e.g., Table II). The similarities in amino acid composition at structurally homologous positions are intriguing, especially in the S domains of tomato bushy stunt virus (Tombusvirus group) and southern bean mosaic virus (Sobemovirus group) (Hopper *et al.*, 1984), and apparently also of tomato bushy stunt virus and turnip crinkle virus (proposed carnation mottle virus group) (Hogle *et al.*, 1986; for review see Chapters 2 and 3 of this volume). A common evolutionary origin of southern bean mosaic and tomato bushy stunt viruses has been suspected on the basis of these observations (Hopper *et al.*, 1984). Genome organization and translational strategies may turn out to be similar for the viruses of the proposed carnation mottle virus group and tobacco necrosis virus (Morris and Carrington, Chapter 3 of this volume), but they are different for Tombusviruses (Martelli *et al.*, Chapter 2; Morris and Carrington, Chapter 3; see Table IV for review).

Tymoviruses (Chapter 6) and maize rayado fino virus (Chapter 7) differ from most other viruses described in this volume (with the exception of parsnip yellow fleck virus, Chapter 10) in having two major types of particles: "full particles" containing genomic and sometimes also subgenomic RNAs and "empty shells," which may or may not encapsidate subgenomic or host RNAs. There are also other similarities between these viruses (Tables I and III; see also Francki *et al.*, 1985), but also many differences (Tables II, V and VI). Maize rayado fino (Chapter 7), maize chlorotic dwarf (Chapter 9), and parsnip yellow fleck and dandelion yellow mosaic viruses (Chapter 10) differ from the other viruses described in this volume by having more than one coat protein species (Table II). Luteoviruses (Chapter 8) are unique among the viruses described here by being transmitted by aphids in the persistent manner; by being restricted mainly to the phloem; and by having somewhat smaller particles. Maize chlorotic dwarf virus has particles with a sedimentation constant S_{20w} around 180 S, which is unusually high for an isometric virus. Its buoyant density is also very high, and a unique feature compared to the other viruses described in this volume is its transmissibility by leafhoppers in a semipersistent manner. Parsnip yellow fleck virus is transmitted by aphids in a semipersistent manner, but only in association with a helper virus. Maize rayado fino virus and the viruses related to it are transmitted by leafhoppers in a persistent manner, and are the only small plant viruses known to replicate in leafhoppers.

TABLE I. Particle Properties I. Electron Microscopical Appearance, Sedimentation Behavior, and Particle Weight

Virus group	Electron microscopical appearance of particles after staining with uranyl acetate	Particle diameter (nm)	Number of components sedimenting in sucrose density gradient centrifugation	S_{20w}	Buoyant density in CsCl (g/cm³)	Particle weight ($M_r \times 10^6$)
Tombusvirus	Rounded to slightly angular with knobby surface	ca. 30	1	ca 135	1.34–1.36	8.9
Carnation mottle and viruses with similar properties	Rounded to slightly angular with knobby surface	ca. 30	1	112–143	1.34–136	8–9
Sobemovirus	Rounded to slightly angular, no clear surface structure	25–30	1	110–120	1.34–1.39	6–7
Tobacco necrosis virus	Angular, sometimes rounded, no clear surface structure	28	1	113	1.4	7.2
Tobacco necrosis satellite virus	Angular, sometimes rounded, no clear surface structure	17	1	50		1.7
Tymovirus	Rounded, morphological subunits clearly visible	29	2 (T and B)	T ca 55 B ca 115	1.29 1.42 additional minor bands	3.6 5.6
Maize rayado fino and viruses with similar properties	Rounded, morphological subunits clearly visible	ca. 30	2 (T and B) (only 1 with oat blue dwarf virus)	T 50 B 120	1.3 1.43	4.5 6.5
Luteovirus	Angular, no clear surface structure	24–27	1	104–127		
Maize chlorotic dwarf virus	Rounded with somewhat knobby surface	30	1	175–183	1.5	8.8
Parsnip yellow fleck virus	Rounded to angular, no clear surface structure	30	2 (T and B)	T 60 B 152	1.29 1.49	8.3

TABLE II. Particle Properties II. Number, Size, and Arrangement of Coat Protein Subunits, Stabilizing Forces in Particles

Virus group	Number of coat protein species	$M_r \times 10^3$	Number of coat protein subunits per particle	Arrangement of coat protein subunits in the capsids	Properties and location of N-termini	Stabilizing forces in particles
Tombusvirus	1	ca. 40	178 + 1 covalently linked dimer	Protein subunits have an internally located R-domain connected by an arm to a shell-forming S-domain and a protrusion-forming P-domain. The P-domains of 2 subunits each form 90 dimers	Basic, internal	Protein–RNA interactions, in addition to protein–protein interactions
Carnation mottle and viruses with similar properties	1 (sometimes additional bands due to proteolysis)	ca. 38	178 + 1 covalently linked dimer	Carnation mottle and turnip crinkle viruses have apparently the same arrangement of protein subunits as tomato bushy stunt virus	Internal	As with tombusviruses
Sobemovirus	1	ca. 30 (24–37)	178 + 1 covalently linked dimer	Southern bean mosaic virus has a similar arrangement of protein subunits as tomato bushy stunt virus with significant amino acid homologies in the S-domains. P-domains, however, are missing.	Highly basic, acetylated, internal	Ca^{2+} protein–protein interactions; pH dependent protein–protein interactions; protein–RNA interactions

Tobacco necrosis virus	1	32	180		Probably acetylated in some strains and butyrylated in others, basic, internal
Tobacco necrosis satellite virus	1	22	60	Similar folding of polypeptide as with southern bean mosaic virus	
Tymovirus	1	20	180	Coat protein subunits form 20 hexamers and 12 pentamers	Internal or external (diverging views)
Maize rayado fino virus	2	L 22 H 26–29 (molar ratio 9 : 1)	180 L 20 H	180 L subunits from a shell, 20 H subunits form projections	Protein–protein interactions, especially hydrophobic ones
Luteovirus	1	26 (22–32)			Protein–protein interactions
Maize chlorotic dwarf virus	2 or 3	18 to 30			
Parsnip yellow fleck virus	3	22;26;31			

TABLE III. Genome Properties I. Size of Genomic RNA, Base Composition, and Structures at 5' and 3' ends

Virus group	Percentage of RNA in particles	Size of RNA in Kb	Base ratios (G;A,C;U)	Structure at 5' end	Structure at 3' end
Tombusvirus	17%	4.7	ca. 28;23–27;21;23–26	No VPg	No Poly(A) in encapsidated RNA, 1% of total viral RNA in infected tissue may be polyadenylated
Carnation mottle and viruses with similar properties	14–23%	4.0	Differs with different viruses	Probably cap structure (carnation mottle virus)	No poly(A) (carnation mottle and turnip crinkle viruses)
Sobemovirus	21%	4.2 (4.0–4.8)	25–29;21–26;22–27;23–27	VPg needed for infection	No poly(A)
Tobacco necrosis	19%	4.2		ppAGU—, no VPg	No poly(A)
Tobacco necrosis satellite	24%	1.239		(p)ppAGU—, no VPg	No poly(A)
Tymovirus	ca. 35% of B particles	5.8	15–18;17–24,32–42;22–29	$m^7G^{5'}$ pp$^{5'}$ G—	tRNA-like structure accepting valine; encapsidated genomic RNA not charged with valine, but approximately one molecule in four of free RNA may be charged with valine in infected tissue
Maize rayado fino	25–30%	6.1	23;17;31;29		Probably no poly(A)
Luteovirus		5.8–7.2		VPg not needed for infection	No poly(A)
Maize chlorotic dwarf	36%	9.7	24;30;17;29		
Parsnip yellow fleck	42%	9.7		Probably VPg needed for infection	Probably poly(A)

TABLE IV. Genome properties II. Arrangement of Coding Regions in Genomic RNA, Number and Size of Subgenomic RNAs, *in vitro* Translation Products

Virus group (information based mainly on virus in parentheses)	Mapping of major open reading frames (ORFs) including coat protein (CP) cistron, from 5' to 3' terminus of genomic RNA	Number (and size in parenthesis) of subgenomic RNAs	Major *in vitro* translation products (G, genomic; S, subgenomic RNAs)
Tombusvirus (Petunia asteroid mosaic)	ORFs at 5' unknown; CP cistron terminates ca 1000 nt before 3' terminus; the ca 1000 3' terminal nts contain several possible ORFs	2 (2.2 and 1.0 Kb); Infected tissues contain dsRNAs in corresponding numbers and sizes	43 K (S, 2.2 Kb) = CP; 40 K (G); 34 K (S) – probably partially contained in 40 K, 22 K (S). Apparently no readthroughs
Carnation mottle and viruses with similar properties (carnation mottle virus)	69 nt 5' leader sequence; ORF 1 (nt 70-2677, punctuated by 2 amber termination codons at nt 805 and 2359, ORF 2 (nt 2334-2515, entirely within ORF 1, but out of phase); ORF 3 (nt 2669-3713, partially overlapping, but out of phase with ORF 1) = CP cistron; 209 nt 3' noncoding region (carnation mottle virus)	2 (1.7; 1.5 Kb), apparently 3' coterminal with genomic RNA. Infected tissues contain dsRNAs in corresponding number and sizes	38 K (S, 1.5 Kb) = CP; 27 K (G); 86 K (G) – probably readthrough of 27 K; 7 K (S, 1.7 Kb)
Sobemovirus (southern bean mosaic)	Coat protein cistron near 3'end; 3'terminal ca 130 nt noncoding region (southern bean mosaic virus)	At least 1 (ca 1 – 2 Kb), no corresponding dsRNA found suggesting that subgenomic RNA is derived from internal initiation in the genomic minus strand	3 to 5 major products, the two largest of ca 100 K (G) and ca 70 K (G) are related by readthrough or processing, ca 30 K (S) = CP.

(continued)

TABLE IV. (*Continued*)

Virus group (information based mainly on virus in parentheses)	Mapping of major open reading frames (ORFs) including coat protein (CP) cistron, from 5' to 3' terminus of genomic RNA	Number (and size in parenthesis) of subgenomic RNAs	Major *in vitro* translation products (G, genomic; S, subgenomic RNAs)
Tobacco necrosis virus		Infected tissues contain 3 3' coterminal dsRNAs (ca 3.9, 1.6 and 1.4 Kbp), the two smaller ones probably corresponding to subgenomic RNAs	Major product of 32 K = CP; minor proteins of 63, 43 and 26 K
Tobacco necrosis satellite virus	29 nt 5' leader sequence; CP cistron nt 30-617; 622 nt 3' noncoding region		Only product of 19.5 K = CP
Tymovirus (turnip yellow mosaic)	95 nt 5' leader sequence; number and size of 5' ORFs unknown; coat protein cistron followed by 109 nt tRNA-like structure at 3' terminus	1 (0.695 Kb), coterminal with genomic RNA. Subgenomic RNA encapsidated in bottom or top component particles	195 K (G), 150 K (G), 120 K (G) and 78 K (G) related by readthrough or processing? 20 K (S) = CP
Maize rayado fino virus			160 K product (G) is not related to the viral coat proteins which are probably translated from nonencapsidated subgenomic RNAs
Luteovirus (potato leaf roll)	No information		71 K product (G?) unrelated to CP
Maize chlorotic dwarf virus			
Parsnip yellow fleck virus	No information		

TABLE V. Biological and Serological Properties

Virus group	Host range (HR)	Mechanical transmission	Transmission by vectors/mode of transmission	Other forms of transmission	Serological relationships among the majority of group members
Tombusvirus	Natural HR narrow, experimental HR wide	+	Apparently none	Vectorless in soil; (seed)*	Yes
Carnation mottle and viruses with similar properties	Narrow	+	(Beetles; Olpidium radicale)*	(Vectorless in soil?; seed)*	No
Sobemovirus	Narrow	+ (blueberry shoe-string −)*	(Chrysomelid beetles; aphids; other insects)*	(Seed)*	No
Tobacco necrosis and tobacco necrosis satellite virus	Wide	+	Olpidium brassicae		No
Tymovirus	Usually narrow	+	Beetles, mainly chrysomelids	(Seed)*	Yes
Maize rayado fino virus	Narrow	−	Cicadellid leafhopper/persistent, propagative. The only small plant viruses known to multiply in leafhoppers		Yes
Luteovirus	Some narrow, some wide	−	Aphids/persistent, circulative		Yes
Maize chlorotic dwarf virus	Narrow	−	Leafhoppers/semipersistent		
Parsnip yellow fleck virus	Narrow	+	Aphids/semipersistent only in association with helper virus		

*Statements in parentheses are valid only for one or a few group members.

TABLE VI. Relations with Tissues and Cells

Virus group	Tissues preferentially infected	Specific cytopathic effects
Tombusvirus	All tissues and cells	Multivesicular bodies usually originating from peroxisomes; cytoplasmic accumulations of virus particles in extensions of the tonoplast protruding into the vacuoles; virus particles in nuclei and cytoplasm including chloroplasts and mitochondria
Carnation mottle and viruses with similar properties	All tissues	High concentrations of virus particles in cytoplasm, but usually not in crystalline arrays, usually not in nuclei; some viruses induce the formation of multivesicular bodies in the cytoplasm
Sobemovirus	All tissues	High concentrations of virus particles in the cytoplasm, sometimes in crystalline arrays; some viruses occur also in nuclei, but usually not in chloroplasts or mitochondria; fibrils in the cytoplasm which may represent dsRNA
Tobacco necrosis and tobacco necrosis satellite virus	Preferentially root tissues	Particles occur in the cytoplasm, those of the satellite usually in crystalline arrays
Tymovirus	All tissues	Double-membrane bounded invaginations of the chloroplasts; "empty" protein shells accumulating in the nuclei; clumping of chloroplasts
Maize rayado fino virus	Epidermis, parenchyma and phloem cells (oat blue dwarf virus only phloem)	Aggregates of virus particles in vacuoles or in membrane-bounded cytoplasmic vesicles
Luteoviruses	Phloem and adjacent cells	Virus particles occur in the cytoplasm, with some viruses also in nuclei; double-membraned vesicles in the cytoplasm and sometimes in the perinuclear space
Maize chlorotic dwarf virus	Mainly vascular tissues, occasionally mesophyll	Dense granular inclusions with embedded viruslike particles in cytoplasm; striated inclusions
Parsnip yellow fleck virus	All leaf tissues	Characteristic inclusion bodies usually adjacent to nucleus; cell wall outgrowths ensheathing cytoplasmic tubules which contain a single row of viruslike particles; such tubules may also be found free in the cytoplasm

TABLE VII. Satellite Viruses and Satellite RNAs in Different Virus Groups

Virus group	Helper viruses	Size of RNA of satellites in Kb	Do satellites encode their own coat protein (CP) or are they encapsidated in helper virus CP?	Influence on symptoms produced by helper virus	Remarks
Tombusviruses	All members of the group	ca. 0.4–0.7	Helper virus CP	Attenuation	Appearance of satellites is host dependent, extensive sequence homologies between satellites of different viruses[a]
Carnation mottle and viruses with similar properties	Turnip crinkle virus	Major form 0.35; additional minor forms	Helper virus CP	Enhancement	3' Terminal sequence (ca 150 nt) of major form has extensive sequence homology with helper virus RNA; minor form has sequence homologies with host DNA
Sobemoviruses	1. Panicum mosaic virus		Own CP	Apparently no influence	
	2. Lucerne transient streak, solanum nodiflorum mottle and others	0.4	Helper virus CP	Enhancement or no influence	
Tobacco necrosis virus	Tobacco necrosis virus	1.239	Own CP	Attenuation	
Tymoviruses Maize rayado fino virus Luteoviruses Maize chlorotic dwarf virus Parsnip yellow fleck virus	No reports				

[a] The satellite RNA of PAMV has been found to be a chimeric, defective-interfering RNA containing sequences derived from the 5'-proximal, internal, and 3'-proximal regions of the helper virus genome (T. J. Morris, personal communication).

ACKNOWLEDGMENTS. I am grateful to my colleagues D.-E. Lesemann, H.-L. Paul, and W. Burgermeister for helpful discussions, and to the Deutsche Forschungsgemeinschaft for financially supporting my work on Tombusviruses, Tymoviruses, and viruses of the proposed carnation mottle virus group.

REFERENCES

Francki, R. I. B., Milne, R. G., and Hatta, T., 1985, *Atlas of Plant Viruses*, Volumes I and II, CRC Press, Boca Raton, FL.

Harrison, S. C., 1984, Multiple modes of subunit association in the structures of simple spherical viruses, *Trends Biochem. Sci* **9**:345.

Hogle, J. M., Maeda, A., and Harrison, S. C., 1986, Structure and assembly of turnip crinkle virus I. X-ray crystallographic structure analysis at 3.2 Å resolution, *J. Mol. Biol.* **191**:625.

Hopper, P., Harrison, S. C., and Sauer, R. T., 1984, Structure of tomato bushy stunt virus V. Coat protein sequence determination and its structural implications, *J. Mol. Biol.* **177**:701.

Kurstak, E. (ed.), 1981 *Handbook of Plant Virus Infections, Comparative Diagnosis*, Elsevier, North Holland Biomedical Press.

Matthews, R. E. F., 1982, Classification and nomenclature of viruses, *Intervirology* **17**:1.

CHAPTER 2

Tombusviruses

G. P. Martelli, D. Gallitelli, and M. Russo

I. INTRODUCTION

The Tombusvirus group was established in 1971 (Harrison *et al.*, 1971). It derives its name from the sigla "tombus," originating from the name of the type member of the group, tomato bushy stunt virus (TBSV).

Tombusviruses have isometric particles approximately 30 nm in diameter, made up of 180 protein subunits, each with a M_r of approximately 40 kd. Their genome is monopartite, consisting of a molecule of positive sense single-stranded RNA made up of 4700 nucleotides, which constitutes approximately 17% of the particle weight (Gallitelli *et al.*, 1985). Accessory smaller RNAs 700 nucleotides in length (satellite RNAs) are associated with all members of the group whose presence is required for their replication (Gallitelli and Hull, 1985b).

The virions are very stable, especially in the acidic range (pH 5–6.5), are not appreciably affected by organic solvents during purification, reach remarkably high concentrations in host tissues (dilution end point of infectivity in expressed sap up to or over 10^{-6}), retain infectivity in expressed sap for four to five weeks or longer, and withstand heating *in vitro* at 90–95°C (see Martelli *et al.*, 1971).

Tombusviruses are readily transmitted by inoculation of sap, are disseminated with propagating material in vegetatively propagated hosts or, in some cases, through seeds. Soil transmission is common and is not mediated by soil organisms.

When first established, the Tombusvirus group was made up of five members: TBSV, artichoke mottled crinkle (AMCV), carnation Italian ringspot (CIRV), pelargonium leaf curl (PLCV), and petunia asteroid mo-

G. P. MARTELLI, D. GALLITELLI, and M. RUSSO • Department of Plant Pathology, University of Bari, Bari, Italy.

saic (PAMV) viruses (Harrison *et al.*, 1971). In 1982, the list had grown to seven definitive members: TBSV, AMCV, PLCV, PAMV, cymbidium ringspot (CyRSV), eggplant mottled crinkle (EMCV) viruses, and two possible members: turnip crinkle (TCV) and saguaro cactus (SCV) viruses (Matthews, 1982). In the latest suggested composition, the group included an eighth definitive member (glycine mottle virus, GMoV) and four more possibles: TCV, galinsoga mosaic (GaMV), hibiscus chlorotic ringspot (HCRSV), and cucumber necrosis (CNV) viruses (Francki *et al.*, 1985).

A virus originally reported from Morocco as the "pepper strain" of TBSV (Fischer and Lockhart, 1974), was recently reinvestigated serologically and found to differ sufficiently from TBSV to warrant a name of its own, i.e., Moroccan pepper virus (MPV) (Makkouk *et al.*, 1981; Vetten and Koenig, 1983). Tombusvirus Neckar (TVN), a virus isolated in Germany from the waters of the river Neckar (Koenig and Lesemann, 1985), and grapevine Algerian latent virus (GALV), a virus recovered from a grapevine in Algeria (G. P. Martelli, unpublished information) appear as additional *bona fide* members of the group because of their biological, physicochemical, ultrastructural, and serological properties (Koenig and Lesemann, 1985; Koenig and Gibbs, 1986; G. P. Martelli, M. Russo, D. Gallitelli, and A. Di Franco, unpublished information).

Therefore, in this presentation, the Tombusvirus group is considered to consist of the ten definitive members listed in Table I.

TBSV-type and TBSV-BS3 can be distinguished from each other serologically and electrophoretically (see among others Koenig and Gibbs, 1986) but cause similar diseases in tomato (*Lycopersicon esculentum* Mill.). In this review, TBSV-type and TBSV-BS3 are treated as strains of the same virus rather than as separate members of the group.

Despite the intriguing similarities with Tombusviruses in some physicochemical properties and cytopathological effects (Behncken and Dale, 1984), GMoV may not be a member of the group. Major differences lie in: (1) a very narrow host range that is restricted to *Leguminosae* (Behncken and Dale, 1984) and does not not include basil (*Ocymum basilicum* L.) and globe amaranth (*Gomphrena globosa* L.), two hosts common to all Tombusviruses; and (2) size and constitution of genomic RNA, which is smaller than that of Tombusviruses (3.9 kb vs. 4.7 kb) and has no sequence homology with any of the members of the group (Gallitelli *et al.*, 1985).

Because of significant differences with Tombusviruses in biological behavior (i.e., host range responses and epidemiology) and strategy of replication (Henriques and Morris, 1979), it had already pointed out that TCV and SCV were very doubtful members of the group (Martelli, 1981). Now it seems that these viruses, together with GaMV, fit best in a group of their own which also comprises carnation mottle virus, as discussed in Chapter 3, this volume.

Recent serological investigations using an indirect form of enzyme-linked immunosorbent assay (ELISA), have shown that TCV is serologically related to five Tombusviruses (TBSV, AMCV, CIRV, PAMV, PLCV)

TABLE I. Members of the Tombusvirus Group, Their Hosts and Geographical Distribution

Virus	Hosts		Geographical distribution	Reference (virus description)
	Cultivated	Wild		
Tomato bushy stunt (TBSV) (type, BS3 and other untypified strains)	Tomato, pepper, eggplant, lettuce, spinach, tulip, piggyback, apple, pear	*Solanum nigrum, Aster scamatus, Poa annua, Urtica urens, Polygonum persicaria, Stellaria media*	Europe, Mediterranean, N. and S. America	Smith (1935)
Artichoke mottled crinkle (AMCV)	Artichoke	None reported	Mediterranean (S. Italy, Morocco, Tunisia, Malta, Greece)	Martelli (1965), Quacquarelli and Martelli (1966)
Eggplant mottled crinkle (EMCV)	Eggplant	None reported	Lebanon	Makkouk *et al.* (1981)
Carnation Italian ringspot (CIRV)	Carnation	None reported	Great Britain (Italy, USA), W. Germany	Hollings *et al.* (1970)
Cymbidium ringspot (CyRSV)	Cymbidium, white clover	None reported	Great Britain	Hollings *et al.* (1977)
Grapevine Algerian latent (GALV)	Grapevine	None reported	Algeria	Unpublished information
Moroccan pepper virus (MPV)	Pepper, tomato, eggplant, pelargonium	None reported	Morocco, W. Germany	Fischer and Lockhart (1974), Vetten and Koenig (1983)
Tombusvirus Neckar (TVN)	Unknown	Unknown	W. Germany	Koenig and Lesemann (1985)
Pelargonium leaf curl (PLCV)	Pelargonium	None reported	Europe, Mediterranean, USA (probably worldwide)	Pape (1927), Hollings (1962)
Petunia asteroid mosaic (PAMV)	Petunia, cherry, grapevine, hop, pepper, plum, privet	*Chenopodium album, Cucumis melo, Plantago major, Stellaria media*	Central Europe (N. Italy, Switzerland, Czechoslovakia, W. and E. Germany), Canada	Lovisolo (1956), Koenig and Kunze (1982)

with a serological differentiation index (SDI) of 10 to 13 (Jaegle and Van Regenmortel, 1985). The significance of such weak serological cross-re-actions as evidence for supporting inclusion of TCV in the Tombusvirus group remains to be established. In fact, cross-reactivity with SDIs of the same order of magnitude, or even lower, was reported between viruses with isometric and rod-shaped particles belonging to quite different tax-onomic groupings (Bercks and Querfurth, 1971; Maat *et al.*, 1978; Paul *et al.*, 1980; Koenig, 1981). Moreover, sodium dodecyl sulfate (SDS)-dis-rupted particles of some Potex- and Carlaviruses were reported to cross-react in Western blot analyses with antisera to several Tombusviruses (Burgermeister and Koenig, 1984; see also T.J. Morris and J.C. Carrington, Chapter 3).

In light of the above considerations, the results of a classification based on the amino acid contents of the coat proteins, whereby TBSV was found to group with CaMV, TCV, SCV and CNV, are similarly in-conclusive (Fauquet *et al.*, 1986). These findings, however, support the notion of a close relationship between the Tombusvirus and the carnation mottle virus groups (see Chapter 3, this volume).

II. DISEASES, HOST RANGES, AND GEOGRAPHICAL DISTRIBUTION

A. General Features

As shown in Table I, only two Tombusviruses (TBSV and PAMV) are reported to infect a number of different cultivated and wild plants in nature, all remaining members of the group being more or less specialized. TBSV and PAMV also enjoy the widest geographical distribution and may cause (TBSV in particular) diseases of economic relevance in an epidemic form.

Except for PLCV, which was likely spread internationally by infected geraniums, other Tombusviruses have a somewhat restricted geographical distribution and infect naturally very few or only a single host. Con-versely, the artificial host range of all Tombusviruses is wide and diver-sified, comprising as a whole well over 100 plant species in 20 different monocotyledonous and dicotlyledonous families (see Martelli *et al.*, 1971).

As a rule, Tombusviruses systemically invade naturally infected plants but remain localized in the majority of the experimental hosts. However, cases of infection to the root system without subsequent spread of the virus to the above-ground parts of the plant have been reported, for in-stance with PAMV (Lovisolo *et al.*, 1965), MPV (Vetten and Koenig, 1983), and TBSV (Tomlinson and Faithfull, 1984).

Symptoms induced by Tombusviruses in infected crops consist mainly of diffuse mottling and malformation of the leaves, and stunting. Seldom are these symptoms specific enough to serve for diagnosis.

Among experimental hosts, the reactions of basil (*Ocymum basilicum* L.), globe amaranth (*Gomphrena globosa* L.), and *Chenopodium quinoa* Willd. may assist in the preliminary identification of Tombusviruses at the group level. All members of the group induce in basil dark brown necrotic lesions with a lighter center four to five days after inoculation; in *G. globosa*, pale necrotic local lesions in two to three days, which enlarge to give red-rimmed whitish rings or irregular spots; and in *C. quinoa*, chlorotic/necrotic lesions within two days (Martelli *et al.*, 1971).

The possibility of identifying individual Tombusviruses using host responses was discussed by Martelli (1981), who concluded that test plants alone cannot provide a reliable basis for a differential diagnosis.

B. Diseases

1. Tomato Bushy Stunt

Two well-characterized, serologically distinct strains of tomato bushy stunt virus (TBSV) are known: (1) the type strain, originally obtained from infected tomato plants in England (Smith, 1935); and (2) the BS3 strain, which was derived from the type strain by serial passages through differential hosts (Steere, 1953). Both these strains are pathogenic to cultivated plants.

In naturally infected tomato plants, TBSV-type induces stunting and bushy growth, accompanied by chlorotic spots, crinkling and deformation of young leaves and, sometimes, by yellowing and purpling of lower leaves (Smith, 1935; Vetten and Koenig, 1983). Oddly enough, in recent years no natural outbreaks of TBSV-type were recorded. All new records have been attributed to TBSV-BS3 (see Koenig and Avgelis, 1983).

TBSV-BS3 epidemics have been observed in field- or glasshouse-grown tomatoes in Morocco (Fischer and Lockhart, 1977), Portugal (Borges *et al.*, 1979), Tunisia (Cherif, 1981; Cherif and Spire, 1983), and Peru (R. Fribourg in Koenig and Avgelis, 1983). Foliar symptoms are comparable to those described by Smith (1935), fruit setting may be drastically reduced, and fruits are smaller than normal and show chlorotic blotching, rings, and line patterns that lower the economic value of the crop or render it unmarketable. Symptom expression, especially in protected crops, is strongly influenced by photoperiod and temperature (Martinez *et al.*, 1974; Borges *et al.*, 1979; see also Hillman *et al.*, 1985)

In Morocco, TBSV represents an economic threat to tomato crops grown for export (Fischer and Lockhart, 1977). In the same country and in Tunisia, infection incidence up to 40% and very heavy crop losses have been recently recorded (M. El Hamry and H. U. Fischer in Martelli, 1981; Cherif and Spire, 1983). Field observations and artificial inoculation tests have shown that 5 tomato cultivars were resistant, 10 were tolerant, and 11 were susceptible to the Tunisian isolate of TBSV-BS3 (Cherif and Spire, 1983).

Natural infection of sweet pepper by TBSV-BS3 was reported from Tunisia. None of the native cultivars was resistant to the virus, which induced generalized stunting, mottling, deformation of the leaves, and unfruitfulness (Cherif, 1981; Cherif and Spire, 1983).

In eggplant, TBSV-BS3 elicits stunting, mottling and crinkling of the leaves, poor fruit setting, chlorotic spots, and deformations of the fruits (Cherif and Spire, 1983; Koenig and Avgelis, 1983).

Additional natural hosts of TBSV are tulip (*Tulipa fosteriana* L.), in which the virus causes a disease characterized by extensive necrosis of leaves and petals (Mowat, 1972), and piggyback (*Tolmiea menziesii* T. et G.), in which it induces stunting and mild mottling of the leaves (Henriques and Schlegel, 1978). These isolates of TBSV differ from the type strain but have not been thoroughly characterized serologically. Likewise, serological characterization is lacking for "TBSV strains" that were isolated frequently (28% incidence) from apple trees (*Malus sylvestris* L.) and occasionally from pear trees (*Pyrus communis* L.) in East Germany (Kegler and Kegler, 1980).

The artificial host range of TBSV is wide. The type strain was reported to infect 52 out of 157 species (Schmelzer, 1958) and the BS3 strain, 33 out of 48 species (Cherif and Spire, 1983).

TBSV is readily inactivated *in vivo* by moderately high temperatures. Whole plants can be freed from virus by exposure at 36°C for three to four weeks (Kassanis, 1954; Kassanis and Lebeurier, 1969; McCarthy, 1983).

2. Artichoke Mottled Crinkle Virus

In artichoke (*Cynara scolymus* L.), artichoke mottled crinkle virus (AMCV) causes stunting, large chlorotic to pale-green blotches, distortion, puckering and, sometimes, lacing of the leaves. Flower heads may be variously distorted, show color breaking of the scales, and are fewer than normal. The yield is reduced but no specific assessment of crop losses has been made. Severely affected plants may wither and die. The intensity of symptoms is stronger during cooler months but apparent recovery occurs when temperature and light intensity increase (Martelli, 1965; Quacquarelli and Martelli, 1966). Severe cases of infection to artichoke have been reported from Morocco (Fischer and Lockhart, 1974) and Tunisia (Rana and Cherif, 1981), but the virus was also isolated from plants with very mild or no apparent symptoms (Martelli *et al.*, 1976, 1981; Rana and Kyriakopoulou, 1982).

A disease of lilac (*Syringa vulgaris* L.) characterized by chlorotic to bright yellow spots and rings on the leaves has been attributed by Novak and Lanzova (1977a) to a virus serologically similar to AMCV. However, the extent of the serological relatedness of the lilac virus with authentic AMCV has not been established.

The experimental host range of AMCV includes 24 species in seven different botanical families (Quacquarelli and Martelli, 1966; Fischer and Lockhart, 1974). Except for *Nicotiana clevelandii* Gray and *Nicotiana benthamiana* Domin., all hosts are infected locally.

Italian, Maltese, Moroccan, and Tunisian AMCV isolates seem to be serologically very close or indistinguishable. A serological variant of this virus was identified in Greece (Rana and Kyriakopoulou, 1982).

3. Eggplant Mottled Crinkle Virus

Eggplant mottled crinkle virus (EMCV) has only been reported from Lebanon, where it induces mottling and malformation of eggplant (*Solanum melongena* L.) leaves, severe stunting of the plants, and poor fruit set (Makkouk *et al.*, 1981). The field syndrome was reproduced by mechanical inoculation of healthy eggplant seedlings.

Artificial host range studies were limited to 12 plant species in five botanical families, most of which were infected locally. Systemic invasion was observed in *N. clevelandii* and *Physalis floridana* (Makkouk *et al.*, 1981). *N. benthamiana* is also infected systemically (Gallitelli *et al.*, 1985). No serological variants are known.

4. Carnation Italian Ringspot Virus

Carnation Italian ringspot virus (CIRV) was described by Hollings *et al.*, (1970) in Britain, where it was isolated only twice in 15 years from carnations (*Dianthus caryophyllus* L.) imported from Italy and the United States. The virus does not cause a serious disease, symptoms consisting of transient chlorotic spots and oval rings in the leaves and a slight reduction of growth.

A more recent record of CIRV, the first in 16 years since the original description, refers to its recovery from the waters of a creek in West German woodlands (Büttner *et al.*, 1987).

The experimental host range is wide, comprising 62 out of 104 plants tested. Sixteen hosts were invaded systemically, including *N. clevelandii* and *N. benthamiana*.

CIRV is readily inactivated *in vivo* by heat treatment. Systemically infected *N. clevelandii* plants were freed from virus when grown for less than two months at 36°C (Hollings *et al.*, 1970).

Two isolates of CIRV of Italian and North American origin, respectively, are known, which are serologically indistinguishable (Hollings *et al.*, 1970).

5. Cymbidium Ringspot Virus

Cymbidium ringspot virus (CyRSV) is another Tombusvirus with a restricted geographical distribution. It is recorded only from Britain, where it has been extensively studied by Hollings *et al.* (1977). In cymbidium orchids, the virus induces a mild chlorotic mottling and rings; in white clover (*Trifolium repens* L.), flecks and mottling of the leaves and slight stunting (Hollings and Stone, 1965; Hollings *et al.*, 1977).

Out of 101 plants tested, the cymbidium strain of cyRSV infected 61 (13 systemically) in 23 families, and the white clover strain infected 64 species (18 systemically) in 22 families. Both strains infected nearly all species tested in the families *Amaranthaceae, Chenopodiaceae, Compositae, Leguminosae,* and *Solanaceae,* but none of those of *Cruciferae* and *Gramineae.*

CyRSV was eliminated from nearly all (96%) explants from systemically infected *N. clevelandii* grown for two months at 35–37°C.

The cymbidium and the white clover strain of the virus show a very similar biological behavior, but can be distinguished serologically (Hollings *et al.*, 1977).

6. Grapevine Algerian Latent Virus

Grapevine Algerian latent virus (GALV) was isolated in 1985 from a grapevine plant (*Vitis vinifera* L.) of Western Algeria (G. P. Martelli, unpublished information).

The reactions of experimental hosts (13 in 6 different botanical families) were comparable to those typically elicited by Tombusviruses. Four *Nicotiana* species (*N. clevelandii, N. benthamiana, N. megalosiphon,* and *N. tabacum* cv. White Burley) and *Chenopodium amaranticolor* Coste et Reyn. were infected systemically.

7. Moroccan Pepper Virus

Moroccan pepper virus (MPV) was originally described in Morocco as a serious pathogen of sweet pepper (*Capsicum annuum* L.) inducing generalized stunting, deformation of the leaves, leaf and flower abscission, and lethal systemic necrosis (Fischer and Lockhart, 1977). This virus was at that time identified as a separate strain of TBSV (Fischer and Lockhart, 1977), but shortly afterward it was recognized as a possibly different entity that was provisionally called MPV (Makkouk *et al.*, 1981). More recently, MPV has been isolated in Germany from glasshouse-grown tomato plants (*Lycopersicon esculentum* Mill.) with mottling and malformation of the upper leaves and stunting of the shoot tips (Vetten and Koenig, 1983). Also in Germany, MPV was recovered from glasshouse-grown geranium plants (*Pelargonium zonale* L.) with leaf malformations and stunting (Vetten and Koenig, 1983).

A trial in which 15 different tomato cultivars were inoculated with
MPV showed that the virus, although capable of infecting a large proportion of plants (ca. 90%), was mildly pathogenic, inducing severe symptoms in only a few individuals of a single cultivar (cv. Rentita) (Vetten
and Koenig, 1983). MPV has also been recovered from naturally infected
eggplants (M. El Maataoui and R. Koenig in Koenig and Avgelis, 1983).

The experimental host range of MPV includes over 30 species in seven
botanical families. Sixteen species are invaded systemically, four of which
(*N. clevelandii*, *N. benthamiana*, *Chenopodium foetidum* L., *Spinacea
oleracea*) may be killed by the infection (Fischer and Lockhart, 1977;
Vetten and Koenig, 1983). Except for an untypified "TBSV strain" infecting spinach in Yugoslavia (Stefanac, 1978), MPV seems to be the only
Tombusvirus liable to invade systemically *S. oleracea*.

Moroccan and German MPV isolates are serologically indistinguishable (Vetten and Koenig, 1983).

8. Petunia Asteroid Mosaic Virus

After the first record from Italy (Lovisolo, 1957), petunia asteroid
mosaic virus (PAMV) has been reported from Canada and several European
countries, being often identified as a strain of TBSV (Allen and Davidson,
1967; Albrechtova *et al.*, 1975; Richter *et al.*, 1977).

The petunia isolate of PAMV induces in *Petunia hybrida* Vilm. color
breaking of the flowers, stunting, and stellate yellowish spots in the
leaves, accompanied by puckering and malformations of the leaf blades
(Lovisolo, 1957).

In cherry (*Prunus avium* L.), PAMV is associated with a very serious
disease in Canada (Allen and Davidson, 1967) and Central Europe (Albrechtova *et al.*, 1975; Kegler and Kegler, 1980; Koenig and Kunze, 1982),
which is known as "detrimental canker." In infected trees, leaves occur
in tight clusters or tufts, and are strongly deformed and twisted because
of veinal necrosis. Shoots are stunted and bent sideways and exhibit
splitting of the bark. Fruits are few and show rounded sunken pits. The
possible agent of this disease was originally identified as a strain of TBSV
serologically distinguishable from the type strain (Allen, 1968), but was
later found to be serologically very close to PAMV (Hollings and Stone,
1975). Similar conclusions were reached by Koenig and Kunze (1982), who
also suggested that Tombusviruses recovered in Switzerland (G. Stouffer
in Koenig and Kunze, 1982), Czechoslovakia (Albrechtova *et al.*, 1975),
and East Germany (Richter *et al.*, 1977) could represent isolates of PAMV.

PAMV was isolated in Czechoslovakia from plum trees (*Prunus do-
mestica* L.) with a poxlike pitting of the fruits (Novak and Lanzova, 1977b)
and hop (*Humulus lupulus* L.) with yellow mottling and malformations
of the leaves (Novak and Lanzova, 1976).

PAMV infection of grapevine has been reported from Germany (Bercks,
1967), Czechoslovakia (Novak and Lanzova, 1976), Canada (H. F. Dias

in Martelli, 1981), and Italy (U. Prota and M. Conti in Martelli, 1981). In all these instances, however, PAMV was in mixed infections with other viruses so that the symptoms induced, if any, could not be identified.

A recent report from Yugoslavia (Erić *et al.*, 1986) indicates that natural PAMV infection to spinach (*Spinacea oleracea* L.) induces clearing of the veins, mottling, curling and twisting of the leaves, and severe stunting.

The experimental host range of PAMV comprises about 30 species in eight families, most of which are infected locally (Lovisolo, 1957; Allen and Davidson, 1967). Virus isolates of diverse geographical origin that have been investigated serologically do not seem to differ from one another (Koenig and Kunze, 1982), but behave differently in some experimental hosts. Petunia, for instance, was invaded systemically by the petunia isolate (Lovisolo, 1957) and two grapevine isolates from Czechoslovakia and Italy (Novak and Lanzova, 1976; U. Prota, personal communication). It reacted with chlorotic or necrotic local lesions to Canadian and West German cherry isolates (Allen and Davidson, 1967; Koenig and Kunze, 1982) and was apparently not infected by the German grapevine isolate (Bercks, 1967).

9. Pelargonium Leaf Curl Virus

Pelargonium leaf curl virus (PLCV) is the agent of the leaf curl disease of *P. zonale*, a disorder that was first described in Germany (Pape, 1927) and was recorded afterwards in several European countries, the United States, and the Mediterranean basin (see Hollings, 1962). Symptoms consisting of yellow stellate spots with necrotic centers appear on the top leaves of diseased plants in winter. As foliar tissues expand, leaves become crinkled and curled downward, and may remain smaller than normal. Leaves produced in spring and autumn are symptomless, so that when the symptomatic leaves are shed, infected plants may appear healthy.

Whereas no natural host of PLCV other than pelargonium is known, the artificial host range of the virus is wide, including 62 out of over 110 plant species tested, 10 of which are invaded systemically (Hollings, 1962; Hollings and Stone, 1965).

In infected *N. clevelandii* and pelargonium plants, PLCV was totally inactivated after exposure for a month at 37°C (Hollings, 1962).

Several PLCV isolates studied by Hollings and Stone (1975) proved to be serologically rather uniform, forming a homogeneous group of closely related strains except for one Welsh isolate that was clearly distinguishable from all others.

Biologically, however, at least three strains were distinguished by the symptom severity in *N. clevelandii* and the type of lesions in *N. glutinosa* (Hollings, 1962).

10. Tombusvirus Neckar

Tombusvirus Neckar (TVN) was recovered from the waters of the river Neckar in West Germany. No natural host is yet known, but the few artificial hosts that were mechanically inoculated (seven plant species in three botanical families) reacted in a manner comparable to that of other Tombusviruses (Koenig and Lesemann, 1985).

III. TRANSMISSION, ECOLOGY, AND CONTROL

A. Transmission

The geographical distribution of Tombusviruses and the increasing number of records of their outbreaks in different crops indicate that in nature many members of the group are efficiently spread and become established in diverse environments.

The mechanisms for natural dissemination and survival of these viruses, largely unknown until recently (reviewed by Martelli, 1981), are now beginning to be unraveled by investigations involving primarily TBSV. In the past, transmission studies had followed diverse experimental approaches, not all of which were successful.

1. Transmission by Contact

Tombusviruses, because of their stable, highly infectious particles that reach large concentrations in host tissues, seem to behave like Potexviruses or Tobamoviruses, which spread readily in crops by mechanical contact between diseased and healthy plants. Transmission by foliage contact has been ascertained experimentally for CyRSV (Hollings et al., 1977) and hypothesized for TBSV and other members of the group in annual crops (R. N. Campbell, in Martelli, 1981). The ecological relevance of this means of dispersal remains, however, to be determined.

2. Transmission through Propagation Material

Tombusviruses that infect clonally propagated herbaceous or woody crops perennate in the propagating material which, therefore, becomes a powerful means for virus dissemination, especially over medium and long distances. There is evidence for international spread of CIRV and PLCV with infected natural hosts (Hollings et al., 1970; Martelli, 1981), and for local spread of AMCV in artichoke (Martelli, 1981), PAMV in cherry (Allen and Davidson, 1967) and TBSV in piggyback (A. H. McCain and M. I. C. Henriques, in Martelli, 1981). Systemically infected apples and grapevines are additional likely candidates for dispersal of TBSV (Allen, 1969; Kegler and Schimanski, 1982) and PAMV (Bercks, 1967).

3. Transmission through Seeds

Reports on seed transmission of Tombusviruses are conflicting. Whereas no transmission of TBSV, PLCV, PAMV, and CyRSV was found in seeds of naturally or artificially infected hosts (Smith, 1935; Gigante, 1955; Hollings, 1962; Lovisolo et al., 1965; Hollings et al., 1977), seed-borne infection of TBSV was demonstrated in apple (up to 7%) (Allen, 1969; Kegler and Schimanski, 1982) and in pepper and tomato, where it ranged from 4% to 65% (Cherif, 1981; Tomlinson and Faithfull, 1984). Furthermore, it was reported by Allen and Davidson (1967) that a "high percentage" of sweet cherry seeds were infected by PAMV. Hence, for some members of the group, seeds of different hosts may constitute a significant route for dissemination and perennation.

4. Transmission through Pollen

The Canadian PAMV isolate was suspected to spread through pollen in sweet cherries. The virus was recovered from dehisced anthers and washed pollen grains, but no transmission was obtained in trees that were hand-pollinated with pollen from infected sources (Allen and Davidson, 1967). Occurrence of TBSV (possibly PAMV, vide Koenig and Kunze, 1982) in cherry pollen was more recently confirmed by Kegler and Kegler (1980).

5. Transmission by Vectors

All attempts to transmit Tombusviruses by means of aerial or soil-borne vectors have failed so far. No success was obtained in any of the several experiments made using different aphid species for transmitting PAMV (Lovisolo, 1957), PLCV (Hollings, 1962), AMCV (Fischer and Lockhart, 1974), CyRSV (Hollings et al., 1977), or TBSV (Smith, 1935; Koenig and Avgelis, 1983). Equally negative were tests with whiteflies for the transmission of PLCV (Hollings, 1962) and mites for the transmission of TBSV (Orlob, 1968).

Although strong circumstantial evidence indicated the possible involvement of a Chytrid fungus in the transmission of PAMV through the soil (Lovisolo, 1966; Lovisolo et al., 1965), subsequent experimental trials failed to demonstrate unequivocally that soil-inhabiting fungi like *Olpidium brassicae* Wor., *Lagenocystis radicicola* Want. et Led., *Fusarium oxysporum* f.sp. *lycopersici* Snyd. et Hans., and *Phytophora capsici* Leon, could act as vectors for PAMV or TBSV (Teakle and Gold, 1963; Allen and Davidson, 1967; Campbell, 1968; Campbell et al., 1975; Cherif, 1981).

6. Transmission through Soil

In spite of the apparent lack of a soil-inhibiting vector, there is little doubt that several Tombusviruses spread through and persist in the soil, from which they can be acquired by the hosts.

Lovisolo et al. (1965) were among the first to obtain indications of the soil-borne nature of a Tombusvirus, demonstrating that plants grown in soil that had hosted PAMV-infected petunias picked up the virus through their roots. Likewise, TBSV appears to be firmly established in some British (Hollings and Stone, 1975) and Canadian (W. R. Allen, in Martelli, 1981) soils from which it is readily acquired by bait plants.

Field spread of PAMV, PLCV, and TBSV in the respective natural hosts has been recorded from several countries (Lovisolo et al., 1965; Hollings and Stone, 1965; Cherif, 1981; M. El Hamry and H. U. Fischer in Martelli, 1981). Moreover, transplanting healthy seedlings in soils that had hosted infected plants or were made infective by the addition of plant sap or concentrated particle suspensions of CyRSV, PAMV, MPV, or TBSV resulted in 10–100% infection, depending on soil type and virus–host combination (Roberts, 1950; Lovisolo et al., 1965; Hollings et al., 1977; Kegler and Kegler, 1981; Kegler et al., 1980; Cherif, 1981; Vetten and Koenig, 1983; Tomlinson and Faithfull, 1984). In particular, infection through roots of Celosia argentea seedlings grown undisturbed in sterilized media to which a purified suspension of an East German TBSV strain was added, was 9.3% in nutrient solution, 10.6% in nutrient agar, and 24.1% in quartz sand. The higher rate of transmission observed in quartz sand was interpreted as being consequent to root injuries that had favored entry of the virus (Kegler and Kegler, 1981; Kleinhempel and Kegler, 1982). The same TBSV strain as above was found to persist in an infective form for up to three months in a naturally infected soil or when added as purified suspension to sterilized soil samples kept outdoors (Kegler and Kegler, 1981). PAMV and TBSV-BS3, on the other hand, were shown to survive in clay soil for over five months (Lovisolo et al., 1965; Cherif, 1981).

Although no appreciable adsorption of TBSV to soil was detected by Tomlinson and Faithfull (1984), previous studies were strongly indicative that adsorption to soil constituents occurs to various extents according to the type of substrate (Kegler and Kegler, 1981). That some soil fractions exert a protecting action on TBSV is shown by the persistence of infectivity in soils autoclaved for 2 hr at 121°C (Kegler and Kegler, 1981).

B. Ecology

As shown in the preceding section, much circumstantial and experimental evidence supports the notion that Tombusviruses are soil-borne and spread at a site without the intervention of living organisms.

The ecological success of these viruses depends, then, on simple mechanisms whereby a consistent presence of sufficient inoculum in the environment must be assured, to allow for contact with the roots and subsequent invasion of susceptible hosts whenever the opportunity arises. Infected hosts and contaminated soil represent the major sources of inoculum for Tombusviruses.

Wild plants (weeds and shrubs) may constitute natural reservoirs for some members of the group, but their relevance in the ecology of these viruses is not yet fully known. Records of Tombusvirus infections in wild plants are relatively few and refer mainly to TBSV, several strains of which were recovered from weeds in East Germany (Richter et al., 1977; Kegler and Kegler, 1980) and Tunisia (Cherif, 1981), and from shrubs in Britain and Czechoslovakia (Martelli, 1981).

Cultivated plants seem to play a much greater role in the perpetuation of Tombusviruses. Several of these viruses, for instance AMCV and PLCV, were never found infecting in nature hosts other than artichoke and pelargonium, respectively. This suggests that Tombusviruses, similarly to Tobamoviruses, with which they seem to share the same ecological niche, may be regarded primarily as cultivated plant-adapted viruses (CULPAD viruses, sensu Harrison, 1981).

Many members of the group (e.g., PAMV, TBSV-type and BS3 strain, CyRSV) have been recovered from leachates of soils in which infected plants were growing (Lovisolo et al., 1965; Campbell et al., 1975; Hollings et al., 1977; Cherif, 1981; M. El Hamry and H. U. Fischer, in Martelli, 1981; Tomlinson and Faithful, 1984).

Therefore, routes for virus entry into the environment are:

1. Crop debris (i.e., roots and leaves falling to the ground) containing high concentrations of virus particles. AMCV, for example, was recovered in large quantities from soil of a severely infected artichoke plot by differential centrifugation of aqueous soil suspensions (Martelli et al., 1981).
2. Living roots of infected hosts, from which infective virus is released in the soil, as shown for PAMV, CyRSV, and TBSV (Lovisolo et al., 1965; Smith et al., 1969; Hollings et al., 1977; Kegler and Kleger, 1981; Cherif, 1981).
3. Water used for irrigation. Infective TBSV was isolated from lakes and rivers whose water is used for irrigation in Britain and many other countries (Tomlinson and Faithfull, 1984; Koenig and Lesemann, 1985), TVN from the river Neckar in Germany (Koenig and Lesemann, 1985), an as-yet-unidentified Tombusvirus from rivers of southern Italy (Piazzolla et al., 1986; Vovlas and Di Franco, 1987), and CIRV from a creek in West Germany (Büttner et al., 1987).

Tomlinson and Faithful (1984) suggested that TBSV may enter rivers either from drainage water of soils where infected crops are grown or from sewage water, as the virus passes unharmed through the alimentary tract of humans (Tomlinson *et al.*, 1982) following consumption of raw vegetables.

Whatever the origin, it becomes increasingly clear that epidemiological significance is to be attached to the presence of Tombusviruses in irrigation water.

The above findings on the epidemiology and ecology of Tombusviruses are consistent with both the endemic nature of some members of the group in certain geographical areas, such as TBSV in Mexico (Martinez *et al.*, 1974), Morocco (Fischer and Lockhart, 1977), and Tunisia (Cherif, 1981; Cherif and Spire, 1983); AMCV in southern Italy (Quacquarelli and Martelli, 1966), Morocco (Fischer and Lockhart, 1974), Malta (Martelli *et al.*, 1976), Tunisia (Rana and Cherif, 1981), and Greece (Rana and Kyriakopoulou, 1982); PAMV in Canada and Central Europe (see Koenig and Kunze, 1982), and with records of sudden and severe outbreaks of diseases in protected crops, for instance TBSV in tomato in Portugal (Borges *et al.*, 1979), TBSV-BS3 in eggplant in Crete (Koenig and Avgelis, 1983), and MPV in tomato in Germany (Vetten and Koenig, 1983).

C. Control

Combating Tombusvirus infections either in the field or in protected environments is, at the present state of knowledge, an almost impossible task. The polyphagy of those members of the group that cause economically relevant diseases (e.g., TBSV and MPV in solanaceous crops, PAMV in stone fruits) and their entrenchment in soils, constitute formidable obstacles to any successful control strategy, for which no handy solution is yet available.

However, prophylactic measures such as sanitary selection and heat therapy can assist in the prevention of indiscriminate dissemination of viruses through propagation material.

Roguing and propagation strictly from healthy mother plants were suggested for restraining dispersal of PLCV in pelargonium (Hollings, 1962) and AMCV in artichoke (Martelli and Rana, 1975).

Thermotherapy, on the other hand, has been successfully applied, especially to ornamentals, for freeing plants from PLCV (Hollings, 1962), CIRV (Hollings *et al.*, 1970), and CyRSV (Hollings *et al.*, 1977).

Tombusviruses have a thermal inactivation point *in vitro* above 90°C, but are surprisingly labile *in vivo*. They are inactivated when infected hosts are exposed for relatively short periods of time at 36°C, a temperature at which these viruses do not multiply. With TBSV, this phenomenon was observed well over 30 years ago (Kassanis, 1954), but its origin and kinetics *in vivo* were investigated more recently (Kassanis and Lebeurier, 1969; McCarthy, 1983). These studies have shown that at nonpermissive temperatures (36°C), the specific infectivity of TBSV decreases very rapidly (to approximately half in five days), whereas the total nucleoprotein content decreases much more slowly, requiring about 22 days to be reduced to half. The viral capsid is apparently not affected (hence the serological properties remain unchanged), not are the stability and physical properties of virus particles affected. The decrease of infectivity was found associated with fragmentation of viral RNA within the particles, possibly through a process involving degradation by free radicals (McCarthy, 1983).

IV. PURIFICATION

Tombusviruses have stable particles that can be extracted readily and in large quantities from infected tissues. Both locally (*G. globosa*) and systemically (*N. clevelandii, N. benthamiana*) infected hosts have been utilized for culturing viruses for purification, but *N. clevelandii* seems to be the most widely used.

An array of different purification schedules have been more or less successfully applied. These include simple differential centrifugation of crude leaf sap extracts (Steere, 1959; Ambrosino *et al.*, 1967); or use of 1.0 to 1.5 vol. of extraction media consisting of distilled water (Quacquarelli *et al.*, 1966) or diverse buffered solutions (50 to 200 mM phosphate, citrate, borate, TRIS-HCl) at pH values close to 7.2 or above (pH 7.6–8.0) neutrality, with the addition of mild reducing agents (1% ascorbic acid, 0.1% 2-mercaptoethanol, 0.05 M sodium sulfite, 0.1% thioglycollic acid) and, occasionally, 1 mM EDTA (Hollings, 1962; Quacquarelli *et al.*, 1966; Hollings and Stone, 1970, 1975; Hollings *et al.*, 1977; Makkouk *et al.*, 1981; Koenig and Kunze, 1982; Hayes *et al.*, 1984). Extraction was followed by clarification with organic solvents or Mg^{2+}-activated bentonite, concentration by cycles of differential centrifugation, and final separation from unwanted host components by sucrose density gradient centrifugation.

Extraction media at pH values above neutrality and containing chelating agents like EDTA are now known to be detrimental to Tombusvirus particles which, under these conditions, destabilize and swell (Krüse *et al.*, 1982a). A purification procedure has therefore been devised that takes into account the requirements for preserving particle integrity (Gallitelli *et al.*, 1985; Hillman *et al.*, 1985). Infected tissues are homogenized in 2 vol. of 50 mM sodium acetate buffer, pH 5.0, containing 1% ascorbic acid. The homogenate is filtered through muslin, kept for 1 hr at 4°C, and centrifuged at 10,000 rpm for 10 min. The pH of the clear supernatant is adjusted to 5.5, 0.2M NaCl and polyethylene glycol (mol. wt. 6000) to the final concentration of 10% are added and dissolved by stirring. After 1 hr at 4° C, the virus-containing precipitate is collected by low-speed centrifugation and resuspended overnight in 50 mM acetate buffer, pH 5.5. Insoluble material is removed by further low-speed centrifugation, the virus is sedimented by centrifuging at 36,000 rpm, and pellets are resuspended in acetate buffer as above. Further purification is achieved through density gradient centrifugation in CsCl at equilibrium: 2.65 g of CsCl are dissolved in 5 ml of virus suspension and centrifuged overnight at 36,000 rpm in a fixed-angle rotor at 10°C. The sharp opalescent virus band is removed with a syringe and dialyzed against 50 mM NaCl, pH 5.5.

This protocol is applicable to all Tombusviruses, giving consistently reproducible results. Virus yields range between 10 and 60 mg/100 g of infected tissues.

V. PROPERTIES OF VIRUS PARTICLES

A. Particle Morphology

As with other plant viruses whose structure is primarily stabilized by protein–RNA interactions, mounting Tombusvirus preparations in neutral phosphotungstate (PTA) may damage the particles, which are penetrated by the stain and disrupted to variable extents (Quacquarelli et al., 1966). Satisfactory preservation is obtained by fixing the preparations prior to PTA staining, or by using ammonium molybdate or un-buffered uranyl acetate (Quacquarelli et al., 1966; Hollings et al., 1977; Francki et al., 1985).

When negatively stained, Tombusvirus particles appear as isometric structures with a diameter ranging from 28 to 30 nm (see Martelli, 1981) or of ca. 34 nm, according to recent estimates (Francki et al., 1985). In PTA mounts, virions may show an angular outline (Quacquarelli et al., 1966; Lovisolo et al., 1967) but, when mounted in uranyl acetate, exhibit a rounded outline and somewhat knobby surface and edges (Fig. 1) (see also Francki et al., 1985). Details of the surface structure, which are poorly resolved in the electron microscope, are clearly seen in three-dimensional image reconstructions from electron micrographs (Crowther and Amos, 1971) or computerized tracings (Fig. 4B) (Robinson and Harrison, 1982; Harrison, 1984b).

The appearance of the particles under the electron microscope does not allow one to distinguish between members of the Tombusvirus and the proposed carnation mottle virus (CarMV) groups (Chapter 3, this volume).

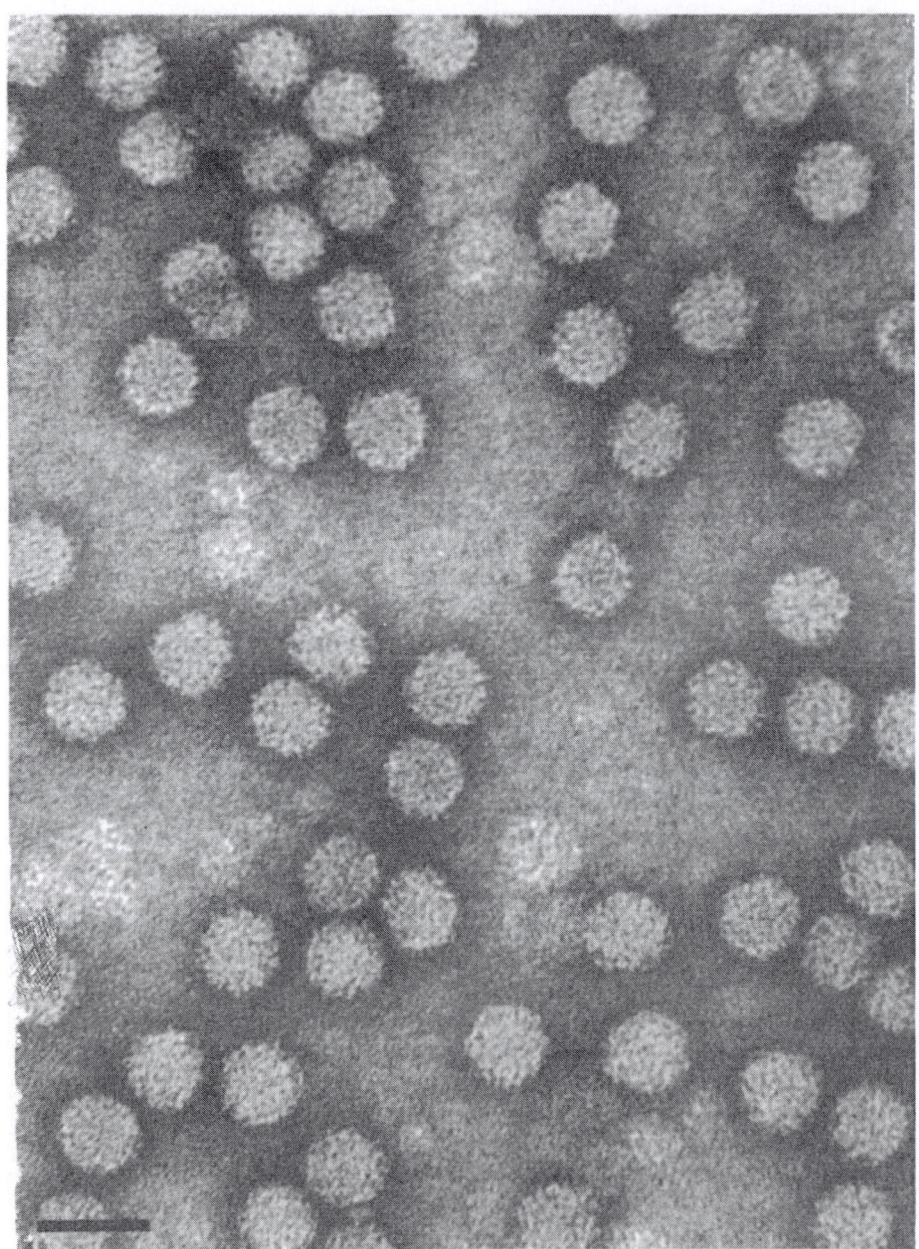

FIGURE 1. Particles of TBSV-type negatively stained with uranyl acetate. Scale bar 50 nm.

TABLE II. Some Properties of Tombusviruses

Virus	Sedimentation coefficient ($S_{20,w}$)	Buoyant density in CsCl (g/cm^3)	M_r of coat protein (kd)	Size of genomic RNA (kb)	Estimated percent sequence homology with TBSV-type	Serological relationship with TBSV-type
TBSV	130[a]	1.35[b]	42[a]	4.7[c]	100[c]	+ + + +[d]
AMCV	132[e]	1.35[e]	43[a]	4.7[c]	7[c]	+ + +[d]
EMCV	132[f]	1.34[a]	41[f]	4.7[c]	7[c]	+[d]
CIRV	135[g]	1.36[a]	37[a]	4.7[c]	11[c]	+[d]
CyRSV	127[i]	1.36[i]	43[a]	4.7[c]	10[c]	±[d]
GALV	128[a]	1.34[a]	37[a]	4.7[a]	15[a]	−[a]
MPV	140[a]	1.36[a]	43[a]	4.7[a]	10[a]	+ +[d]
TVN	133[a]	1.36[a]	45[a]	4.7[a]	15[a]	+[d]
PAMV	134[h]	1.35[i]	41[a]	4.7[c]	11[c]	+ + +[d]
PLCV	136[j]	1.34[a]	41[a]	4.7[c]	24[c]	+ + +[d]

[a] D. Gallitelli and M. Russo (1986 and unpublished information).
[b] Mayo and Jones (1973).
[c] Gallitelli et al. (1985).
[d] Koenig and Gibbs (1986).
[e] Quacquarelli et al. (1966).
[f] Makkouk et al. (1981).
[g] Hollings et al. (1970).
[h] Ambrosino et al. (1967).
[i] Hollings and Stone (1965).
[j] Hollings et al. (1977).

B. Sedimentation

Purified preparations of Tombusviruses yield a single light-scattering band when centrifuged in sucrose density gradient columns (Quacquarelli *et al.*, 1966; Hollings *et al.*, 1977). However, multiple bands were observed with PLCV and CIRV preparations, which were attributed to aggregation or, more likely, partial degradation of virus particles (Hollings and Stone, 1965; Hollings *et al.*, 1970).

In the analytical ultracentrifuge, Tombusvirus particles sediment as a single component at 127–140 S. The particles are isodense and stable in CsCl, in which they band as a single buoyant density class with values ranging from 1.34 to 1.36 g/cm^3 for different members of the group (see references in Table II). The buoyant density of TBSV in Cs_2SO_4 is 1.29 g/cm^3 (McCarthy, 1983).

Heating *in vivo* at 36°C does not change the sedimentation behavior of TBSV particles in rate-zonal density gradient centrifugation nor in moving boundary or isopycnic centrifugation in the analytical ultracentrifuge (McCarthy, 1983).

C. Electrophoretic Properties

Preparations of all Tombusviruses are electrophoretically homogeneous, migrating as a single species. The rate and direction of migration, however, vary with individual members.

As shown in Table III, some viruses (e.g., AMCV, CIRV, PLCV) migrate either toward the anode or the cathode according to the conditions

TABLE III. Electrophoretic Behavior of Tombusviruses

Electrophoresis in 0.8-1% agar, 0.03M phosphate buffer, pH 7.6 or 1N Na-Veronal buffer, pH 8.6		Electrophoresis in 1% agarose, 0.02-0.05M phosphate buffer, pH 7.0	
Migration to		Migration to	
Anode (+)	Cathode (−)	Anode (+)	Cathode (−)
TBSV untypified strain[e]	TBSV-type[e]	TBSV-BS3[g,i,k]	TBSV-type[i]
TBSV-BS3[a,b,c]	TBSV-tulip[e]	AMCV[k]	CyRSV[k]
PAMV[a,b]	AMCV[c,e]	EMCV[g]	
	CIRV[d,e]	CIRV[k]	
	CyRSV[f]	MPV[g,i,k]	
	PLCV[b,c,e]	PAMV[g,h,k]	
		PLCV[k]	
		TVN[k]	

[a] Lovisolo *et al.* (1964). [b] Bercks and Lovisolo (1965). [c] Martelli and Quacquarelli (1966). [d] Hollings *et al.* (1970). [e] Hollings and Stone (1975). [f] Hollings *et al.* (1977). [g] Makkouk *et al.* (1981). [h] Koenig and Kunze (1982). [i] Koenig and Avgelis (1983). [j] Vetten and Koenig (1983). [k] Koenig and Gibbs (1986).

of electrophoresis (i.e., type of medium, type and molarity of buffer, pH). Other viruses seem less influenced by these conditions, showing consistent anodic (TBSV-BS3, PAMV, EMCV, MPV) or cathodic (TBSV-type, CyRSV) migration (see references in Table III). Different migration rates toward either pole have been observed among serologically indistinguishable isolates of the same virus (Hollings and Stone, 1975; Koenig and Kunze, 1982; Koenig and Avgelis, 1983; Koenig and Gibbs, 1986).

D. Coat Protein Subunits

Coat protein subunits of Tombusviruses are made up of a single polypeptide with M_r ranging from 37 to 45 kd (Table II). These differences in M_r value derive from differential migration rates consistently shown by subunits of diverse members of the group when coelectrophoresed in polyacrylamide slabgels (Fig. 2).

Minor polypeptides with M_r of 28 kd and 87kd, respectively, have been found in TBSV coat protein preparations in some laboratories (Butler, 1970; Ziegler et al., 1974). These components are not always detected, especially if freshly made preparations are analyzed (McCarthy, 1983; M. Russo and D. Gallitelli, unpublished information). The smaller polypeptide (28 kd) may be a cleavage product of the major coat protein (ca. 40 kd) (Ziegler et al., 1974) whereas the 87 kd molecule used to be somewhat of a mystery.

Ziegler et al. (1974) indicated that this large polypeptide occurs in the proportion of one chain per virion and is located inside the particles. Butler (1970) suggested that it could be a polymerase attached to viral RNA. This hypothesis, however, conflicted with the notion that highly purified TBSV RNA preparations are infectious and that their infectivity is not affected by treatments with proteinase K or pronase (Mayo et al., 1982). It has now been ascertained that the 87 kd protein is a stable

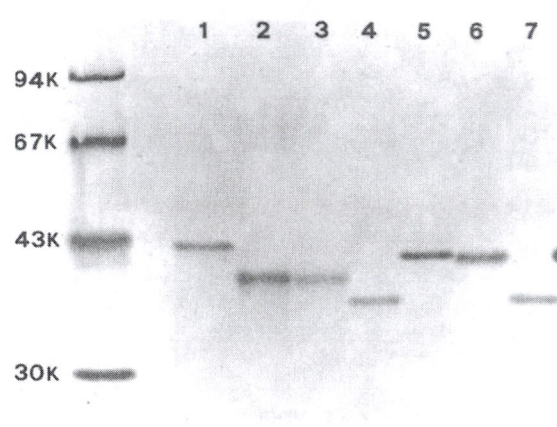

FIGURE 2. Gel electrophoresis of coat protein subunits of 7 different Tombusviruses. Lane 1, NRV; 2, PAMV; 3, PLCV; 4, CIRV; 5, CyRSV; 6, AMCV; 7, GALV. Reference markers on the extreme left lane.

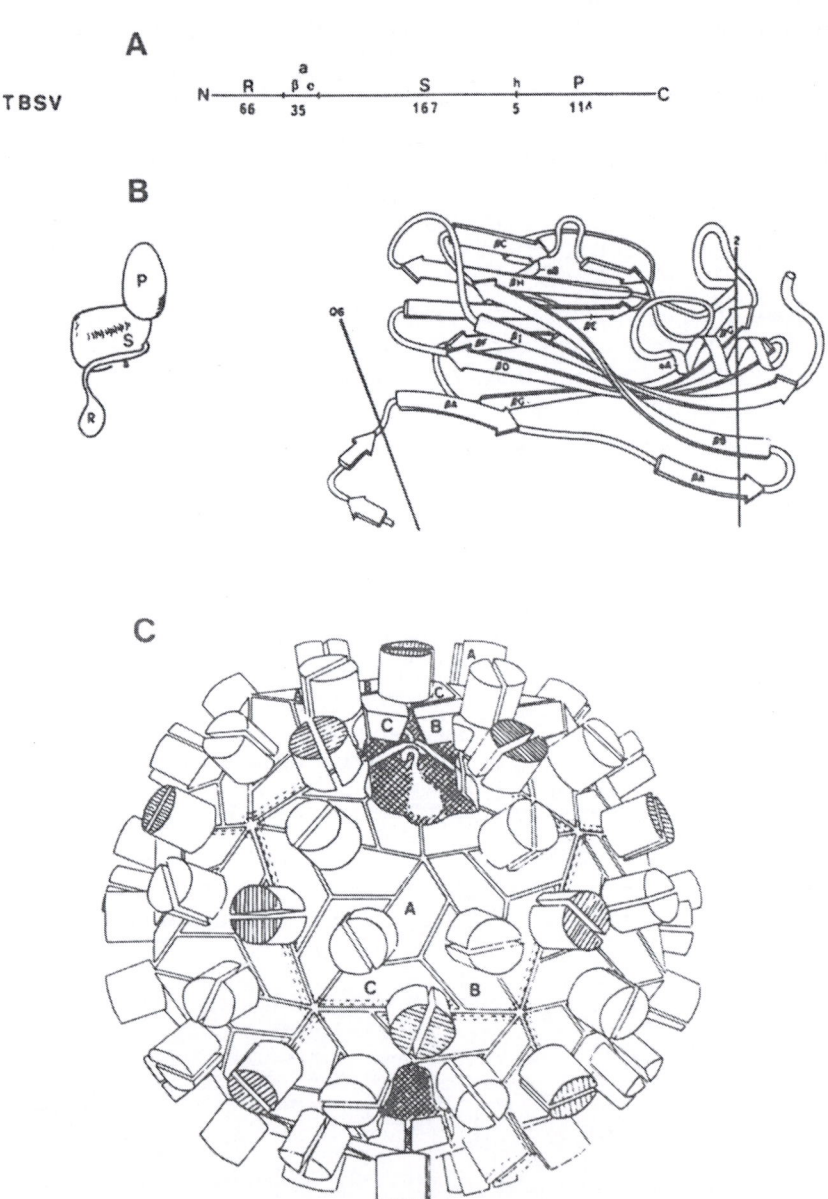

FIGURE 3. Structure of TBSV according to Hopper *et al.* (1984). (A) Linear map of the coat protein subunit. (B) Left: diagrammatic representation of the protein subunit showing the spatial arrangement of the three domains (P, S, and R) and connecting arm (a); right: schematic view of the folded polypeptide. (C) Arrangement of the subunits in the virus particle, showing the three packing environments A, B, and C. Courtesy of the authors and permission of Academic Press Ltd., London.

covalently linked dimer of two protein coat subunits (Stockley *et al.*, 1986).

The coat protein subunits of TBSV contain 387 amino acid residues, whose sequence has been almost entirely determined by chemical analysis (Hopper *et al.*, 1984). Interestingly enough, there is extensive sequence homology between structurally corresponding residues of the shell-forming tract (S domains) of TBSV and southern bean mosaic virus subunits (Hopper *et al.*, 1984). A similar situation seems to exist between TBSV and TCV (Chapter 3, this volume), which may account for the distant serological relationship found between these two viruses (Jaegle and Van Regenmortel, 1985). Similarities in the amino acid composition of coat protein occur also between TBSV, SCV, CMV, and CaMV, which has led Fauquet *et al.* (1986) to suggest a common grouping of these viruses.

The backbone folding of the TBSV subunit polypeptide and its spatial configuration have been reconstructed (Fig. 3A) and the tertiary architecture of the capsid has been determined to 2.9 Å resolution by X-ray crystallographic analysis (Harrison *et al.*, 1978; Olson *et al.*, 1983; Hopper *et al.*, 1984).

E. Structure of Virus Particles

The protein shell of TBSV and other tombusviruses has a cubic symmetry and an architectural organization based on dimer clustering of structural subunits located on the two fold position of a $T = 3$ icosahedral surface lattice (Finch *et al.*, 1970; Crowther and Amos, 1971; Harrison *et al.*, 1978). There are 180 identical structural subunits in the shell which, because of their dimer clustering, give rise to 90 morphological subunits forming rings of five and six about the fivefold and threefold symmetry axes, respectively. Each subunit (Fig. 3B) is folded into four distinct modules: an internal positively charged domain (R), comprising 66 residues, which is connected to the other regions of the molecule by an arm 35 residues long (Fig. 3A). Arm and R domains form the entire N-terminal region of the polypeptide (101 residues altogether). This region is flexibly linked to two distinct globular modules, one of which (domain S) forms the shell of the virus particle, and the other (domain P) protrudes on the particle surface. P and S domains are in turn connected to one another by a flexible hinge 5 residues long. The P (protruding) domains, which are thought to be the result of gene duplication (Argos *et al.*, 1980), are clustered in pairs, thus providing the dimer contacts that help stabilize the particle structure. Because of their flexibility at the level of the hinge between the S and P domains, coat polypeptides can accommodate up to three distinct packing positions denoted A, B, and C (for details see Harrison *et al.*, 1978; Olson *et al.*, 1983; Harrison, 1984a,b). The connecting arms of the 60 C-type subunits are folded in an orderly manner at the bottom of the respective S domains, interlocking to form a cagelike internal framework, which determines particle size (Olson *et al.*, 1983). The N-termini of the 120 remaining A- and B-type subunits lie toward the interior of the particle and are less ordered than in the C configuration. These terminal regions, together with positively charged residues of the inward-facing S domain surfaces, are potential sites for RNA binding (Harrison *et al.*, 1978; Chauvin *et al.*, 1978; Hopper *et al.*, 1984).

Viral RNA is tightly packed within the protein shell, most of it

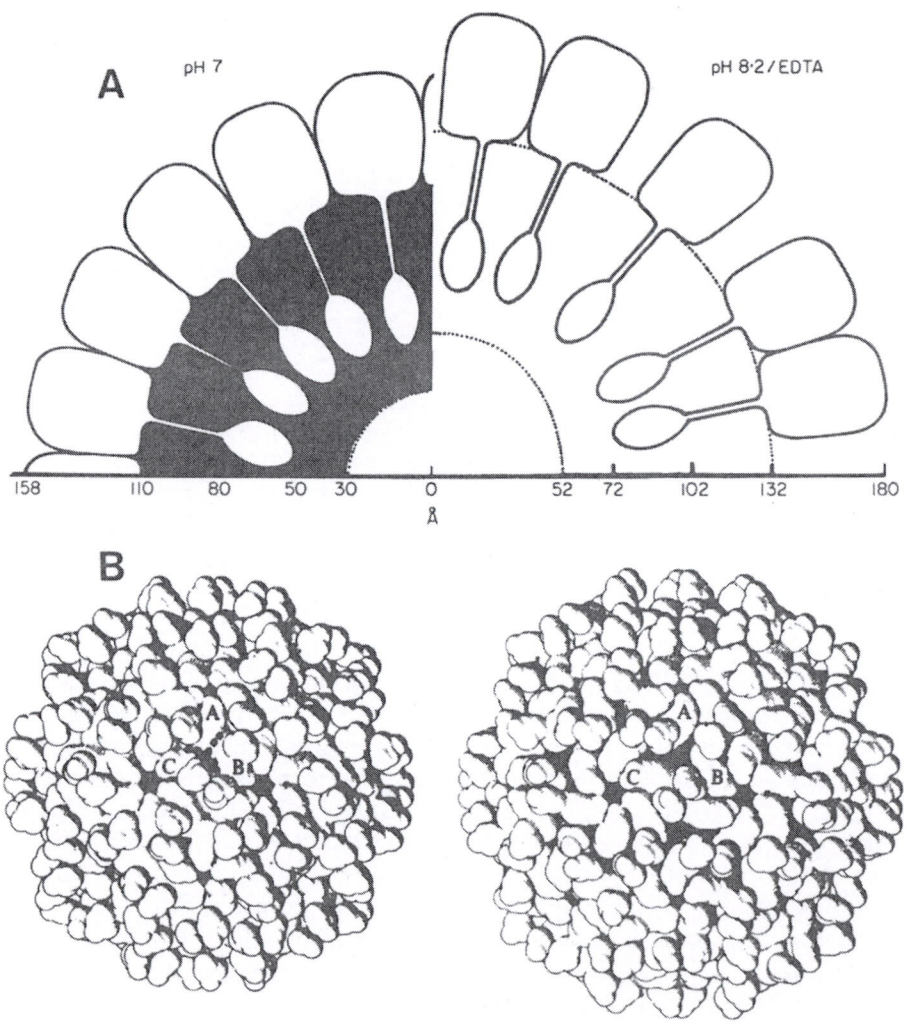

FIGURE 4. (A) Schematic view of a sector of a TBSV particle in compacted (pH 7) and swollen (pH 8.2/EDTA) state according to Krüse *et al.* (1982a). (Courtesy of the authors and permission of Academic Press Ltd., London.) (B) Pictorial view according to Robinson and Harrison (1982) of compact and expanded TBSV particles showing cation-binding sites between type-A, -B, and -C subunits. Courtesy of the authors and permission of McMillan Journals Ltd., London.

probably located in the narrow space (ca. 3 nm) between the S domains and the internal concentric shell made up of N-terminal regions (Fig. 4A) (Chauvin et al., 1978). ^{31}P nuclear magnetic resonance studies have shown that the RNA is highly structured and immobilized in native but not in swollen particles (Munowitz et al., 1980). Although most of the mobility of RNA is lost in recompacted particles, these do not resume the full stability of the native virions (Krüse et al., 1982b).

That RNA-protein linkages stabilize TBSV particle structure had been shown years ago (Kaper, 1975; Piazzolla et al., 1977). However, the extent to which these interactions operate is not fully known (Harrison, 1984a). In this connection it is worth mentioning that heat-induced fragmentation of TBSV-RNA in situ does not seem to impair particle stability, as though the interactions responsible for capsid stabilization were not affected by the RNA degradation process (McCarthy, 1983).

There is evidence that TBSV particles stability depends also on protein–protein interactions (Boatman and Kaper, 1976), i.e., on P domain dimer contacts, which are very stable, and on trimer interactions between S domains of A, B, and C subunits (Harrison et al., 1978). The strength of these latter contacts is enhanced by Ca^{2+} but not Mg^{2+} ions (Robinson and Harrison, 1982; Krüse et al., 1982a; Hogle et al., 1983). In native TBSV preparations Hogle et al. (1983) have located two cation binding sites for each pair of interacting A, B, and C subunits, where EDTA-chelatable Ca^{2+} ions are bound (Fig. 4B). When these cations are removed and the pH is raised above 7.0, the particles undergo reversible expansion. Expanded particles are approximately 12% larger than the compact ones, do not suffer changes in domain conformation, and retain the same radial distribution of RNA and protein (Fig. 4A). Dimer clustering is also preserved upon swelling (Krüse et al., 1982a; Robinson and Harrison, 1982). Addition of Ca^{2+} or lowering the pH reverts the swelling, thus recompacting the particles (Krüse et al., 1982b).

Expanded TBSV is not sensitive to RNase but is partially sensitive to protease. The terminal regions of some subunits undergo proteolytic attack, possibly because they are extruded through the gaps in the expanded shell (Krüse et al., 1982; Robinson and Harrison, 1982). From hydrogen-ion titration curves, it has been established that the protein shell of TBSV is semipermeable, thus having the same ion-handling properties as lipid membranes (Durham et al., 1984).

VI. SEROLOGY AND IMMUNOCHEMISTRY

A. General Features

Tombusviruses are good immunogens. Antisera with titers of 1:512–1:2048 and up to 1:16 384 are readily obtained (see among others Hollings and Stone, 1975) by injecting purified virus suspensions into rabbits, as most investigators do, or hens (Jaegle and Van Regenmortel, 1985). Virus particles do not require stabilization with fixatives (e.g., formaldehyde, glutaraldehyde) prior to injections as do, for instance, members of virus groups with tripartite genome (reviewed by Rybicki and Von Wechmar, 1985).

Immunization schedules used in different laboratories vary considerably. Although as little as 0.025 mg of antigen given as a single dose proved sufficient for a satisfactory immune response (Allen, 1968), customarily, 1 to 2 mg of purified virus for each injection are used, with peaks of 5–6 mg/injection (Hollings and Stone, 1975; Hollings et al., 1977; Tomlinson and Faithfull, 1984).

The number and type of injections are also variable. Multiple injections (three or more), at least one of which was intramuscular and emulsified with Freund's complete or incomplete adjuvant, followed by two or more intravenous injections over a period of 3–4 weeks, have been used, especially in the past (Lovisolo et al., 1964; Bercks and Lovisolo, 1965; Martelli and Quacquarelli, 1966; Wetter and Luisoni, 1969; Hollings et al., 1970, 1977; Hollings and Stone, 1975; Rana and Kyriakopoulou, 1982; Cherif and Spire, 1983). More recently, Koenig and coworkers (Makkouk et al., 1981; Koenig and Kunze, 1982; Koenig and Lesemann, 1985; Koenig and Gibbs, 1986) have adopted an immunization schedule based on two intramuscular injections a week apart, the first one emulsified in Freund's complete adjuvant and the second in Freund's incomplete adjuvant. A comparable schedule was used by Hollings and Stone (1975) to prepare narrow-spectrum (i.e., strain-specific) antisera to TBSV-type and CIRV.

Immunized rabbits are usually bled a couple of weeks after the last injection for a period of 2–3 months, sometimes longer (up to two years, Koenig and Gibbs, 1986). The antisera can be stored in lyophylized form, frozen at −20°C, or kept cold (2–3°C) with the addition of an equal volume of glycerol or 0.025% sodium azide as preservatives.

B. Serology in Virus Identification

Symptomatological responses of test plants and morphology of virus particles in crude sap extracts, although liable to indicate the presence of a Tombusvirus in the sample under examination, are no substitute for serology in identification. Serology, in fact, still represents the best single means for a correct diagnosis at the virus level.

Since the majority of Tombusviruses are serologically interrelated, broad spectrum antisera to any of them can recognize many members of the group, with few exceptions. For reliable identification it is advisable to make tests with antisera specific to each member of the group, and include homologous antigens for comparison.

The following are the serological techniques that have been most widely used for Tombusvirus detection and identification.

1. Double Diffusion in Gels

A simple technique such as the Ouchterlony's double diffusion test in agar or agarose gels is quite suitable for routine diagnostic purposes and, as discussed below, for establishing relationships between group members. In agar plates Tombusviruses give a single sharp precipitin line that begins to form 4 to 6 hr after loading the plate wells with reactants.

The high concentration of Tombusvirus particles in glasshouse-grown experimentally inoculated hosts either with localized (*G. globosa* for instance) or systemic infections, allows for the production of clear-cut precipitin reactions using undiluted crude sap. By means of this technique, AMCV and PLCV were detected in sap expressed from field-grown naturally infected artichoke and pelargonium plants during winter, when symptoms are strong and virus concentration is high (Martelli, 1981). No additional records seem to exist of direct identification of other Tombusviruses in natural hosts by double diffusion in gel.

2. Latex Agglutination Test

Agglutination consists of the clumping that antigen- or antibody-coated inert carrier particles of relatively large size undergo when they are exposed to homologous reactants (antibodies or antigens, respectively). In the latex agglutination (LA) test, the carrier particles are polystyrene spheres sensitized directly with antibodies (Bercks and Querfurth, 1969). The technique is simple, quick (visible precipitates form within 5–10 min) and 25 to 1000 times more sensitive than precipitin tests (reviewed by Van Regenmortel, 1982).

LA has been used for detecting PAMV in grapevine and cherry. In particular, PAMV was identified in leaves of 23 out of 132 vines from the major viticultural areas of West Germany (Bercks, 1967) and in young leaves and petals of 7 out of 54 cherry trees from East Germany (Fuchs *et al.*, 1979). Whereas LA proved more sensitive than biological tests with grapevine (Bercks, 1967), the reverse was apparently true with cherry, as PAMV was recovered by mechanical inoculation from ten trees, but detected serologically in seven trees only (Fuchs *et al.*, 1979).

3. Enzyme-Linked Immunosorbent Assay

This test, universally known as ELISA, has gained progressive popularity since its first application in plant virology (Clark and Adams, 1977) and is now widely used, especially for large-scale surveys. Several ELISA procedures are known, whose methodology has recently been reviewed by Clark and Bar-Joseph (1984).

With EMCV and TBSV-BS3, the sensitivity of ELISA proved to be 5000 to 10,000 times higher than that of gel double diffusion tests (Makkouk *et al.*, 1981; Cherif, 1981). Direct double antibody sandwich ELISA was used for detecting PAMV in young leaves and petals of cherry trees in East Germany. In these tests, 12 trees out of 54 were ELISA positive as compared with 7 and 10 trees found to be infected by LA test and mechanical inoculation, respectively (Fuchs et al., 1979).

The same ELISA techniques were extensively used in a study of TBSV-BS3 infecting solanaceous crops in Tunisia, whereby the virus was identified in natural weed hosts, soil extracts, and symptomless cultivated hosts. ELISA was also utilized to assess the degree of susceptibility of tomato cultivars to TBSV-BS3 infection, the rate of transmission through seeds, and whether the virus was located in the seed coats or germinated embryos (Cherif, 1981).

Additional applications of ELISA were reported by Vetten and Koenig (1983), who used this procedure for assessing the level of infection of tomato cultivars inoculated with TBSV-type and MPV and the percentage of natural root infection by soil-transmitted MPV. In the same study, Vetten and Koenig failed to demonstrate the presence of MPV in leaves and roots of infected pelargonium plants, probably because virus concentration was below the level of detection by ELISA.

4. Dot-Immunobinding Assay

Although the applicability of methods for detecting Tombusviruses spotted on nitrocellulose membranes (Hawkes et al., 1982) as drops of crude sap extracts has not been investigated in detail, there are indications that dot-immunobinding assays can be useful. CyRSV, for instance, was readily and reliably identified in sap of infected N. clevelandii with a detection limit of approximately 5 ng, using a procedure similar to that of Hibi and Saito (1985) (M. Russo, unpublished information).

5. Immunosorbent Electron Microscopy

The merits and procedures of immunosorbent electron microscopy (ISEM) have recently been reviewed by Milne and Lesemann (1984).

Except for a study by Appiano and D'Agostino (1985), who used ISEM for determining the distribution of TBSV-BS3 in the root tissues of G. globosa, not much is reported in the literature on the application of this technique to Tombusvirus detection. However, ISEM is routinely utilized in our laboratory for the identification of different members of the group in both naturally and artificially infected hosts (M.A. Castellano and A. Di Franco, unpublished information).

ISEM has also been used to investigate the degree of serological relatedness of EMCV to other Tombusviruses. In these studies, serological binding of EMCV particles was highest with homologous antiserum but decreased progressively with antisera to PLCV, TBSV-BS3, and AMCV (Makkouk et al., 1981).

C. Serology in Virus Classification

The current taxonomy of Tombusviruses is essentially based on serology. Serological comparisons, often between no more than two or three distinct members of the group, have been reported from many laboratories (Lovisolo *et al.*, 1964; Bercks and Lovisolo, 1965; Hollings and Stone, 1965; Martelli and Quacquarelli, 1966; Allen, 1968; Wetter and Luisoni, 1969; Makkouk *et al.*, 1981; Vetten and Koenig, 1983; Cherif and Spire, 1983; Jaegle and Van Regenmortal, 1985). These studies, together with more exhaustive comparative investigations by Hollings and Stone (1975) and Koenig and Gibbs (1986), have contributed to laying the basis for the present classification.

An array of different techniques have been used in the above investigations: (1) tube precipitin tests (Hollings and Stone, 1975); (2) double diffusion in gels (see among others, Hollings and Stone, 1975; Koenig and Gibbs, 1986); (3) intragel absorption (Wetter and Luisoni, 1969); (4) immunoelectrophoresis (see among others, Bercks and Lovisolo, 1965; Martelli and Quacquarelli, 1966; Hollings and Stone, 1975; Makkouk *et al.*, 1981); (5) ISEM (Makkouk *et al.*, 1981); (6) indirect ELISA (Jaegle and Van Regenmortel, 1985); and (7) electro-blot immunoassay (Western blot) (Burgermeister and Koenig, 1984).

Tube precipitin tests, being greatly influenced by the characteristics of individual antisera (i.e., whether they are broad-spectrum or strain-specific or originate from early or late bleedings) are more useful for assessing antigen and antibody concentrations than for quantitative differentiations between viruses. This is better achieved by gel diffusion and intragel absorption tests (Wetter and Luisoni, 1969; Hollings and Stone, 1975; Koenig and Gibbs, 1986). Indirect ELISA is also suitable to this effect (Jaegle and Van Regenmortal, 1985). Conversely, Western blot appears of little use because, with tombusviruses, it was found that the antisera cross-reacted with coat proteins of viruses belonging to quite different taxonomic groups (i.e., Carlavirus and Potyvirus) and that the SDS-dissociated coat protein subunits did bind aspecifically with alkaline phosphatase-labeled antibodies (Burgermeister and Koenig, 1984).

The results of recent double diffusion test studies by Koenig and Gibbs (1986), based on more than 2000 titer determinations with 222 antisera obtained from 36 rabbits, have shown that all Tombusviruses are serologically interrelated, although to variable extents (Fig. 5). GALV, which was not included in the above investigations, is no exception, for in gel double diffusion tests it proved to be serologically distantly related to MPV (SDI = 5) but apparently unrelated to all remaining members of the group (G. P. Martelli, M. Russo, and D. Gallitelli, unpublished information).

The relationship among Tombusviruses range from close (e.g., PAMV and AMCV, SDI = 1), to intermediate (e.g., TBSV-type and AMCV or PLCV, SDI = 3), to distant (e.g., TBSV-type and EMCV or CIRV, SDI =

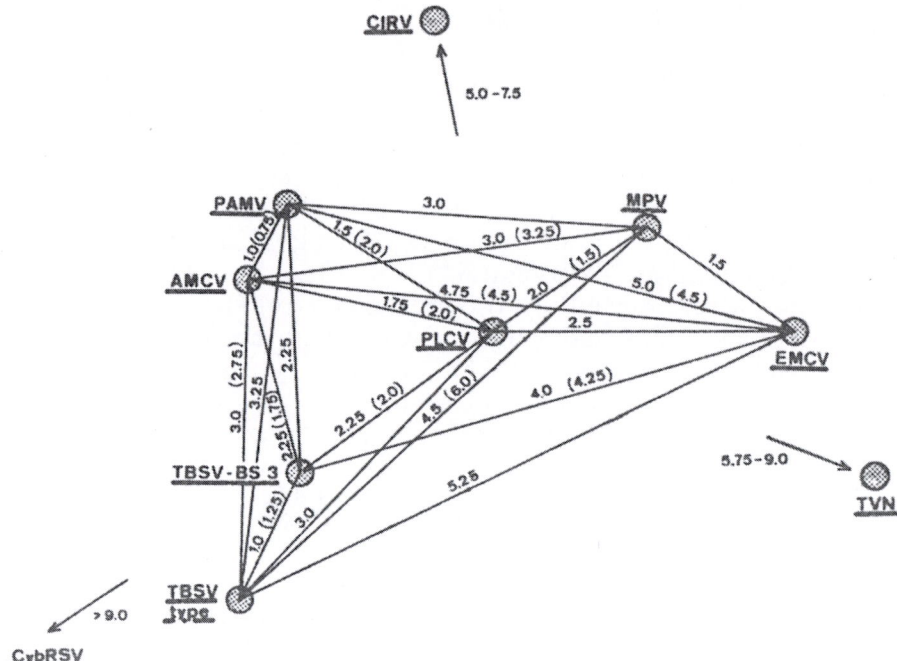

FIGURE 5. Serological classification of Tombusviruses according to Koenig and Gibbs (1986), based on average serological differentiation indices from reciprocal tests depicted as length units. This scheme does not include GALV. Courtesy of the authors and permission of the Society for General Microbiology, Reading.

5–7), to nearly undetectable (e.g., TBSV-type and CyRSV, SDI > 9) to undetectable (e.g., TBSV-type and GALV or CyRSV and NRV). Unfortunately, serological variations are also found among isolates (or strains) of individual members as with TBSV (Martelli *et al.*, 1971, 1977; Koenig and Gibbs, 1986), PLCV (Hollings and Stone, 1975), CyRSV (Hollings *et al.*, 1977), and AMCV (Rana and Kyriakopoulou, 1982). This situation may complicate the correct identification at the "species" level of viruses like TBSV, PAMV, PLCV, and AMCV, which fall within a cluster characterized by close relationships. Despite these difficulties, the consensus is that the viruses included in this chapter continue to be considered as separate members of the Tombusvirus group.

D. Immunochemical Studies

To our knowledge, the antigenic constitution of Tombusviruses has been little investigated.

Allen (1968) studied the kinetics of the specificity of the immune response to PAMV-cherry and TBSV-type. He found that the degree of relationship between these two viruses depended very much upon the time of bleeding during the primary response. In particular, the percentage of cross-reactivity of antisera decreased after the first titration (four days after injection) until the highest homologous titer was reached (3–4 weeks after injection). Then, cross-reactivity increased gradually to reach a maximum about three months after injection. These changes were associated with a differential synthesis and/or release of specific antibodies.

Specific and cross-reactive antibodies were present among both light (7S) and heavy (19S) immunoglobulins. Although both types of immunoglobulins were already present in the earliest bleedings (four days after injection), the greatest part of total cross-reactivity was associated with 19S antibodies and did not change over a five-week period. Conversely, cross-reacting 7S antibodies were very few initially, but increased in number concomitantly with the increase of total 7S and whole antiserum titers.

A comparative antigenic analysis of TBSV-BS3, PAMV, and PLCV was made by Wetter and Luisoni (1969) who, by means of intragel absorption tests, identified the following antibody fractions: PAMV = G,a,d; PLCV = G,c,e; and TBSV-BS3 = G,b,d,e. Thus, these three viruses were shown to have a single antigenic determinant in common (G). This determinant was also the only one shared by PAMV and PLCV. Each of these two latter viruses, however, shared an additional determinant ("d" for PAMV and "e" for PLCV) with TBSV-BS3, whose presence accounted for the closer relationship existing between either PAMV or PLCV and TBSV-BS3, than between PAMV and PLCV (Wetter and Luisoni, 1969).

VII. GENOME PROPERTIES AND REPLICATION

A. Structure of Genomic RNA

All Tombusviruses contain a monopartite genome made up of a molecule of positive-sense single-stranded RNA whose size, according to estimates made in different laboratories under nondenaturing conditions, was reported to vary between 4.1 and 5 kb (Martelli, 1981; Francki et al., 1985). More recent determinations indicated that the genomic RNA of all members of the group, including MPV, NRV, and GALV, is a molecule with an apparent length of 4.7 kb (Gallitelli et al., 1985; D. Gallitelli and M. Russo, unpublished information), a value very close to that determined for TBSV RNA (4.8 kb) by Hayes et al. (1984) and Morris et al. (1987). Thus, the consensus is that the genome of Tombusviruses is significantly larger than that of viruses with similar properties that have been grouped together with carnation mottle virus (CarMV) (Chapter 3, this volume).

Molar base ratio of RNA has been determined for four members of the group: (1) TBSV-BS3, G28.6; A26.3; C21.2; U26.3; (2) AMCV, G28.6;

A23.5; C21.1; U26.8; (3) CyRSV, G27.8; A24.9; C21.3; U26.1; and (4) PAMV, G28.2; A26.8; C21.5; U23.4 (Martelli, 1981).

No proteinase K-sensitive structures needed for infectivity are present at the 5' end of TBSV-type and CyRSV RNA (Mayo et al., 1982; Burgyan et al., 1986).

Encapsidated RNA of TBSV-type, CIRV, and CyRSV appeared to be deprived of poly(A) sequences (Gallitelli et al., 1985), but indirect evidence was obtained that about 1% of TBSV-type genomic RNA extracted from infected plants contains poly(A) sequences (Gallitelli and Hull, 1985a).

Nucleic acid hybridization studies using cDNAs prepared by the random primer method to genomic RNAs of Tombusviruses, have demonstrated that all members of the group, including MPV, NRV, and GALV, have nucleotide sequences in common with TBSV-type and with one another (Gallitelli et al., 1985; D. Gallitelli and M. Russo, unpublished information). The percentage of sequence homology varies with different viruses (Gallitelli et al., 1985; see also Table II). No homology, however, has been detected between TBSV-type (and some other Tombusviruses) and RNAs of some CarMV-like viruses (Gallitelli et al., 1985; Chapter 3, this volume). This may be taken as an indication that molecular hybridization techniques using RNAs immobilized on solid supports (Northern blot or dot blot) could prove useful for distinguishing Tombusviruses from other isometric viruses with monopartite genomes.

Incidentally, AMCV RNA sequences were successfully detected in naturally infected artichoke plants using the spot hybridization method by Maule et al. (1983) (D. Gallitelli, unpublished information).

B. Expression of the Genome

The existence of less than genomic-size virus-specific RNAs has been reported for most Tombusviruses (Gallitelli et al., 1985; Gallitelli and Hull, 1985b). These RNAs show a great variability in length, but some are consistently present either in purified virus preparations or in total RNA extracts from infected tissues. TBSV genomic RNA-specific bands, approximately 2.1, 1.8, and 1.2 kb in size, were found associated with polyribosomes extracted from infected plants (Hayes et al., 1984; Gallitelli and Hull, 1985b). CyRSV genomic RNA-specific bands with mobilities similar to those reported above were found in infected cowpea protoplasts (Russo and Gallitelli, 1985). In a more recent study, besides the full-length genomic RNA, only two major RNA species were found in RNA preparations from CyRSV-infected tissues. By using clones made in different regions of cloned complementary DNA, the two possible subgenomic RNA species were shown to reside in the 3' half of the viral genome (M. Russo and J. C. Carrington, unpublished information).

A greater number of possible subgenomic RNAs have been identified as dsRNAs in TBSV-infected tissues (Morris, 1983; Hayes *et al.*, 1984; Hillman *et al.*, 1985).

In vitro translation studies have been made with CyRSV (Burgyan *et al.*, 1986). In rabbit reticulocyte lysates programmed with unfractionated RNA, four major polypeptides were produced with an apparent size of 43,000 (43K), 40,000 (40K), 34,000 (34K), and 22,000 (22K). A similar pattern of *in vitro* products was observed in preliminary experiments with AMCV, TVN, and PAMV (M. Russo and D. Gallitelli, unpublished information). In cowpea protoplasts infected with CyRSV RNA, only the 43K and 22K products were found.

The 43K polypeptide was identified as the virus coat protein. Peptide mapping showed that the 34K protein is partially contained in the 40K product. Translation experiments with size-fractionated RNAs indicated that the translation strategy of CyRSV RNA uses both genomic-size and subgenomic species as templates, the 40K protein being the product coded directly by the genomic RNA (Burgyan *et al.*, 1986).

The lack of the complete sequence of CyRSV RNA does not allow one to give its genome map. However, because of the messenger activity of genomic RNA, it can be postulated that the initiation codon for the synthesis of the 40K polypeptide occurs near the 5' end. Peptide mapping and the absence of the 34K product in lysates programmed with genomic RNA only, indicate that this protein is partially contained in the 40K and is expressed via a subgenomic species. Another subgenomic RNA codes for the 22K protein. By analogy with other monopartite RNA viruses, the gene for coat protein would be expected to map near the 3' end and be expressed via a third subgenomic species about 2 kb in length. However, Hillman and Morris (1986a) have shown that in TBSV genomic RNA the coat protein coding region terminates about 1000 nucleotides from the 3' end, and this distal sequence contains several possible open reading frames, one of which might encode a polypeptide of 112 amino acids. At the extreme 3' terminus there is a region sharing extensive sequence homology with two different symptom-modulating RNA species with properties of defective interfering RNAs (Hillman and Morris, 1986b).

In conclusion, the results of translation and sequencing studies carried out so far, although not thoroughly conclusive, indicate that Tombusviruses differ from other monopartite RNA viruses as they lack a readthrough mechanism and do not have the coat protein coding region near the 3' end.

C. Viral Replication

The events taking place during replication of Tombusviruses have been little investigated. With CyRSV, it was shown that poly-*L*-ornithine is not essential for infecting cowpea protoplasts, probably because the virus itself possesses a net positive charge even at moderately alkaline pH (7.6) while cowpea protoplasts have a relatively small negative charge (Russo and Gallitelli, 1985). These authors also showed that 70S ribosomes are not involved in the synthesis of virus protein, and that host transcription is required both in the early phase of the infective process and at a later stage of virus multiplication. Cowpea protoplasts support CyRSV multiplication to an extent comparable to that of host plant tissues. Both virus particles and genomic RNA are detected as early as 12 hr after inoculation, together with putative subgenomic RNAs (2.1, 1.8, and 1.2 kb) and the satellite RNA (0.7 kb) (Russo and Gallitelli, 1985).

A different approach has been used to study the replicative strategy of TBSV, through the analysis of dsRNAs extracted from infected plants. In these studies (Morris, 1983; Hayes *et al.*, 1984; Hillman and Morris, 1986a) many dsRNA species were detected in infected plant extracts, some of them having a size compatible with that of genomic RNA and at least three possible subgenomic RNA species. It is worth mentioning that Hillman *et al.* (1985) found a wide range of dsRNA species with a lower size (0.2–0.4 kb) depending on the conditions under which host plants were grown and on whether or not a set of low-molecular-weight RNAs was present.

VIII. RELATIONS WITH CELLS AND TISSUES

A. General Cytology of Infected Cells

1. Localized Infections

As discussed in Section II, Tombusvirus infections remain localized in the majority of the experimental hosts because of the plant's hypersensitive response. In these cases, viruses are confined within necrotic lesions developing upon inoculation.

Appiano and D'Agostino (1980) investigated the cytopathological events taking place in epidermal cells of *G. globosa* from a few minutes up to 4 hr after mechanical inoculation with TBSV-BS3. Although the mechanisms underlying virus entry into cells were not elucidated, interesting observations were made on the effect of rubbing on the cell structure. Severely injured cells that exhibited large breaches in the cell walls were extensively plasmolyzed 15–30 min after inoculation and turned necrotic within 1 hr. Cells with minor injuries were not killed, and reacted with production of callose deposits in the plasmalemma–cell wall interface underneath the wound. Callose deposits were detected 2 and 4 hr after inoculation by light and electron microscopy, respectively. These healing cells were regarded as possible infection sites (Appiano and D'Agostino, 1980).

TBSV and CyRSV, like all known Tombusviruses, induce lesions in *G. globosa*, which become visible within 40 hr from inoculation. These lesions have a necrotic center, surrounded by darkened tissues with cells in various stages of necrosis and a translucent yellowish halo, which becomes progressively and increasingly reddish due to betacyanin formation, from the fourth to the fifth day after inoculation. By the fifth to the sixth day, the lesions cease to grow, attaining a maximum diameter of approximately 2 mm (Pennazio *et al.*, 1978; Russo and Martelli, 1981).

Changes occurring in inoculated *G. globosa* leaves in advance of the appearance of TBSV-BS3-induced local lesions were studied by Pennazio *et al.* (1979). The ultrastructural aspects of local lesion formation in the same host were investigated by Appiano and coworkers (Appiano *et al.*, 1977, 1981; Pennazio *et al.*, 1978) with TBSV-BS3, and by Russo and Martelli (1981) with CyRSV. Both viruses induced amazingly similar alterations. The time sequence of the development of cytological modifications was the same, as well as their type and distribution in cells of different areas of the lesion.

The center of local lesions was made up of collapsed necrotic cells, filled with starch grains and containing many virus particles. In the inner part of the halo, surrounding the central zone, necrotic and necrosing cells were interspersed with apparently active cells. These had a much deranged cytology and exhibited cell walls heavily thickened by callose and suberin depositions (Fig. 6) and intercellular spaces filled with lignin and suberin.

The cells of the central part of the halo were severely altered. The ground cytoplasm contained many virus particles, altered organelles (especially mitochondria and chloroplasts), and the membranous inclusions known as multivesicular bodies (Fig. 7), whose nature and significance will be discussed later in this section.

The cytology of cells of the outer part of the halo was little affected and cell walls were also normal. Only chloroplasts were swollen and misshapen because of the massive accumulation of starch grains. Virus particles were not seen in these cells. Likewise, no cytological alterations were detected in the cells of green tissues bordering the lesions. Plasmodesmata were not plugged by callose deposits, nor were other barrier substances, like suberin and lignin, seen as in the innermost part of the lesion.

As shown by Pennazio and Redolfi (1977), tissues surrounding TBSV-BS3-induced lesions possess a high level of localized acquired resistance, which decreases progressively with increasing distance from the lesion. Thus, restriction of virus movement may be ascribed to physiological rather than structural mechanisms (Pennazio *et al.*, 1978; Appiano *et al.*, 1981; Russo and Martelli, 1981).

The factor responsible for TBSV-induced hypersensitive reaction is not known. A flavanoid accumulating within and around the lesions in a specific response to viral infection was suggested to play a role in the

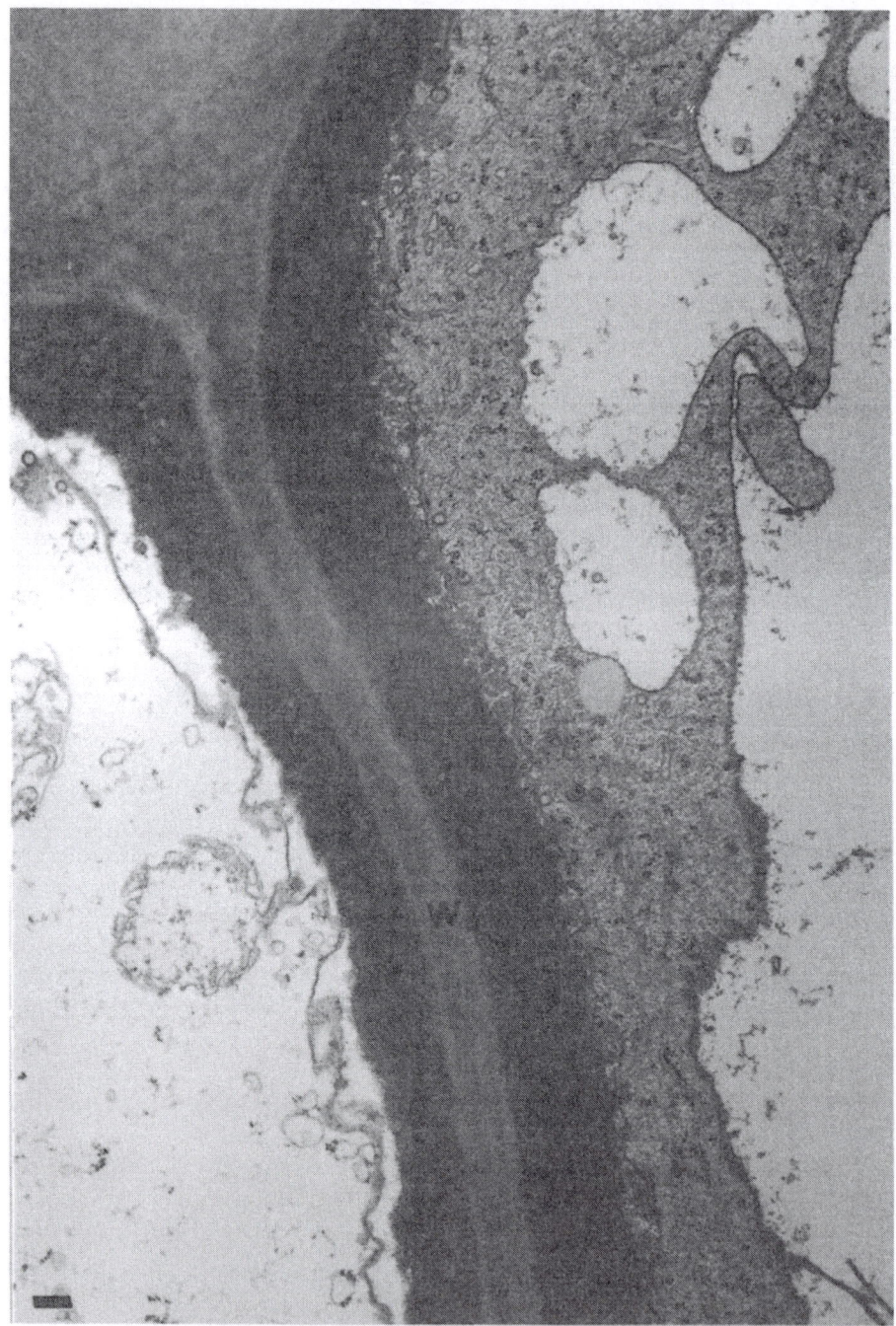

FIGURE 6. Adjacent cells from a CyRSV-induced local lesion in *G. globosa* whose cell walls (W) are heavily thickened because of the apposition of densely staining material, possibly suberin. Scale bar, 200 nm.

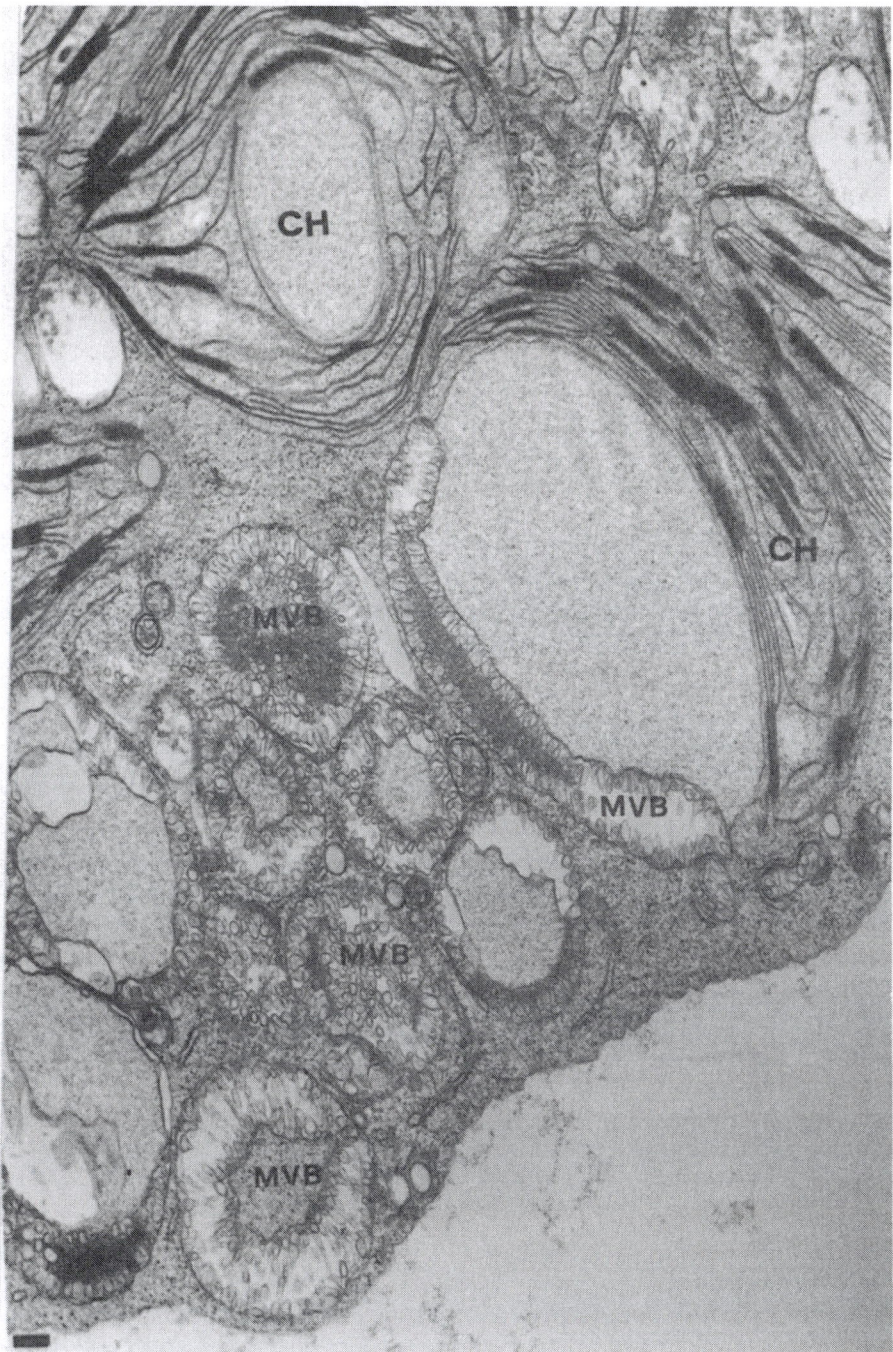

FIGURE 7. A mesophyll cell of *G. globosa* infected with TBSV-type. The cytoplasm contains altered chloroplasts (CH) and many multivesicular bodies (MVB). Scale bar, 200 nm.

localization process (Redolfi *et al.*, 1978), but this idea was not substantiated by subsequent work (Pennazio *et al.*, 1979).

2. Systemic Infections

All definitive members of the Tombusvirus group have been studied ultrastructurally in systemically infected artificial hosts. AMCV and PLCV were also investigated in naturally infected artichoke (Russo *et al.*, 1968) and pelargonium (Rubio-Huertos and Garcia-Hidalgo, 1971; Martelli and Russo, 1972a) plants and CyRSV in protoplasts of *C. quinoa* and cowpea (De Varennes *et al.*, 1984; Russo and Gallitelli, 1985).

The cytopathology of individual cells of systemically infected tissues does not differ from that of cells in local lesions. A variety of abnormalities of the cell constituents have been observed, whose type and intensity depend on the virus, the host, and, perhaps, on whether or not satellite RNAs are present in the inoculum.

Ordinarily, the general architecture of infected cells is deranged to some extent, the ground cytoplasm being intermingled with virus particles, which may occur in massive accumulations, organelles in various stages of degeneration, and different types of inclusion bodies.

Among major organelles, nuclei seem to be the least affected by Tombusvirus infections. The nuclei of PLCV-infected pelargonium plants had enlarged nucleoli, little chromatic material, and a conspicuously vesiculated nucleoplasm (Martelli and Castellano, 1969; Martelli and Russo, 1972a). Likewise, the nuclei of *C. quinoa* infected with AMCV exhibited prominent inclusions made up of bundles of membranous cisternae (Martelli and Russo, 1973). PLCV and AMCV, however, did not alter the nuclei of any of the other hosts in which they were studied (e.g., *G. globosa*, *N. clevelandii*, *D. stramonium*). Likewise, none of the remaining Tombusviruses was found to elicit nuclear modifications of any consequence (Appiano *et al.*, 1978; Martelli and Russo, 1981; Francki *et al.*, 1985; A. Di Franco, M. Russo, and G.P. Martelli, unpublished information).

In most virus-host combinations, the chloroplasts of infected cells show mild to severe signs of degeneration such as swelling, vesiculation, bending, abnormalities of the lamellar system, disruption of the membranes, and fragmentation of the body. A detailed account of the modifications induced by TBSV-BS3 in *D. stramonium* chloroplasts was recently given by Bassi *et al.* (1985). In *D. stramonium*, virus infection causes a metabolic derangement of the membranes, which gives rise to an impressive variety of modifications of the thylakoids and formation of prolamellarlike bodies. Severe modifications of the chloroplast's structure were also reported in AMCV-infected *C. quinoa* (Martelli and Russo, 1973), locally and systemically TBSV-infected *G. globosa* (Appiano *et al.*, 1977, 1978, 1981; Martelli *et al.*, 1984), and CyRSV-infected *G. globosa*, *N. benthamiana*, and *N. clevelandii* (Martelli and Russo, 1981).

In pelargonium plants naturally infected with PLCV, chloroplasts exhibited a budding activity (Martelli and Russo, 1972a), which was not seen in hosts artificially infected with the same virus nor in any other Tombusvirus-host system investigated so far (A. Di Franco, M. Russo, and G.P. Martelli, unpublished information).

In *G. globosa* inoculated with TBSV-BS3, localized swelling and separation of thylakoids are the very early signs of infection that becomes visible 24 hr after inoculation, i.e., well in advance of the appearance of the lesions (Appiano *et al.*, 1981). With time, alterations of the chloroplasts become progressively severe, leading to profound modifications of the shape and internal organization and to clumping of the organelles (Appiano *et al.*, 1981; Martelli *et al.*, 1984).

Mitochondria can also be affected in various ways by Tombusviruses. Besides their transformation in multivesicular bodies, as discussed later in this section, the alterations most commonly observed consist of the dilation and reduction of the number of cristae, increased electron opacity of the matrix, and presence of internal inclusions such as the convoluted myelinlike profiles seen in TBSV-infected *D. stramonium* (Russo and Martelli, 1972) or the clumps of amorphous electron-dense material resembling protein present in plants infected with several members of the group (Fig. 9C) (A. Di Franco, M. Russo, and G.P. Martelli, unpublished information).

As to dictyosomes, increased vesiculating activity was observed in cells of *D. stramonium* and *N. clevelandii* infected with TBSV-BS3 and CyRSV, respectively (Russo and Martelli, 1972; Martelli and Russo, 1981).

Modifications of the cell wall-plasmalemma interface, consisting of localized wall thickenings, detachments of plasmalemma from the cell wall that contain paramural bodies, and callose deposits, were observed in systemic infections by CyRSV (Martelli and Russo, 1981) and, occasionally, in other members of the group (A. Di Franco, M. Russo, and G. P. Martelli, unpublished information).

B. Appearance and Intracellular Distribution of Virus Particles

Cells infected by Tombusviruses contain variable amounts of virus particles. These appear as electron-dense rounded bodies with a smooth and regular outline, sometimes with a hexagonal contour and a diameter of approximately 26 nm. Although RNase treatment has been used for clearing tissues of ribosomes (Hatta and Francki, 1981), no need for such treatment seems necessary since virus particles stand up clearly and can be readily identified even when they occur in very low concentrations and in a scattered fashion.

All tissues and cell types can be invaded by Tombusviruses and possibly support their multiplication. For instance, AMCV particles were observed in leaf-hair cells, epidermis, mesophyll cells, and vascular bun-

dles, including differentiating and mature tracheary elements (Russo
et al., 1967, 1968). Likewise, PLCV was found in parenchymatous and
conducting tissues of infected leaves (Martelli and Russo, 1972a),
and TBSV-BS3 was observed in leaf veins and foliar parenchymas
(Martelli and Russo, 1972b) as well as in most cell types of G. globosa
roots, including mature xylem elements, cortex, and epidermis (Appiano
and D'Agostino, 1985).

In infected cells, virus particles occur in the cytoplasm and the central
vacuole. Cytoplasmic virions may be either scattered at random or ap-
pressed in a disorderly fashion into more or less compact accumulations,
or regularly arranged in a crystalline lattice.

With most members of the group, virus particles also accumulate
within extensions of the tonoplast that give rise to bubblelike vesicles
protruding into the vacuole (Fig. 8) (Russo et al., 1968; Martelli and Russo,
1972a, 1981; Russo and Martelli, 1972; Makkouk et al., 1981; Koenig and
Lesemann, 1985). These protuberances, as originally suggested by Russo
et al. (1968), become detached from the tonoplast and move into the
vacuole where virions are released.

Cell organelles such as nuclei, chloroplasts, and mitochondria are
additional sites of virus accumulation. For instance, particles of TBSV-
BS3, AMCV, PLCV, PAMV and CIRV were seen in the nucleoplasm (Fig.
9A), the nucleolus, or both, of nuclei of several hosts (Martelli and Castellano, 1969; Martelli and Russo, 1972a, 1972b; Russo and Martelli,
1972). However, virus-containing nuclei did not always suffer obvious
structural modifications even when particles were plentiful (Russo and
Martelli, 1972).

Small groups of scattered virions were observed in the stroma of
mitochondria of different hosts infected with TBSV-type, TBSV-BS3, EMCV,
CIRV, TVM, GALV, and CyRSV (Hatta and Francki, 1981; Martelli and
Russo, 1981; Russo and Martelli, 1981; Koenig and Lesemann, 1985; A.
Di Franco, M. Russo, and G. P. Martelli, unpublished information). The
fine structure of these mitochondria was relatively well preserved and
their outer envelope was apparently intact (Fig.9D). These features and
the extremely frequent and consistent association of virus particles of
some members of the group (e.g., CyRSV) with intact mitochondria, has
led to suggestion that intramitochondrial synthesis or assembly of virions
could take place (Martelli and Russo, 1981).

The association of Tombusvirus particles with chloroplasts is much
less frequent and, in most cases, seems to be accidental. With the exception of PLCV, whose particles have been seen inside apparently intact
pelargonium chloroplasts (Fig. 9B) (Martelli and Russo, 1972a), there are
reasons to believe that the intraplastidial presence of AMCV and EMCV
derives from entry of whole particles through a broken bounding membrane (Martelli and Russo, 1973; Makkouk et al., 1981; A. Di Franco,
personal communication).

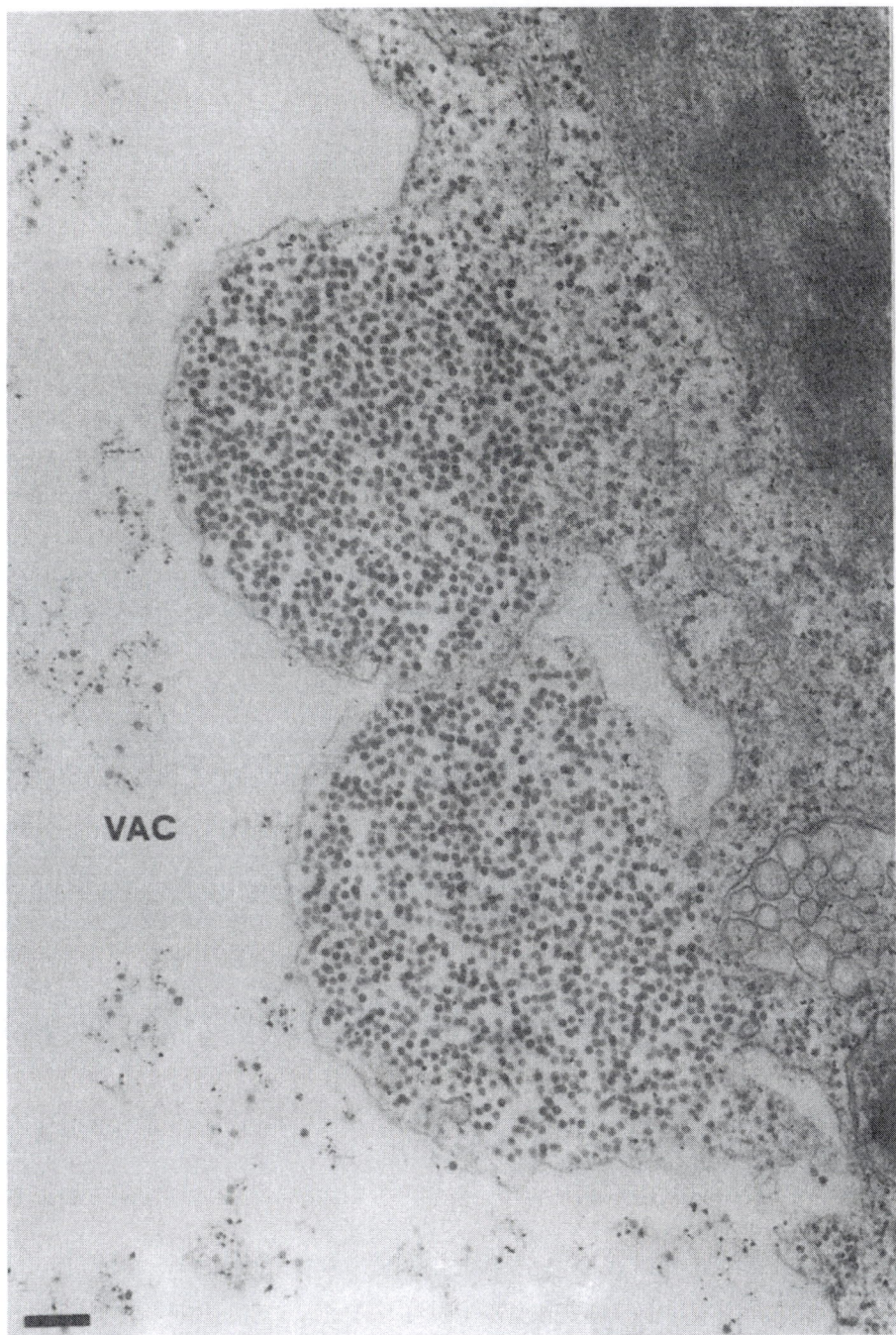

FIGURE 8. Two bubblelike extensions of the tonoplast filled with virus particles, protruding into the central vacuole (VAC) of an AMCV-infected cell. Scale bar, 200 nm.

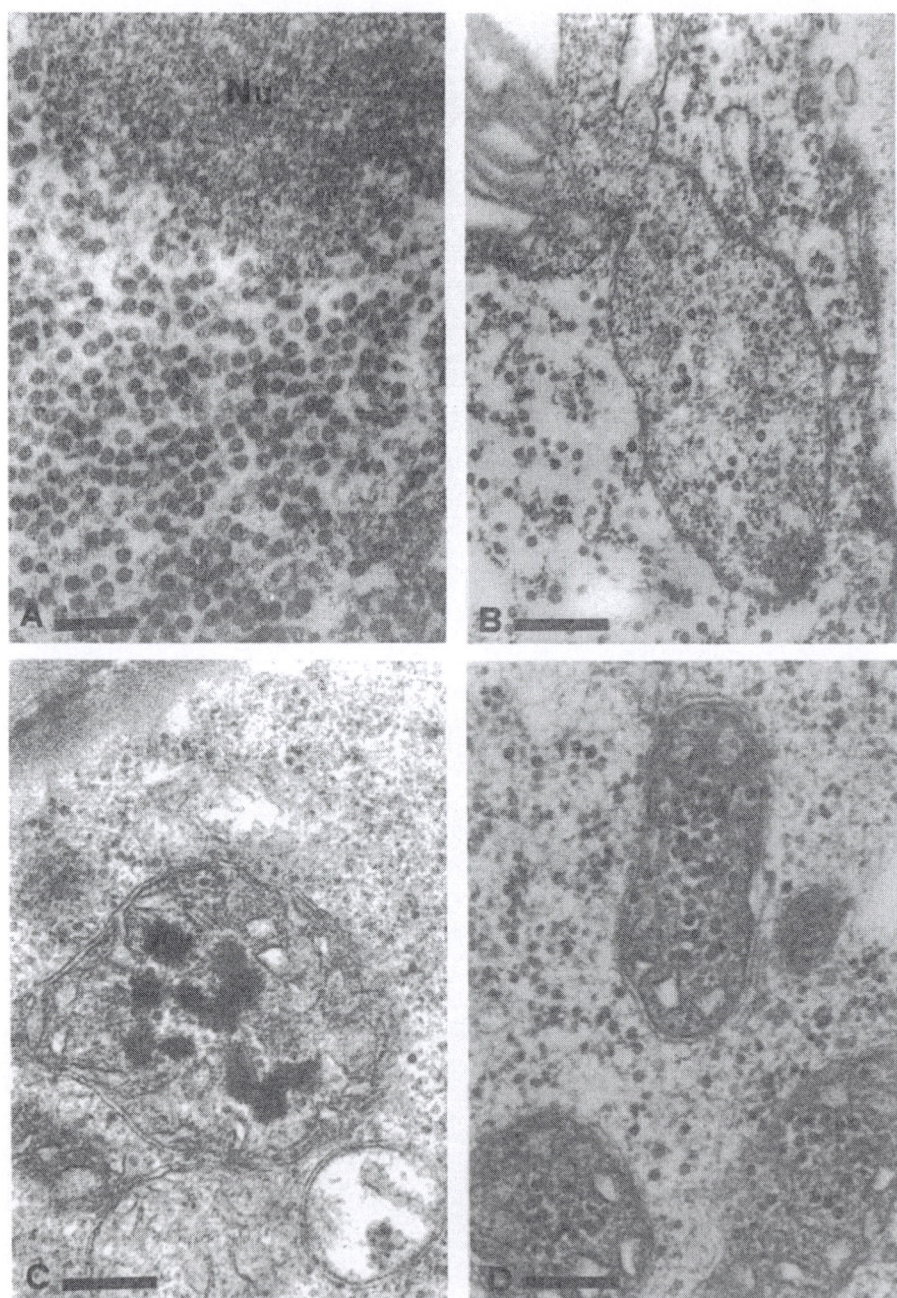

FIGURE 9. (A) Intranuclear TBSV-BS3 particles accumulating next to the nucleolus (Nu). Scale bar, 100 nm. (B) PLCV particles in a plastidial bud. Scale bar, 200 nm. (C) Intramitochondrial patches of electron-dense amorphous material in a CyRSV-infected cell. Scale bar, 200 nm. (D) Virus particles within seemingly intact mitochondria in a CyRSV-infected cell. Scale bar, 200 nm.

C. Inclusion Bodies

Cytoplasmic and nuclear inclusions staining reddish with Azure A were observed with the light microscope in TBSV-infected cells (Christie and Edwardson, 1977). If these inclusions, as stated by these authors, correspond to the large viral aggregates seen in thin sections, they are likely to be elicited also by all other members of the group. This, however, and the diagnostic usefulness of the inclusions, has not been ascertained experimentally.

In thin sections, three types of cytoplasmic inclusions have been recognized: i.e., crystalline bodies, patches of amorphous proteinaceous material, and multivesicular bodies.

Virus crystals exhibiting an apparently cubic close-packed structure are frequent in hosts infected with AMCV, PLCV, and, to a lesser extent, PAMV and TBSV-BS3 (Russo et al., 1968; Martelli and Russo, 1972a; Russo and Martelli, 1972; Appiano et al., 1978; Christie and Edwardson, 1977). Crystalline inclusions can be located in the cytoplasm (Fig. 10) or the vacuole, have a compact solid structure with few or no gaps in the lattice, and may attain a very large size. As discussed by Martelli and Russo (1984), the presence of virus crystals in Tombusvirus infections is limited and too inconsistent to serve as a diagnostic marker at the group level.

The small patches of electron-dense amorphous material originally observed in the cytoplasm of G. globosa cells infected with TBSV-BS3 (Appiano et al., 1977), were also recorded in CyRSV infections (Martelli and Russo, 1981) and, more recently, in infections by TBSV-type (Fig. 11A) AMCV, CIRV, and PAMV (A. Di Franco, M. Russo, and G. P. Martelli, unpublished information).

With TBSV-BS3, this material was digested by pronase (Appiano et al., 1983) and was specifically tagged by colloidal gold (Fig. 11B) in immunolabeling tests with homologous antibodies (Appiano et al., 1985). Hence, it was identified as virus coat protein.

Multivesicular bodies (MVB) are membranous cytoplasmic inclusions that are invariably associated with Tombusvirus infections regardless of the virus, the host, or the tissue examined. MVB can therefore be regarded as prominent and specific ultrastructural features that are useful for the recognition of group members.

There is no doubt that MVB are virus-induced structures. They are absent from tissues of healthy plants used as controls in all ultrastructural studies carried out so far in different laboratories; were not seen in healthy G. globosa leaves with necrotic local lesions elicited by senescence but resembling ultrastructurally TBSV lesions (D'Agostino and Pennazio, 1981); and were formed in protoplasts of two hosts infected with CyRSV nucleoproteins or isolated CyRSV RNA (De Varennes et al., 1984; Russo and Gallitelli, 1985).

The early suggestion that MVB could develop from a rearrangement

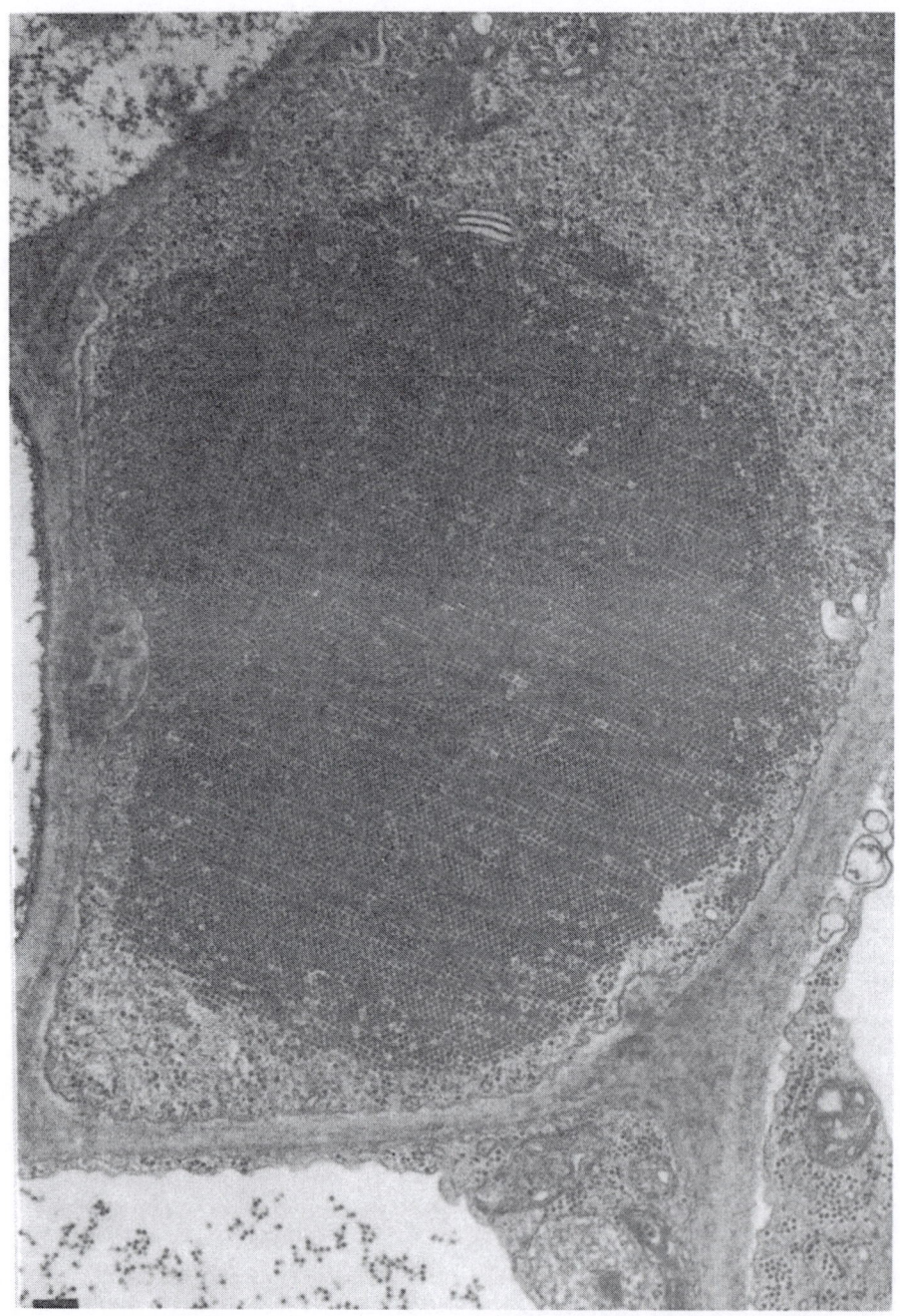

FIGURE 10. A large intracytoplasmic crystal in an AMCV-infected cell. Scale bar, 200 nm.

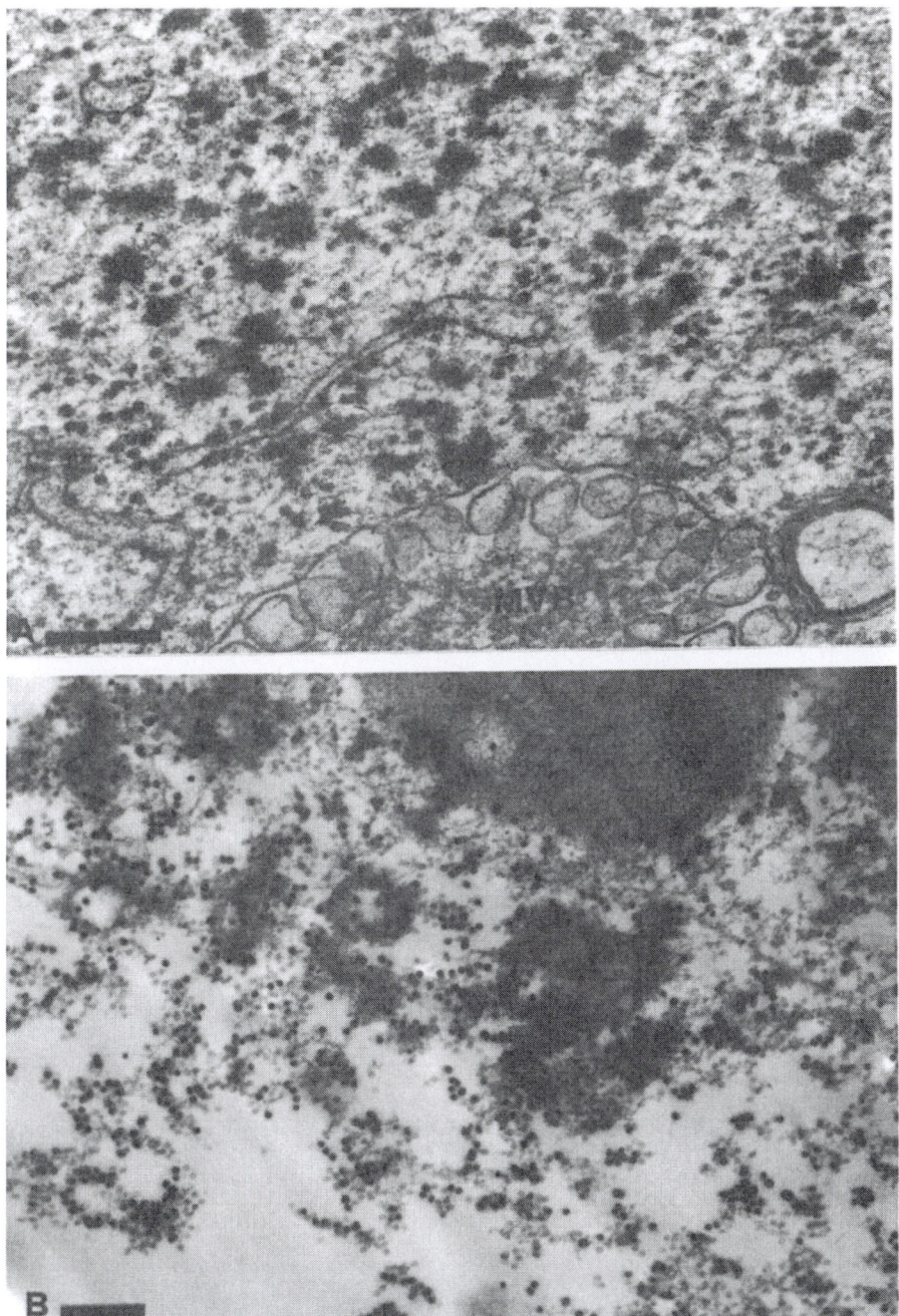

FIGURE 11. (A) Patches of electron-dense amorphous material intermingled with virus particles in the cytoplasm of a cell infected with TBSV-BS3. MVB, multivesicular body. Scale bar, 200 nm. (B) Amorphous material as above and groups of virus particles specifically tagged by colloidal gold in immunolabeling tests. Scale bar, 200 nm. Courtesy of Dr. A. Appiano.

into pseudo-organellar form of normal cell constituents like endoplasmic reticulum strands and dictyosomal vesicles (Russo and Martelli, 1972), or from modifications of chloroplasts (Appiano et al., 1978), do not hold true in the light of more recent investigations.

Cytochemical studies providing evidence that MVB contain glycolate oxidase (Fig. 12, inset), demonstrated that these structures derive from peroxisomes. This was experimentally shown with CyRSV (Russo et al., 1983), TBSV-type and TBSV-BS3 (Martelli et al., 1984), and AMCV and EMCV (M. Russo, unpublished information). Furthermore, peroxidase activity was evidenced in the matrix of MVB in TBSV BS3-infected tissues (A. Appiano, personal communication).

In infected cells, peroxisomes undergo a progressive vesiculation of the bounding membrane through the possible addition of membranous material from the endoplasmic reticulum. Modified peroxisomes become very plastic, engulfing portions of the ground cytoplasm through the invagination of the limiting membrane or through the production of membranous appendages that fold back on the main bodies (Russo et al., 1983; Koenig and Lesemann, 1985). These modifications lead to a novel structure as large as a small plastid, made up of three major components: (1) a peripheral irregularly thickened membrane, sometimes obviously connected with endoplasmic reticulum strands; (2) an electron-dense finely granular or crystalline matrix; and (3) many globose to ovoid vesicles measuring 80–150 nm in diameter, which contain finely fibrillar material (Fig. 12). A recent study of cells of five hosts infected locally (C. quinoa, G. globosa, D. stramonium) or systemically (N. clevelandii, N. benthamiana) with CIRV, a little-investigated member of the group, has shown that with this virus, MVB originate consistently from mitochondria rather than from peroxisomes. Visual and cytochemical evidence (i.e., identification of cytochrome oxidase in developing MVB) supported this finding (Di Franco et al., 1984). As with peroxisomes, mitochondria undergo a progressive peripheral vesiculation (Fig. 13 and insets) leading to a transformation of the organelles into disorderly aggregates of vesicles intermingled with remnants of shredded stroma.

The results of a comparative unpublished study made in our laboratory in which all members of the group were investigated in three hosts (G. globosa, N. clevelandii, and C. quinoa) have shown that the origin of MVB can be influenced by the host. In fact, in C. quinoa cells infected with TBSV-type and TBSV-BS3, MVB developed either from peroxisomes or mitochondria. A similar situation occurs, to a much lesser extent, in GALV infections (Table IV).

An additional source of variation may reside in the presence of satellite RNAs in the inoculum. Preliminary observations relative to C. quinoa infected with CIRV indicate that when virus cultures also contain satellite RNA, MVB originate from mitochondria and peroxisomes rather than from mitochondria alone as when the inoculum is satellite-free.

Peripheral vesiculation of some chloroplasts was observed in cells

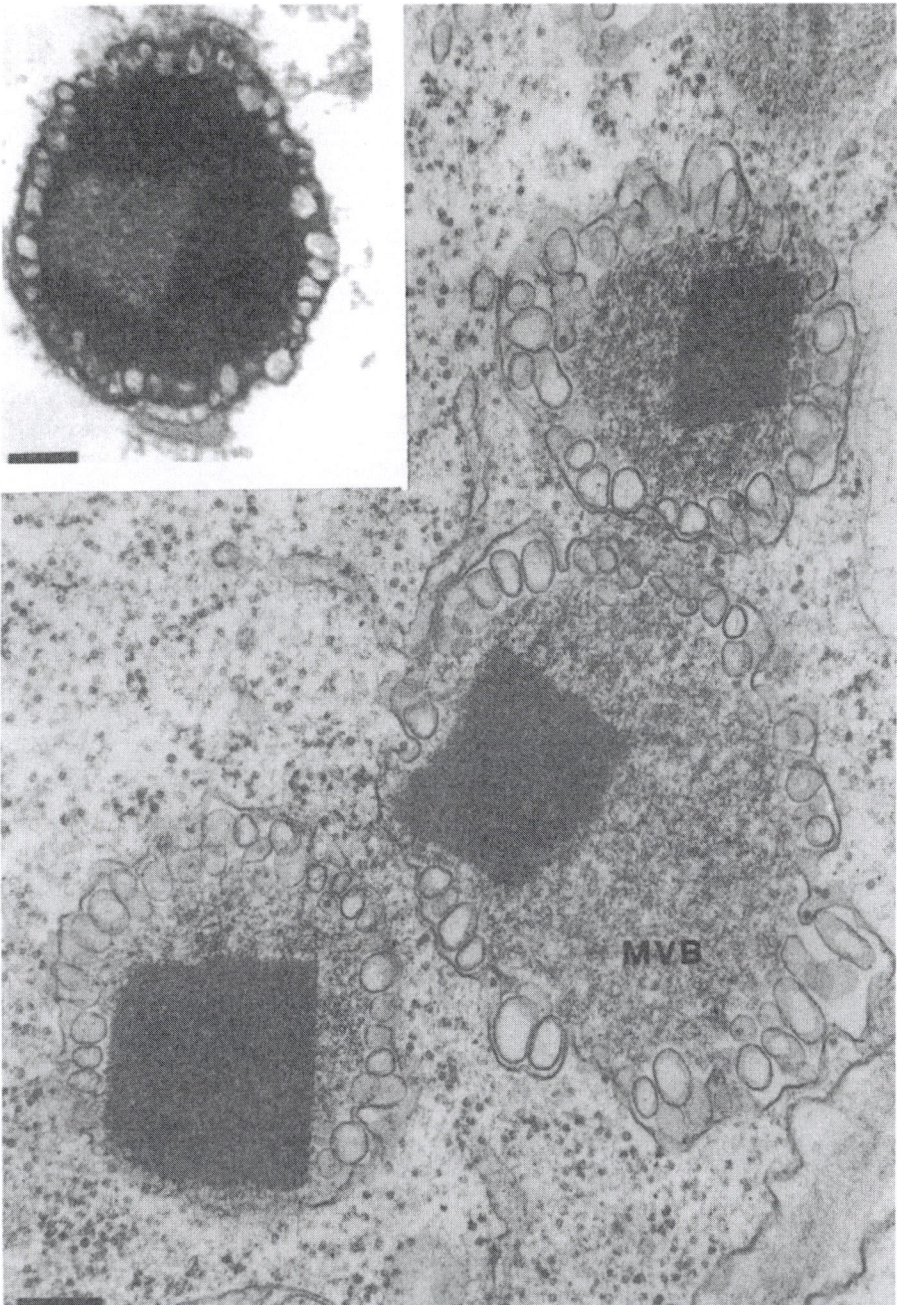

FIGURE 12. Three multivesicular bodies (MVB) obviously originating from peroxisomes in a cell infected with TBSV-BS3. Inset shows an MVB with positive reaction for glycolate oxidase in the matrix but not in the crystalline core or peripheral vesicles. Scale bars, 200 nm.

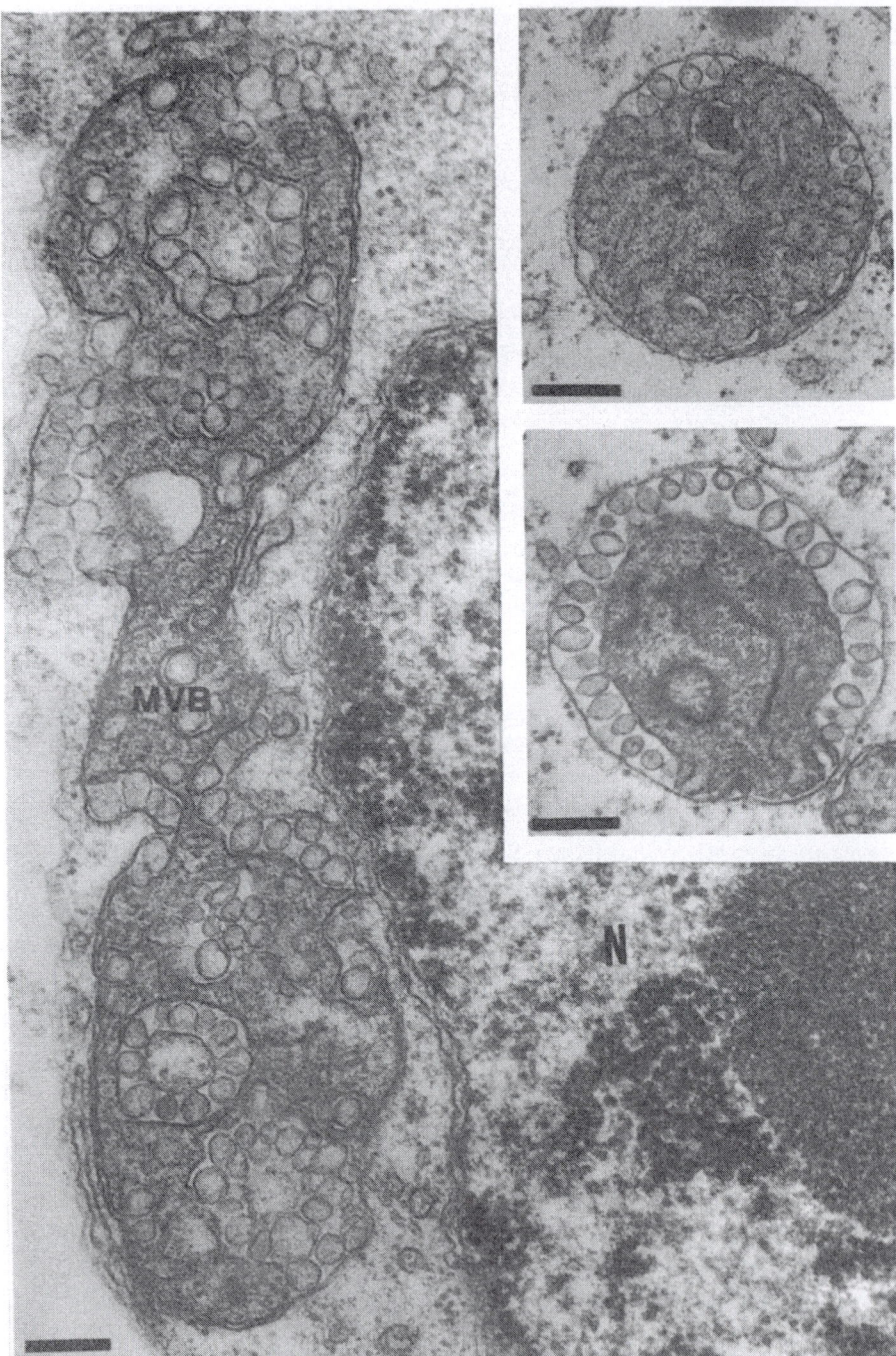

FIGURE 13. A large MVB derived from a modified mitochondrion in a CIRV-infected cell. Insets show mitochondria in intermediate stages of transformation into MVB. N, nucleus. Scale bars, 200 nm.

TABLE IV. Origin of Multivesicular Bodies in Cells of Different Hosts Infected
by Tombusviruses[a]

Virus	Nicotiana clevelandii		Gomphrena globosa		Chenopodium quinoa	
	Peroxisomes	Mitochondria	Peroxisomes	Mitochondria	Peroxisomes	Mitochondria
TBSV-type	+	−	+	−	+	+
TBSV-BS3	+	−	+	−	+	+
AMCV	+	−	+	−	+	−
EMCV[b]	+	−	+	−	+	−
CIRV	−	+	−	+	−	+
CyRSV	+	−	+	−	+	−
GALV[c]	+	−	+	−	+	±
MPV	+	−	+	−	+	−
TVN[c]	+	−	+	−	+	±
PAMV	+	−	+	−	+	−
PLCV	+	−	+	−	+	−

[a] Based on unpublished comparative studies by A. Di Franco, M. Russo, and G. P. Martelli.
[b] Chloroplasts of C. quinoa exhibit peripheral vesiculation.
[c] Chloroplasts and, occasionally, mitochondria of C. quinoa exhibit peripheral vesiculation.

infected with EMCV and GALV. Rows of vesicles were formed between
the outer and inner lamella of the plastid envelope (Fig. 14), but in no
case did the vesiculating activity progress to the point of transforming
chloroplasts into veritable MVB. As mentioned previously, chloroplasts
were suspected to participate in the genesis of MVB (Appiano et al., 1978;
Appiano and Pennazio, 1982), primarily because of the close connection
detected between these organelles and MVB in TBSV-type and TBSV-BS3
infections. In both instances, chloroplast abnormalities consisting of
clumping, peripheral vacuolization, crescentlike bending, and clustering
around MVB were seen, but no direct correlation between chloroplasts
and MVB could be established (Martelli et al., 1984). The peroxisomal
origin of TBSV-induced MVB was demonstrated by cytochemical reactions whereby the presence of glycolate oxidase in MVB was ascertained
(Martelli et al., 1984), as well as a difference in the composition of the
membranes of MVB and chloroplasts (Appiano et al., 1983). Further support of the nonplastidial nature of TBSV-induced MVB was provided by
Appiano and D'Agostino (1985), who recorded the presence of MVB in
root tissues where no chloroplasts occur.

As to the function of MVB in the economy of virus multiplication,
there is mounting evidence that these structures may be directly implicated in the replicative process. Double-stranded RNA supposed to be of
viral origin was detected in the vesicles of MVB of cells infected with
CyRSV and CIRV (Russo et al., 1983; Di Franco et al., 1984). In addition,
MVB of TBSV-BS3-infected cells were shown to incorporate [³H]uridine
(Appiano et al., 1983). These findings and the notion that MVB appear in
inoculated tissues before progeny virus is detected (Appiano et al., 1981),

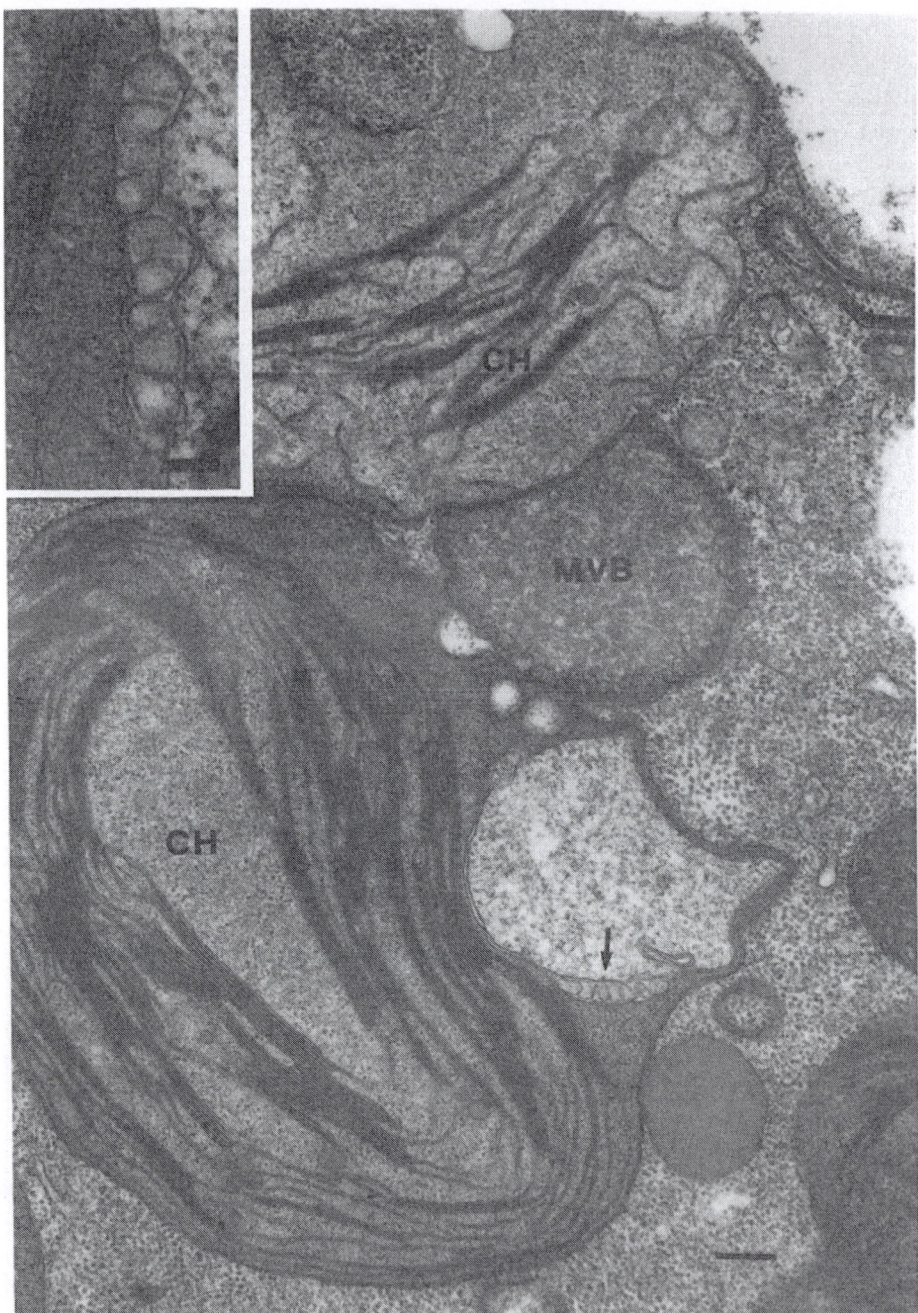

FIGURE 14. A GALV-infected cell showing a peroxisome-derived MVB and a deranged chloroplast (CH) with peripheral vesiculation. (8mm = 200 nm.) Inset shows a close-up view of a row of membranous vesicles between the lamellae of the plastidial envelope. (7mm = 100 nm.)

strongly support the likelihood that MVB are the site of viral RNA replication.

Inclusions seemingly identical to MVB are induced by three viruses (GaMV, TCV, and GMoV) that are close to Tombusviruses in many respects but are not regarded as members of the group (Chapter 3, this volume). MVB of GaMV and TCV originate from mitochondria (Russo and Martelli, 1982; Hatta *et al.*, 1983), and those of GMoV from peroxisomes (Behncken and Dale, 1984). It is therefore evident that, contrary to previous views (Martelli and Russo, 1984), MVB should no longer be regarded as ultrastructural features strictly specific to the Tombusvirus group.

IX. SATELLITES

Experimental evidence that satellite RNAs (S-RNA) are associated with Tombusviruses was provided by Gallitelli and Hull (1985b), who reported the occurrence of RNA species with properties of true S-RNA and determined some of their characteristics.

S-RNAs are noninfectious, probably linear molecules with an apparent size of 0.7 kb, which lack appreciable sequence homology with genomic RNAs of helper viruses, and need the assistance of helper virus RNA for their replication but are not necessary for the replication of viral RNA (Gallitelli and Hull, 1985b).

The amount of S-RNA in preparations of different Tombusviruses is variable and may depend on the helper virus. For example, in unpublished recent studies in our laboratory, it was observed that AMCV S-RNA reaches very high concentrations after serial passages through *N. benthamiana*, whereas CyRSV S-RNA behaves in the opposite way, its concentration decreasing rapidly after a few passages through the same host. In this respect, CyRSV S-RNA seem to differ from S-RNA of other members of the group.

It has not yet been ascertained whether S-RNA is packaged in viral capsids by itself or together with less than genomic-size RNAs. Centrifuging to equilibrium in CsCl CyRSV preparation with or without S-RNA, showed that both particle populations were isodense, thus suggesting that S-RNA is not packaged with genomic RNA. This possibility is also supported by the fact that the virus preparations containing the satellite also have variable amounts of RNAs with a size intermediate between genomic and S-RNA (D. Gallitelli and M. Russo, unpublished information).

S-RNAs of different Tombusviruses have extensive sequence homology with one another, the lowest being that of CyRSV S-RNA. Their replication can be supported by heterologous Tombusviruses but not by other isometric viruses with monopartite genomes. For example, TCV, SCV, GaMV, and GMoV failed to act as helper viruses for Tombusvirus S-RNA (Gallitelli and Hull, 1985b). These observations indicate that specific functions coded only by Tombusvirus genomic RNA are required for S-RNA replication, and that all S-RNAs may have a common origin.

Gallitelli and Hull (1985b) have reported that when a satellite-free preparation of TBSV-type RNA is passaged through *N. benthamiana*, it acquires S-RNA. This raises intriguing questions as to the origin of S-RNA. Purified virus preparations do not seem to contain multimeric forms of S-RNA comigrating with genomic RNA. In fact, satellite-free TBSV isolates are readily obtained by inoculating genomic RNA alone in hosts other than *N. benthamiana*. In total RNA extracts from TBSV-type infected tissues, only dimer and trimer molecules of S-RNA were detected, but in low concentration and not in all preparations (D. Gallitelli and M. Russo, unpublished information). Altenbach and Howell (1982) reported that an S-RNA associated with TCV shares some sequence homology with host DNA. TBSV S-RNA does not have any sequence homology with TCV satellite RNA (Gallitelli and Hull, 1985b), but its acquisition from a specific host upon inoculation of simple genomic RNA suggests that it may also originate from the host (*N. benthamiana*) genome. Recent investigations on artichoke plants naturally infected with AMCV failed to detect S-RNA specific sequences in any of 30 individually tested plants (D. Gallitelli, unpublished information).

S-RNA of Tombusviruses may lack messenger activity. This was proven for CyRSV, whose S-RNA in *in vitro* and *in vivo* translation experiments failed to express specific products (Burgyan *et al.*, 1986). In this respect, CyRSV S-RNA resembles the satellite RNAs of tobacco ringspot virus (Owens and Schneider, 1977) and TCV (Altenbach and Howell, 1982).

The biological behavior of S-RNAs is also intriguing. Gallitelli and Hull (1985b) have shown that their presence induces a remarkable attenuation of symptoms as compared with disease expression in hosts infected with satellite-free virus isolates. The mechanism underlying this protective effect is not known. However, this effect may be of potential practical value if chimeric genes originating from S-RNAs were constructed and inserted into susceptible host plants; a possibility that is now under investigation in our laboratory.

Hillman *et al.* (1985) have described a set of low-molecular-weight RNAs (0.4–0.5 kb) associated with TBSV. These RNA species share sequence homology with viral genome and, therefore, strictly speaking, they cannot be regarded as true S-RNAs (Murant and Mayo, 1982). However, these RNAs have a modulating effect on symptoms expression and require genomic RNA for replication, thus acting like defective interfering RNAs (Hillman and Morris, 1986b). Structurally, these RNAs are single-stranded molecules with a linear form in both virions and tissue extracts. The nucleotide sequence homology derives from two distinct regions of the viral genome, one of which seems to be located at the 3' terminus (Hillman and Morris, 1986b).

A satellitelike RNA species apparently different from all the above, named X-RNA, was found associated with EMCV only (Gallitelli and Hull, 1985b). This RNA has not yet been characterized.

ACKNOWLEDGMENTS. The authors wish to express grateful thanks to Dr. R. Koenig for supplying the cultures of MPV and NRV used in this study, Dr. T. J. Morris for providing unpublished information and for helpful discussion, Dr. A. Appiano for supplying an unpublished micrograph, Dr. S. Harrison and Dr. J. Krüse for permission to use illustrations from their papers, and Mr. M. Fanelli for help with the photography.

X. RECENT DEVELOPMENTS

Recent papers presented at the 7th International Congress of Virology (Edmonton, Canada, August 1987) confirm that TBSV (possibly PAMV) and CyRSV have similar genome organization and expression. The coat protein coding region is not located at the 3' end. The 4.7-kb genomic RNA encodes a gene product of c. 40 kd, whereas two 3'-coterminal subgenomic RNA species, c. 2.1- and 1.0-kb, code for the 43-kd coat protein and for a polypteptide of 20-to 22-kd, respectively (Burgyan and Russo, 1987; Bradley et al. 1987).

Cucumber necrosis virus (CNV) was found to have a molecular biology comparable to that of TBSV and CyRSV and extensive sequence homology with TBSV genomic RNA, which suggests its inclusion in the Tombusvirus group (Rochon and Tremaine, 1987).

The Plant Virus Subcommittee of the International Committee on Taxonomy of Viruses has now placed CNV among possible members of the Tombusvirus group together with cucumber leaf spot (CLSV) and cucumber soil-borne (CSBV) viruses. CLSV and CFSV have a genome size and dsRNA pattern similar to that of TBSV (H. J. Vetten, personal communication) but are serologically unrelated to all tombusviruses tested, do not show detectable sequence homology with TBSV RNA (D. Gallitelli, unpublished data), and have a cytopathology distinctively different from that of Tombusviruses (Di Franco and Martelli, 1987).

REFERENCES

Albrechtova, L., Chod, J., and Zimandl, B., 1975, Nachweis des Tomatenzwergbusch-Virus (tomato bushy stunt virus) in Süsskirschen, die mit virösen Zweigkrebs befallen waren, *Phytopath. Z.* **82**:25.

Allen, W. R., 1968, Tomato bushy stunt virus from *Prunus avium*. II. Serological typing and characterization of antibodies, types and activities, *Canad. J. Bot.* **45**:229.

Allen, W. R., 1969, Occurrence and seed transmission of tomato bushy stunt virus in apple, *Canad. J. Plant Sci.* **49**:797.

Allen, W. R., and Davidson, T. R., 1967, Tomato bushy stunt from *Prunus avium* L. I. Field studies and virus characterization, *Canad. J. Bot.* **45**:2375.

Altenbach, S. B., and Howell, S. H., 1982, In vitro translation products of turnip crinkle virus RNA, Virology 118:128.

Ambrosino, C., Appiano, A., Rialdi, G., Papa, G., Redolfi, P., and Carrara, M., 1967, Caratterizzazione chimica e chimicofisica del Petunia asteroid mosaic virus (PAMV), Atti Acc. Sci. Torino 101:301.

Appiano, A., and D'Agostino, G., 1980, Cytological alterations of the epidermal cells of Gomphrena globosa following mechanical inoculations with tomato bushy stunt virus, Phytopath. Medit. 19:129.

Appiano, A., and D'Agostino, G., 1985, Distribution of tomato bushy stunt virus in root tips of systemically infected Gomphrena globosa, J. Ultrastruct. Res. 85:239.

Appiano, A., and Pennazio, S., 1982, Some observations on tomato bushy stunt virus-induced inclusion bodies, Proc. 10th Internat. Congr. Elect. Microsc. (Hamburg) 3:155.

Appiano, A., Bassi, M., and D'Agostino, G., 1983, Cytochemical and autoradiographic observations on tomato bushy stunt virus-induced multivesicular bodies, Ultramicroscopy 12:162.

Appiano, A., Pennazio, S., D'Agostino, G., and Redolfi, P., 1977, Fine structure of necrotic local lesions induced by tomato bushy stunt virus in Gomphrena globosa leaves, Physiol. Pl. Pathol. 11:327.

Appiano, A., Pennazio, S., and Redolfi, P., 1978, Cytological alterations in tissues of Gomphrena globosa plants systemically infected with tomato bushy stunt virus, J. Gen. Virol. 40:277.

Appiano, A., D'Agostino, G., Redolfi, P., and Pennazio, S., 1981, Sequence of cytological events during the process of local lesion formation in the tomato bushy stunt virus– Gomphrena globosa hypersensitive system, J. Ultrastruct. Res. 76:173.

Appiano, A., Barbieri, N., Bassi, M., D'Agostino, G., dell'Orto, P., and Viola, G., 1985, Ulteriori indagini sui corpi d'inclusione indotti da TBSV, Atti 15° Congr. Naz. Microscop. Elettr. A3.

Argos, P., Tsukihara, T., and Rossman, M. G., 1980, A structural comparison of concanavalin A and tomato bushy stunt virus protein, J. Mol. Evol. 15:169.

Bassi, M., Appiano, A., Barbieri, N., and D'Agostino, G., 1985, Chloroplast alterations induced by tomato bushy stunt virus in Datura leaves, Protoplasma 126:233.

Behncken, G. M. and Dale, J. L., 1984, Glycine mottle virus, a possible member of the tombusvirus group, Intervirology 21:159.

Bercks, R., 1967, Über der Nachweis des Tomatenzwergbusch-Virus (tomato bushy stunt virus) in Reben, Phytopath. Z. 60:273.

Bercks, R., and Lovisolo, O., 1965, Serologischer Verleich von Stämmen des tomatenzwergbusch-Virus (tomato bushy stunt virus), Phytopath. Z. 52:96.

Bercks, R., and Querfurth, G., 1969, Weitere methodische Untersuchungen über den Latextest zum serologischen Nachweis pflanzenpathogener Viren, Phytopath. Z. 65:243.

Bercks, R., and Querfurth, G., 1971, Serologische Beziehungen zwischen einem gestreckten (tobacco mosaic) und einem isometrischen (cocksfoot mild mosaic) Viren, Phytopath. Z. 72:354.

Boatman, S., and Kaper, J. M., 1976, Molecular organization and stabilizing forces of simple RNA viruses. IV. Selective interference with protein–RNA interactions by use of sodium dodecyl sulphate, Virology 70:1.

Borges, M. de L., Sequeira, J. C., and Louro, D., 1979, Aparecimento em Portugal do virus do emanjericado do tomateiro (tomato bushy stunt virus). Hospedeiros, morfologia e localização nas celulas de Pimenteiro, Phytopath. Medit. 18:118–122.

Bradley, et al., 1987, Abstr. 7th Internat. Congr. Virol. 84.

Burgermeister, W., and Koenig, R., 1984, Electro-blot immunoassay. A means for studying serological relationships among plant viruses? Phytopath. Z. 111:15.

Burgyan, J., and Russo, M., 1987, Abstr. 7th Internat. Congr. Virol. 82.

Burgyan, J., Russo, M., and Gallitelli, D., 1986, Translation of Cymbidium ringspot virus RNA in cowpea protoplasts and rabbit reticulocyte lysates, J. Gen. Virol. 67:1149.

Butler, P. J. G., 1970, Structure of turnip crinkle and tomato bushy stunt viruses, *J. Mol. Biol.* **52**:589.

Büttner, C., Jacobi, V., and Koenig, R., 1987, Isolation of carnation Italian ringspot virus from a creek in a forested area southwest of Bonn, *J. Phytopathol.* **118**:131.

Campbell, R. N., 1968, Transmission of tomato bushy stunt virus unsuccessful with Olpidium, *Plant Dis. Rept.* **52**:379.

Campbell, R. N., Lovisolo, O., and Lisa, V., 1975, Soil transmission of petunia asteroid mosaic strain of tomato bushy stunt virus, *Phytopath. Medit.* **14**:82.

Chauvin, C., Witz, J., and Jacrot, B., 1978, Structure of tomato bushy stunt virus: a model for protein–RNA interaction, *J. Mol. Biol.* **124**:641.

Cherif, C., 1981, Contribution à la connaissance du virus du rabougrissement buissoneux de la tomate (tomato bushy stunt virus) en Tunisie, *Thèse Diplome Docteur 3eme Cycle, Univ. P. and M. Curie,* Paris.

Cherif, C., and Spire, D., 1983, Identification du virus du rabougrissement buissoneux de la tomate (tomato bushy stunt virus) en Tunisie sur tomate, piment et aubergine: Quelques caractéristiques de la souche tunisienne, *Agronomie* **3**:701.

Christie, R. G., and Edwardson, J. R., 1977, Light and electron microscopy of plant virus inclusions, *Fla Agric. Exp. Stn. Monogr.* Series 9.

Clark, M. F., and Adams, A. M., 1977, Characteristics of the microplate method of enzyme-linked immunosorbent assay for the detection of plant viruses, *J. Gen. Virol.* **34**:457.

Clark, M. F., and Bar-Joseph, M., 1984, Enzyme immunosorbent assay in plant virology, in: *Methods in Virology,* Vol. 8 (K. Maramorosch and H. Koprowski, eds.), 51–85, Academic Press, New York.

Crowther, R. A., and Amos, L. A., 1971, Three dimensional image reconstruction of some spherical viruses, *Cold Spring Harbor Symp. Quant. Biol.* **36**:489.

D'Agostino, G., and Pennazio, S., 1981, An ultrastructural study of the senescence induced in *Gomphrena globosa* leaves by mineral deficiency, *J. Submicrosc. Cytol.* **13**:373.

De Varennes, A., Russo, M., and Maule, A. J., 1984, Infection of protoplasts from *Chenopodium quinoa* with cowpea mosaic and cymbidium ringspot viruses, *J. Gen. Virol.* **65**:1851.

Di Franco, A., and Martelli, G. P., 1987, J. Submicroscop. Cytol. **19**(in press).

Di Franco, A., Russo, M., and Martelli, G. P., 1984, Ultrastructure and origin of cytoplasmic multivesicular bodies induced by carnation Italian ringspot virus, *J. Gen. Virol.* **65**:1233.

Durham, A. C. H., Witz, J., and Bancroft, J. B., 1984, The semipermeability of simple spherical virus capsids, *Virology* **133**:1.

Erić, Z., Štefanac, Z., and Plavsic, B., 1986, Characteristics of the tombusvirus from spinach (*Spinacea oleracea*), *Acta Bot. Croat.* **45**:7.

Fauquet, C., Dejardin, J., and Thouvenel, J. C., 1986, Evidence that the aminoacid composition of the particle proteins of plant viruses is characteristic of the virus group. I. Multidimensional classification of plant viruses, *Intervirology* **25**:1.

Finch, J. T., Klug, A., and Lebermann, R., 1970, The structures of turnip crinkle and tomato bushy stunt viruses. II, *J. Mol. Biol.* **50**:215.

Fischer, H. U., and Lockhart, B. E. L., 1974, Occurrence in Morocco of a virus disease of artichokes related to artichoke mottled crinkle virus, *Plant Dis. Reptr.* **58**:1117.

Fischer, H. U., and Lockhart, B. E. L., 1977, Identification and comparison of two isolates of tomato bushy stunt virus from pepper and tomato in Morocco, *Phytopathology* **67**:1352.

Francki, R. I. B., Milne, R. G., and Hatta, T., 1985, *Atlas of Plant Viruses I.,* p. 222, CRC Press, Boca Raton.

Fuchs, E., Marker, D., and Kegler, G., 1979, Des Nachweis des chlorotischen Blattfleckungs-Virus des Apfels (apple chlorotic leafspot virus), des Stammfurchungs-Virus des Apfels (apple stem grooving virus) und des Tomatenzwergbusch-Virus (tomato bushy stunt virus) mit dem ELISA-test, *Arch. Phytopathol. Pflanzenschutz.* **15**:421.

Gallitelli, D., and Hull, R., 1985a, Preparation of complementary DNA by direct synthesis on plant virus RNA from agarose gels, *J. Virol. Meth.* **11**:141.

Gallitelli, D., and Hull, R., 1985b, Characterization of satellite RNAs associated with tomato bushy stunt virus and five other definitive tombusviruses, *J. Gen. Virol.* **66:**1533.

Gallitelli, D., Hull, R., and Koenig, R., 1985, Relationships among viruses in the tombusvirus group: Nucleic acid hybridisation studies, *J. Gen. Virol.* **66:**1523.

Gallitelli, D., and Russo, M., 1987, Some properties of Moroccan pepper and tombusvirus Neckar viruses, *J. Phytopathol.* **119:**106.

Gigante, R., 1955, Il rachitismo cespuglioso del pomodoro, *Boll. Staz. Pat. Veg. Roma, S III* **12:**43.

Harrison, B. D., 1981, Plant virus ecology: Ingredients, interactions and environmental influences, *Ann. Appl. Biol.* **99:**195.

Harrison, B. D., Finch, J. T., Gibbs, A. J., Hollings, M., Shepherd, R. J., Valenta, V., and Wetter, C., 1971, Sixteen groups of plant viruses, *Virology* **45:**356.

Harrison, S. C., 1984a, Multiple modes of subunit association in the structures of simple spherical viruses, *Trends Biochem. Sci.* **9:**345.

Harrison, S. C., 1984b, Structure of viruses, in: *The Microbes 1984. 1. Viruses* (B. W. J. Mahy and J. R. Pattison, eds.), pp. 29–73, Cambridge University Press, Cambridge.

Harrison, S. C., Olson, A. J., Schutt, C. E., Winkler, F. K., and Bricogne, G., 1978, Tomato bushy stunt virus at 2.9 Å resolution, *Nature (Lond.)* **276:**368.

Hatta, T., and Francki, R. I. B., 1981, Identification of small polyhedral virus particles in thin sections of plant cells by an enzyme cytochemical technique, *J. Ultrastruct. Res.* **74:**116.

Hatta, T., Francki, R. I. B., and Grivell, C. J., 1983, Particle morphology and cytopathology of galinsoga mosaic virus, *J. Gen. Virol.* **64:**687.

Hawkes, R., Niday, E., and Gordon, J., 1982, A dot-immunobinding assay for monoclonal and other antibodies, *Anal. Biochem.* **119:**142.

Hayes, R. J., Buck, K. W., and Brunt, A. A., 1984, Double-stranded and single-stranded subgenomic RNAs from plant tissue infected with tomato bushy stunt virus, *J. Gen. Virol.* **65:**1239.

Henriques, M. I. C., and Schlegel, D. E., 1978, Tomato bushy stunt virus isolated from piggyback plants (abstract 43), *3rd International Congress on Plant Pathology.*

Henriques, M. I. C., and Morris, T. J., 1979, Evidence for different replicative strategies in the plant tombusviruses, *Virology* **99:**66.

Hibi, T., and Saito, Y., 1985, A dot-immunobinding assay for the detection of tobacco mosaic virus in infected tissues, *J. Gen. Virol.* **66:**1191.

Hillman, B. I., and Morris, T. J., 1986a, cDNA cloning and sequencing of the 3' end of tomato bushy stunt virus RNA: Implications for genome organization (abstract), *American Society for Virology*, Santa Barbara, CA.

Hillman, B. I., and Morris, T. J., 1986b, Symptom modulating RNAs associated with tombusviruses (abstract), *Vth International Conference on Comparative Virology.*

Hillman, B. I., Morris, T. J., and Schlegel, D. E., 1985, Effects of low molecular weight RNA and temperature on tomato bushy stunt virus symptom expression, *Phytopathology* **75:**361.

Hogle, J., Kirchhausen, T., and Harrison, S. C., 1983, Divalent cation sites in tomato bushy stunt virus: Difference maps at 2.9 Å resolution, *J. Mol. Biol.* **171:**95.

Hollings, M., 1962, Studies of pelargonium leaf curl virus: I. Host range, transmission and properties *in vitro, Ann. Appl. Biol.* **50:**189.

Hollings, M., and Stone, O. M., 1965, Studies of pelargonium leaf curl virus: II. Relationships to tomato bushy stunt and other viruses, *Ann. Appl. Biol.* **56:**87.

Hollings, M., and Stone, O. M., 1975, Serological and immunoelectrophoretic relationships among viruses in the tombusvirus group, *Ann. Appl. Biol.* **80:**37.

Hollings, M., Stone, O. M., and Bouttell, G., 1970, Carnation Italian ringspot virus, *Ann. Appl. Biol.* **65:**299.

Hollings, M., Stone, O. M., and Barton, R. J., 1977, Pathology, soil transmission and characterisation of cymbidium ringspot, a virus from cymbidium orchids and white clover (*Trifolium repens*), *Ann. Appl. Biol.* **85:**233.

Hopper, P., Harrison, S. C., and Sauer, R. T., 1984, Structure of tomato bushy stunt virus. V. Coat protein sequence determination and its structural implications, *J. Mol. Biol.* **177**:701.

Jeagle, M., and Van Regenmortel, M. H. V., 1985, Use of ELISA for measuring the extent of serological cross-reactivity between plant viruses, *J. Virol. Methods* **11**:189.

Kaper, J. M., 1975, The chemical basis of virus structure, in: *Dissociation and Reassembly*, p. 236. North-Holland Biomedical Press, Amsterdam.

Kassanis, B., 1954, Heat-therapy of virus-infected plants, *Ann. Appl. Biol.* **41**:470.

Kassanis, B., and Lebeurier, G., 1969, The behaviour of tomato bushy stunt virus at different temperatures *in vivo* and *in vitro*, *J. Gen. Virol.* **4**:385.

Kegler, G., and Kegler, H., 1980, Untersuchungen zur natürlichen Übertragung des tomato bushy stunt virus bei Obstgehölzen, *Tag. Ber. Akad. Landwirtsch.-Wiss. DDR* **184**:297.

Kegler, G., and Kegler H., 1981, Beiträge zur Kenntnis der vektorlosen Übertragung pflanzenpathogener Viren, *Arch. Phytopathol. u. Pflanzenschutz* **17**:307.

Kegler, G., and Schimanski, H. H., 1982, Untersuchungen zur Verbreitung und Samenübertragbarkeit des Tomatenzwergbusch-Virus (tomato bushy stunt virus) an Kern- und Steinobst in der DDR, *Arch. Phytopathol. u. Pflanzenschutz* **16**:73.

Kegler, G., Kleinhempel, H., and Kegler, H., 1980, Untersuchung zur Bodenbürtigkeit des tomato bushy stunt virus, *Arch. Phytopathol. u. Pflanzenschutz* **16**:73.

Kleinhempel, H., and Kegler, G., 1982, Transmission of tomato bushy stunt virus without vectors, *Acta Phytopath. Acad. Sci. Hung.* **17**:17.

Koenig, R., 1981, Indirect ELISA methods for the broad specificity detection of plant viruses, *J. Gen. Virol.* **55**:53.

Koenig, R., and Kunze, L., 1982, Identification of tombusvirus isolates from cherry in southern Germany as petunia asteroid mosaic virus, *Phytopath. Z.* **103**:361.

Koenig, R., and Avgelis, A., 1983, Identification of a virus similar to the BS3 strain of tomato bushy stunt virus in eggplant, *Phytopath. Z.* **106**:349.

Koenig, R., and Lesemann, D.-E., 1985, Plant viruses in German rivers and lakes, I. Tombusviruses, a potexvirus and carnation mottle virus, *Phytopath. Z.* **112**:105.

Koenig, R., and Gibbs, A., 1986, The serological relationships of definitive tombusviruses, *J. Gen. Virol.* **67**:75.

Krüse, J., Krüse, K. M., Witz, J., Chauvin, C., Jacrot, B., and Tardieu, A., 1982a, Divalent ion-dependent reversible swelling of tomato bushy stunt virus and organization of the expanded virion, *J. Mol. Biol.* **162**:393.

Krüse, J., Witz, J., and Spegt, P., 1982b, Rigidity of RNA in recompacted tomato bushy stunt virus, *J. Mol. Biol.* **162**:415.

Lovisolo O., 1957, Petunia: nuovo ospite del virus del rachitismo cespuglioso del pomodoro, *Boll. Staz. Pat. Veg. Roma* **14**:103 (1956).

Lovisolo, O., 1966, Indagini su virosi di piante ornamentali, *Atti 1° Congr. Un. Fitopat. Medit.* (Bari-Napoli) **1966**:574.

Lovisolo, O., Ambrosino, C., Liberatori, J., and Papa, G., 1964, Ricerche sul virus del rachitismo cespuglioso del pomodoro (tomato bushy stunt virus). I. Differenziazione per via elettrocinetica ed immunochimica del ceppo Petunia dal ceppo BS-3, *Atti Accad. Sci. Torino* **98**:391.

Lovisolo, O., Bode, O., and Völk, J., 1965, Preliminary studies on the soil transmission of petunia asteroid mosaic virus (= petunia strain of tomato bushy stunt virus), *Phytopath. Z.* **53**:323.

Lovisolo, O., Ambrosino, C., Luisoni, E., and Bellaudo, M., 1967, Micrografia elettronica di petunia asteroid mosaic virus (PAMV) in comparizione con tomato bushy stunt virus (TBSV), *Atti Accad. Sci. Torino* **101**:229.

Maat, D. Z., Huttinga, H., and Hakkaart, F. A., 1978, Nerine latent virus: Some properties and serological detectability in Nerine bodwenii, *Neth. J. Plant Path.* **84**:47.

Makkouk, K. M., Koenig, R., and Lesemann, D.-E., 1981, Characterization of a tombusvirus isolated from eggplant, *Phytopathology* **71**:572.

Martelli, G. P., 1965, L'arricciamento maculato del carciofo, *Phytopath. Medit.* **4**:58.

Martelli, G. P., 1981, Tombusviruses, in: *Handbook of Plant Virus Infections and Comparative Diagnosis* (E. Kurstak, ed.), pp. 61–90, Elsevier/North-Holland Biomedical Press, Amsterdam.

Martelli, G. P., and Castellano, M. A., 1969, The relation of pelargonium leaf curl virus to the nuclei of host cells, *Virology* **39**:610.

Martelli, G. P., and Quacquarelli, A., 1966, Ricerche sull'aricciamento maculato del carciofo. III. Dimostrazione sierologica della parentela con il virus del rachitismo cespuglioso del pomodoro, *Atti 1° Congr. Un. Phytopath. Medit.* (Bari-Napoli) **1966**:195.

Martelli, G. P., and Rana, G. L., 1975, Viruses and virus diseases of globe artichoke and cardoon, *Atti 2° Congr. Internaz. Studi sul Carciofo* (Bari) **1973**:811.

Martelli, G. P., and Russo, M., 1972a, Pelargonium leaf curl virus in host leaf tissues, *J. Gen. Virol.* **15**:193.

Martelli, G. P., and Russo, M., 1972b, Ultrastructure of tomato bushy stunt virus strains in plant tissues, *Mikrobiologija* **9**:177.

Martelli, G. P., and Russo, M., 1973, Electron microscopy of artichoke mottled crinkle virus in leaves of *Chenopodium quinoa* Wild, *J. Ultrastruct. Res.* **42**:93.

Martelli, G. P., and Russo, M., 1981, The fine structure of Cymbidium ringspot virus in host tissues. I. Electron microscopy of systemic infections, *J. Ultrastruct. Res.* **77**:93.

Martelli, G. P., and Russo, M., 1984, Use of thin sectioning for visualization and identification of plant viruses, in: *Methods in Virology*, Volume 8 (K. Maramorosch and H. Koprowski, eds.), pp. 143–224, Academic Press, New York.

Martelli, G. P., Quacquarelli, A., and Russo, M., 1971, Tomato bushy stunt virus, *CMI/AAB Descriptions of Plant Viruses* No. 69.

Martelli, G. P., Russo, M., and Rana, G. L., 1976, Occurrence of artichoke mottled crinkle virus in Malta, *Pl. Dis. Reptr.* **60**:130.

Martelli, G. P., Russo, M., and Quacquarelli, A., 1977, Tombusvirus (Tomato bushy stunt virus) group, in: *The Atlas of Insect and Plant Viruses* (K. Maramorosch, ed.), pp. 257–279, Academic Press, New York.

Martelli, G. P., Russo, M., and Rana, G. L., 1981, A survey of the virological problems of *Cynara* species, *Atti 3° Congr. Internaz. Studi Carciofo* (Bari-Polignano) **1979**:301.

Martelli, G. P., Di Franco, A., and Russo, M., 1984, The origin of multivesicular bodies in tomato bushy stunt virus-infected *Gomphrena globosa* plants, *J. Ultrastruct. Res.* **88**:275.

Martinez, J., Galindo, J., and Rodriguez, R., 1974, Etudio sobre la enfermedad del "Pinto" del jitomate (*Lycopersicon esculentum* Mill.) en la region de Actopan, Hgo. *Agrociencia* **18**:71.

Matthews, R. E. F., 1982, Classification and nomenclature of viruses, *Intervirology* **17**:4.

Maule, A. J., Hull, R., and Donson, J., 1983, The application of spot hybridization to the detection of DNA and RNA viruses in plant tissues, *J. Virol. Meth.* **6**:215.

Mayo, M. A., and Jones, A. J., 1973, The protein and nucleic acid components of elderberry latent virus, *J. Gen. Virol.* **19**:245.

Mayo, M. A., Barker, H., and Harrison, B. D., 1982, Specificity and properties of the genome-linked proteins of nepoviruses. *J. Gen. Virol.* **59**:149.

McCarthy, D. A., 1983, Inactivation of tomato bushy stunt virus *in vivo*, *Physiol. Pl. Path.* **22**:181.

Milne, R. G., and Lesemann, D. E., 1984, Immunosorbent electron microscopy in plant virus studies, in: *Methods in Virology*, Vol. 8 (K. Maramorosch and H. Koprowski, eds.), pp. 85–101, Academic Press, New York.

Morris, T. J., 1983, Virus-specific double-stranded RNA: functional role in RNA virus infection, in: *Plant Infectious Agents: Viruses, Viroids, Virusoids and Satellites* (H. D. Robertson, S. H. Howell, M. Zaitlin, R. L. Malberg, eds.), pp. 80–83, Cold Spring Harbor Laboratory, Cold Spring Harbor, New York.

Morris, T. J., Hillman, B. I., Young, M., and Carrington, J. C., 1987, Comparison of some tentative members for a proposed carnation mottle virus group, *Intervirology* (in press).

Mowat, W. P., 1972, A necrotic disease of tulip caused by tomato bushy stunt virus, *Plant Path.* **21**:171.

Munowitz, M. G., Dobson, C. M., Griffin, R. G., and Harrison, S. C., 1980, On the rigidity of RNA in tomato bushy stunt virus, *J. Mol. Biol.* **141**:152.

Murant, A. F., and Mayo, M. A., 1982, Satellites of plant viruses, *Ann. Rev. Phytopathol.* **20**:49.

Novak, J. B., and Lanzova, J., 1976, Identification of alfalfa mosaic and tomato bushy stunt virus in hop (*Humulus lupulus* L.) and grapevine (*Vitis vinifera* subsp. *sativa* DC/Hegi) plants in Czechoslovakia, *Biol. Plant.* **18**:152.

Novak, J. B., and Lanzova, J., 1977a, Purification and properties of arabis mosaic and tomato bushy stunt viruses isolated from lilac (*Syringa vulgaris* L.), *Biol. Plant.* **19**:264.

Novak, J. B., and Lanzova, J., 1977b, Identification of tomato bushy stunt virus in cherry and plum trees showing fruit pitting symptoms, *Biol. Plant.* **19**:234.

Olson, A. J., Bricoque, G., and Harrison, S. C., 1983, Structure of tomato bushy stunt virus. IV. The virus particle at 2.9 Å resolution, *J. Mol. Biol.* **171**:61.

Orlob, G. B., 1968, Relationships between *Tetranichus urticae* Koch and some plant viruses, *Virology* **35**:121.

Owens, R. A., and Schneider, I. R., 1977, Satellite of tobacco ringspot virus RNA lacks detectable mRNA activity, *Virology* **90**:222.

Pape, H., 1927, Das verheerende Auftreten der Kräusel–Krankheit bei Pelargonium. I. Die Ansicht der Wissenschaft, *Gartenwelt* **31**:329.

Paul, H. L., Querfurth, G., and Huth, W., 1980, Serological studies on the relationships of some isometric viruses of Gramineae, *J. Gen. Virol.* **47**:67.

Pennazio, S., and Redolfi, P., 1977, Localized and systemic resistance induced in *Gomphrena globosa* leaves by tomato bushy stunt virus, *Riv. Pat. Veg.* (S.IV) **13**:103.

Pennazio, S., Appiano, A., and Redolfi, P., 1979, Changes occurring in *Gomphrena globosa* leaves in advance of the appearance of tomato bushy stunt virus necrotic local lesions, *Physiol. Pl. Pathol.* **15**:177.

Pennazio, S., D'Agostino, G., Appiano, A., and Redolfi, P., 1978, Ultrastructure and histochemistry of the resistant tissue surrounding lesions of tomato bushy stunt virus in *Gomphrena globosa* leaves, *Physiol. Pl. Pathol.* **13**:165.

Pennazio, S., Redolfi, P., and Martin, G., 1976, Hypersensitivity of *Gomphrena globosa* to tomato bushy stunt virus, *Phytopath.* **87**:161.

Piazzolla, P., Castellano, M. A., and De Stradis, A., 1986, Presence of plant viruses in some rivers of Southern Italy, *Phytopath. Z.* (in press).

Piazzolla, P., Gallitelli, D., and Quacquarelli, A., 1977, The behaviour of some isometric plant viruses heated *in vitro* as determined by particle stabilizing forces, *J. Gen. Virol.* **37**:373.

Quacquarelli, A., and Martelli, G. P., 1966, Ricerche sull'arricciamento maculato del carciofo. I. Ospiti differenziali e proprietà, *Atti 1° Congr. Un. Fitopat. Medit.* (Bari-Napoli) **1966**:168.

Rana, G. L., and Cherif, C., 1981, Occurrence of artichoke mottled crinkle virus in Tunisia, *Phytopath. Medit.* **20**:179.

Rana, G. L., and Kyriakopoulou, P. L., 1982, Artichoke Italian latent and artichoke mottled crinkle viruses in artichoke in Greece, *Phytopath. Medit.* **21**:101.

Redolfi, P., Cantisani, A., and Pennazio, S., 1978, Unusual phenolic compounds in the hypersensitive reaction of *Gomphrena globosa* to tomato bushy stunt virus, *Phytopath. Z.* **93**:325.

Richter, J., Kleinhempel, H., Schimanski, H. H., and Kegler, H., 1977, Nachweis des Tomatenzwergbusch-Virus (tomato bushy stunt virus) in Obstgehölzen, *Arch. Phytopathol. u. Pflanzenschutz* **13**:367.

Roberts, F. M., 1950, The infection of plant viruses through roots, *Ann. Appl. Biol.* **37**:385.

Robinson, I. K., and Harrison, S. C., 1982, Structure of the expanded state of tomato bushy stunt virus, *Nature* (London) **297**:563.

Rochon, and Tremaisu, 1987, *Abstr. 7th Internat. Congr. Virol.* 85.

Rubio-Huertos, and Garcia-Hidalgo, F., 1971, Electron microscopy of two *Pelargonium* viruses, *Protoplasma* **72**:449.

Russo, M., Di Franco, A., and Martelli, G. P., 1983, The fine structure of Cymbidium ringspot virus infections in host tissues. III. Role of peroxisomes in the genesis of multivesicular bodies, *J. Ultrastruct. Res.* **82**:52.

Russo, M., and Gallitelli, D., 1985, Infection of cowpea protoplasts with Cymbidium ringspot virus, *J. Gen. Virol.* **66**:2033.

Russo, M., and Martelli, G. P., 1972, Ultrastructural observations on tomato bushy stunt virus in plant cells, *Virology* **49**:122.

Russo, M., and Martelli, G.P., 1981, The fine structure of Cymbidium ringspot virus in host tissues. II. Light and electron microscopy of localized infections, *J. Ultrastruct. Res.* **77**:105.

Russo, M., and Martelli, G. P., 1982, Ultrastructure of turnip crinkle- and saguaro cactus virus-infected tissues, *Virology* **118**:109.

Russo, M., Martelli, G. P., and Quacquarelli, A., 1967, Occurrence of artichoke mottled crinkle virus in leaf vein xylem, *Virology* **33**:555.

Russo, M., Martelli, G. P., and Quacquarelli, A., 1968, Studies on the agent of artichoke mottled crinkle. IV. Intracellular localization of the virus, *Virology* **34**:679.

Rybicki, E. P., and Von Wechmar, M. B., 1985, Serology and immunochemistry, in: *Plant Viruses. I. Polyhedral Viruses with Tripartite Genomes* (R. I. B. Francki, ed.), pp. 207–244, Plenum Press, New York.

Schmelzer, K., 1958, Wirstpflanzen des Tomatenzwergbusch-Virus (*Marmor dodecahedron* Holmes), *Z. Pflanzenschutz* **65**:80.

Smith, K. M., 1935, A new virus disease of the tomato, *Ann. Appl. Biol.* **22**:731.

Smith, K. M., Campbell, R. M., and Fry, R. R., 1969, Root discharge and soil survival of viruses, *Phytopathology* **59**:1678.

Steere, R. C., 1953, Strains of tomato bushy stunt virus, *Phytopathology* **43**:485.

Steere, R. C., 1959, The purification of plant viruses, *Adv. Virus Res.* **6**:1.

Stefanac, Z., 1978, Disease of spinach caused by tomato bushy stunt virus, *3rd Internat. Congr. Pl. Path.* **37** (abstract).

Stockley, P. G., Kirsh, A. L., Chow, E. P., Smart, J. E., and Harrison, S. C., 1986, Structure of turnip crinkle virus. III. Identification of a unique coat protein dimer, *J. Mol. Biol.* (in press).

Teakle, D. S., and Gold, A. H., 1963, Further studies on *Olpidium* as a vector of tobacco necrosis virus, *Virology* **19**:310.

Tomlinson, J. A., and Faithfull, E. M., 1984, Studies on the occurrence of tomato bushy stunt in England's rivers, *Ann. Appl. Biol.* **104**:485.

Tomlinson, J. A., Faithfull, E. M., Flewett, T. H., and Beards, G., 1982, Isolation of infective tomato bushy stunt virus after passage through the human alimentary tract, *Nature* (London) **300**:637.

Van Regenmortel, M. H. V., 1982, *Serology and Immunochemistry of Plant Viruses*, p. 302, Academic Press, New York.

Vetten, H. J., and Koenig, R., 1983, Natural infection of tomato and pelargonium in Germany by a tombusvirus originally described from pepper in Morocco, *Phytopath. Z.* **108**:215.

Vovlas, C., and Di Franco, A., 1987, Nuovi dati sulla presenza di virus delle piante nei corsi d'acqua di Puglia a Basilicata, *Informatore Fitopatol.* **37**:55.

Wetter, C., and Luisoni, E., 1969, Precipitin, agar gel diffusion, and intragel absorption tests with three strains of tomato bushy stunt virus, *Phytopath. Z.* **65**:231.

Ziegler, A., Harrison, S. C., and Leberman, R., 1974, The minor proteins of tomato bushy stunt and turnip crinkle viruses, *Virology* **59**:509.

CHAPTER 3

Carnation Mottle Virus and Viruses with Similar Properties

T. J. MORRIS and J. C. CARRINGTON

I. INTRODUCTION

Carnation mottle virus (CarMV) is a 30-nm, isometric plant virus that contains a single-component, positive-sense genome of 4.0 kb (molecular weight of about 1.4×10^6) and a capsid composed of protein subunits of $M_r = 38,000$. The nucleotide sequence and genetic organization of the viral genome has recently been elucidated, making CarMV one of the few monopartite small RNA plant viruses to be well characterized at the molecular level. A review of CarMV is appropriate at this time, in view of the detailed knowledge we now have about this and some similar viruses such as turnip crinkle virus; the lack of previous significant reviews; and the currently unsettled nature of their classification (Francki et al., 1985a,b).

Carnation mottle virus shares similar morphological and physico-chemical properties with some 20 other viruses in addition to the seven definitive members of the Tombusvirus group. Each has a 30-nm spherical particle, a single sedimenting component of 120–130 S, a single genomic RNA species with a molecular weight between $1.4–1.6 \times 10^6$, and a capsid protein of approximately $M_r = 38,000$. Table I lists the viruses to be considered in this chapter and appropriate abbreviations for their names. Plant viruses have generally been assigned to a defined group on the basis

T. J. MORRIS AND J. C. CARRINGTON • Department of Plant Pathology, University of California, Berkeley, California 94720.

TABLE I. Properties of 30-nm, Icosahedral Plant Viruses with Monopartite Genomic RNAs of $M_r = 1.4–1.6 \times 10^6$ and Capsid Proteins of $M_r = 40,000$

Virus name	S_{20w}	RNA M_r ($\times 10^3$)	Protein M_r ($\times 10^6$)	References
Tombusviruses	131–140	1.4–1.7	38–44	Francki et al. (1985), Martelli (1981)
Carnation mottle (CarMV)	122	1.4	38	Waterworth and Kaper (1972), Nelson and Tremaine (1975)
Bean mild mosaic (BMMV)	127	1.27	ND	Waterworth et al. (1977), Waterworth (1981)
Blackgram mottle (BGMV)	122	1.4	38	Scott and Phatak (1979), Scott and Hoy (1981)
Cowpea mottle (CMeV)	122	1.4	44	Bozarth and Shoyinka (1979)
Cucumber fruit streak (CFSV)	132	1.45	47	Gallitelli et al. (1983)
Cucumber leafspot (CLSV)	—	—	—	Weber et al. (1982)
Cucumber necrosis (CNV)	133	1.4	40.3	Dias and McKeen (1972), Tremaine (1972)
Cucumber soil-borne (CSBV)	120	1.5	41	Koenig et al. (1983)
Elderberry latent (ELV)	112	1.55	40	Jones (1974)
Galinsoga mosaic (GMV)	118	1.55	36.4	Behncken et al. (1982), Hatta et al. (1983)
Glycine mottle (GMeV)	138	1.48	39	Behncken and Dale (1984)
Hibiscus chlorotic ringspot (HCRSV)	118	1.4	38	Waterworth (1980), Hearon (1984)
Melon necrotic spot (MNSV)	134	1 sp.	46	Bos et al. (1984), Gonzalez-Garza et al. (1979)
Narcissus tip necrosis (NTNV)	123	1.6	42	Mowat et al. (1976)
Pelargonium flower break (PFBV)	125	ND	41	Stone and Hollings (1973), Hollings and Stone (1974)
Pelargonium line pattern (PLPV)	115	1 sp.	41	Stone and Hollings (1977), Plese and Stefanac (1980)
Plantain 6 (PIV6)	143	1.6	45	Hammond (1981)
Saguaro cactus (SCV)	118	1.4	39	Nelson et al. (1975), Nelson and Tremaine (1975)
Tephrosia symptomless (TSV)	127	1.5	42	Bock et al. (1981), Bock (1982)
Turnip crinkle (TCV)	129	1.4	38	Hollings and Stone (1972)

of the physicochemical properties mentioned above (Francki, 1983), while data on serological relationships and biological properties such as host range and symptomatology have been used to discriminate among viruses within a group. The problem virologists have had in classifying CarMV and the other viruses listed in Table I is the degree of similarity that the viruses share with each other and with an already established virus group, the Tombusviruses (see Martelli *et al.*, Chapter 2 of this volume). Definitive members of the Tombusvirus group form a monotypic cluster of serologically interrelated viruses that induce characteristic multivesicular bodies in the cytoplasm of infected host plants (Martelli *et al.*, 1977). The viruses listed in Table I, however, are neither serologically related to each other nor to any definitive Tombusvirus, although some of them do induce multivesicularlike bodies in infected cells. This latter property is no longer considered to be taxonomically definitive for Tombusviruses (see Martelli *et al.*, Chapter 2 and page 102 of this chapter). Francki *et al.* (1985a) have provisionally listed four of these viruses as possible Tombusviruses (TCV, GMV, HCRSV, and CNV), but excluded SCV on the basis of a lower sedimentation rate and inability to induce multivesicular bodies. The validity of the tentative Tombusvirus status of TCV and SCV was first queried by Hull (1977a) on the basis of distinctive sedimentation and biological properties. This skepticism has subsequently been reinforced for TCV on the basis of its distinctive translation strategy and cytopathology. More recent studies, to be discussed in greater detail later in the chapter, have further confirmed the distinctive nature of such tentative Tombusviruses as TCV, SCV, GMV, and GMeV on the basis of their genome properties.

Hull (1977a) was the first to propose a distinct group that included CarMV, PFBV, ELV, and NTNV. Koenig *et al.* (1983) subsequently presented an expanded list of 1 viruses which they considered to be potential candidates for the Carnation mottle virus group, but with the proviso that more definitive studies were necessary. Francki *et al.* (1985b) tentatively listed eight viruses (BGMV, CMeV, CSBV, ELV, NTNV, PFBV, PLPV, and SCV) as CarMV relatives. A final decision on the taxonomic affinities of all of these viruses will require more precisely defined classification criteria.

More detailed information on the genetic organization of some of these viruses is beginning to accumulate. It is apparent from these studies that the genome organization of the Tombusviruses differs from those of CarMV and TCV, which have very similar genome organizations. In attempting to resolve the classification enigma, we are focusing our attention in this review on those viruses for which we have good information at the molecular level. The rest of the viruses listed in Table I have been included on the basis of their physicochemical similarities, even though additional studies on their genetic structures are needed. In presenting the chapter in this fashion, we hope to clarify the status of group affinities and provide a direction for future research.

II. GEOGRAPHICAL DISTRIBUTION, HOST RANGE, AND TRANSMISSION

The viruses considered in this review have relatively restricted natural host ranges and most, with the exceptions of carnation mottle and cowpea mottle viruses, have not been reported to be economically important pathogens. There is, therefore, relatively little information available on epidemiology and control for the majority of these viruses. In addition, we wish to emphasize that we are presenting a tentative grouping of physically similar viruses, which lack the serological relationship and the degree of biological cohesiveness generally associated with the more definitive plant virus groups. A common feature shared by all of the viruses is their relative ease of transmission by mechanical inoculation. It is interesting to note that many were fortuitously isolated from asymptomatic or mildly symptomatic, systemically infected natural hosts upon inoculation to a host range. The viruses all appear to be restricted in their ability to systemically infect natural host plants and, except for the four beetle-transmitted viruses, are not insect-vectored. A common feature of the ecology of most of the viruses is that their spread is slow and appears to involve vectorless soil transmission similar to that described for the Tombusviruses (Martelli, 1981). For at least one member, cucumber necrosis, a soil fungus vector has been identified. Other than those that have been distributed worldwide in vegetative propagation stock, the viruses tend to be restricted in their distribution as well. In view of their transmission characteristics and restricted distribution, the individual viruses for the most part cause diseases in only a limited number of plants. To facilitate presentation of the biology and pathology of this somewhat disparate group, we have chosen to review biological properties in the context of the assigned host categories indicated in Table II.

A. Diseases of Ornamental Crops

1. Carnation Mottle Virus

Carnation mottle virus (CarMV) occurs worldwide in cultivated carnations (*Dianthus caryophyllus*). Incidence of the virus in commercial plantings has been the subject of many investigations (e.g., Brierley and Smith, 1957; Smookler and Loebenstein, 1975; Makkouk and Shehab, 1980; Lommel *et al.*, 1982), and it is generally recognized as an important component of virus disease complexes that occur in this crop throughout the world. Although it has been frequently described as a symptomless virus, several studies have demonstrated that infection can have a detrimental effect on cut-flower production. This may be particularly significant when CarMV occurs in mixed infections with other viruses

TABLE II. Natural Distribution, Host Range, and Transmission Properties of CarMV-like Viruses.

Virus name	Natural host	Distribution	Host range	Mode of transmission
		Diseases of ornamentals		
Carnation mottle	Dianthus	Worldwide	15 families	Vegetative propagation
Narcissus tip necrosis	Narcissus	Worldwide	Narcissus	Vegetative propagation
Hibiscus chlorotic ringspot	Hibiscus	Worldwide	Malvaceae	Vegetative propagation
Pelargonium flower break	Pelargonium	England	6 families	Vegetative propagation
Pelargonium line pattern	Pelargonium	England, Yugoslavia	5 families	Vegetative propagation
		Diseases of curcurbits		
Cucumber necrosis	Cucumis	Canada	7 families	Soil, olpidium
Cucumber soil-borne	Cucumis	Lebanon	5 families	Soil, mechanical
Cucumber fruit streak	Cucumis	Crete	7 families	Soil, mechanical
Melon necrotic spot	Cucumis	Japan, USA, Europe	Cucurbitaceae	Seed, soil, beetle
Cucumber leaf spot	Cucumis	Germany	5 families	Mechanical
		Diseases of legumes		
Cowpea mottle	Vigna	Nigeria	Leguminosae	Seed, beetle
Bean mild mosaic	Phaseolus	Central America	Leguminosae	Soil, beetle
Glycine mottle	Glycine	Australia	Leguminosae	Mechanical
Tephrosia symptomless	Tephrosia	Kenya	Leguminosae	Mechanical, soil?
Blackgram mottle	Vigna	Australasia	Lebuminosae	Seed, beetle
		Diseases of other hosts		
Turnip crinkle	Brassica	UK, Yugoslavia	20 families	Beetle
Galinsoga mosaic	Galinsoga	Australia	7 families	Soil, mechanical
Saguaro cactus	Cactus	USA	6 families	Mechanical
Elderberry latent	Sambucus	USA	4 families	Vegetative propagation
Plantain virus 6	Plantago	UK	3 families	Mechanical

(Hakkaart, 1964; Brierley, 1964; Hollings *et al.*, 1977). The natural host range of CarMV is generally restricted to members of the Caryophyllaceae and especially to species of *Dianthus* L., in which it produces either a mild mottle or is symptomless (Hollings and Stone, 1970). It has also been reported to occur in *Daphne* spp., *Lactuca*, and possibly *Ranunculus*, an indication of a wider natural host range than previously recognized (Morris-Krsinich and Milne, 1977). The experimental host range includes over 30 species in 15 plant families. *Chenopodium quinoa* is an acceptable propagation and local lesion host. Despite the advent of heat treatment and meristem culture methods to eliminate it, the virus persists world-wide in commercial plantings, primarily through introduction of infected vegetative propagation stock. A recent survey in California demonstrated that infection incidence was as high as 78% in flower production green-houses (Lommel *et al.*, 1982), while other studies have reported incidence as high as 100% (Hollings *et al.*, 1977; Makkouk and Shehab, 1980). Rapid mechanical spread of the virus can occur from a few foci of infection to new transplants because of the extensive plant handling that is a common practice in commercial operations. Effective control could be achieved through rigorous indexing of propagation stock and decontamination of workers and cutting implements after plant handling (Zandvoort, 1973; Lommel *et al.*, 1983). The recent report of the isolation of CarMV from river water in Germany raises some interesting questions about its ecology and epidemiology (Koenig and Lesemann, 1985).

2. Narcissus Tip Necrosis Virus

Narcissus tip necrosis virus (NTNV) is widespread in some stocks of Narcissus cultivars in the Netherlands and the U.K., and it likely occurs wherever commercial cultivars are grown. The virus is restricted to *Narcissus* species, and failed to infect some 46 other plant species from 14 families in host range studies. Different cultivars differ in symptoms, which may vary from leaf tip necrosis to complete absence of symptoms. Diagnosis is complicated by the occurrence of eight other small isometric viruses that infect narcissus, which can be distinguished from NTNV serologically. There is no information on the mechanism of transmission other than by distribution of infected propagation stock (Mowat *et al.*, 1976).

3. Pelargonium Viruses

Several different isometric viruses with very similar physical properties have been reported in *Pelargonium* species. Among these viruses, pelargonium leaf curl is a recognized member of the Tombusvirus group (see Martelli, this volume). Pelargonium flower break virus (PFBV) was originally reported in England, occasionally in many symptomless cultivars and in a few cultivars with flowerbreaking symptoms (Stone and Hollings, 1973). The experimental host range included 15 species in six different plant families, among which *C. quinoa* was the best local lesion host and the only one in which systemic infection was reported. The virus was not transmitted by the aphid, *Myzus persicae,* nor by dodder or through seed of infected cultivars. The main mechanism of distribution was by mechanical transmission and by vegetative propagation. Thermotherapy was only moderately effective in eliminating the virus from infected cultivars.

Pelargonium line pattern virus (PLPV) has been reported in England (Stone and Hollings, 1977) as well as Yugoslavia (Pleše and Stafanać, 1980). It was recognized as a similar but serologically distinct virus from PFBV (Stone, 1980). The virus was mechanically transmissible from *C. quinoa* and *Vicia faba,* as well as to 18 additional hosts in five plant families (Pleše and Stafanać, 1980). The mechanism of transmission, other than by vegetative propagation, is unknown.

4. Hibiscus Chlorotic Ringspot Virus

Hibiscus chlorotic ringspot virus (HCRSV) occurs naturally in many cultivars of *Hibiscus rosa-sinensis,* in which it causes variable symptoms that range from a generalized mottle to ring and veinbanding patterns. It was first described on a cultivar imported to the U.S. from El Salvador, and likely occurs throughout the world wherever hibiscus is grown (Waterworth, 1980). Infection levels of 50% have been reported in nursery stock in Australia, and it has been found in cultivars from Thailand, Hawaii, the U.S., New Guinea, and possibly Nigeria (Waterworth, 1980; Jones and Behncken, 1980). HCRSV incites systemic mosaics when inoculated to other species of the Malvaceae, but it is not thought to be an important virus disease of other members of this family. It is restricted to the inoculated leaves of only a few members of four other plant families in experimental host range tests. *C. quinoa* and *Cyamopsis tetragonoloba* are suitable local-lesion hosts. The virus is not vectored by the aphid *Myzus persicae* or by the Mexican bean beetle, and no mechanisms other than mechanical transmission and vegetative propagation appear to be important in its distribution.

B. Diseases of Cucurbits

Reports of as many as five viruses with properties similar to those outlined for this group have been described as infecting cucurbit hosts. Although they appear to be distinct, a complete serological comparison of all the isolates has not been published. A common feature with most of these viruses is their association with the soil.

1. Cucumber Necrosis Virus

Cucumber necrosis virus (CNV) has been found in naturally infected greenhouse cucumbers (*Cucumbus sativus*) in Canada (McKeen, 1959). Cucumber was the only host to become systemically infected, although the virus was capable of causing local lesions on a fairly broad range of plants in five different plant families (Dias and McKeen, 1972). The virus was not transmitted through cucumber seed or by the melon aphid (*Aphis gossypii*). Natural spreads of CNV was facilitated in soil by zoospores of the fungus *Olpidium cucurbitacearum*, recently described as *O. radicale* (Stobbs *et al.*, 1982). CNV does not represent a disease threat to other crops because the host range of this vector is limited to species of the Cucurbitaceae (Dias, 1970).

2. Cucumber Fruit Streak and Cucumber Leafspot Viruses

Cucumber fruit streak virus (CFSV) was recently isolated from cucumbers with cracked and chlorotically streaked fruit grown under plastic in Crete (Gallitelli *et al.*, 1983). This virus, which is serologically unrelated to CNV, produced local lesions in 26 of 46 plant species tested in seven plant families, as well as systemic symptoms in *C. sativus* and several *Nicotiana* species. Soil transmission trials suggested that the virus could be acquired by cucumber roots through the soil in the absence of a vector, although the field syndrome could not be reproduced by either mechanical or soil transmission. The disease was not widespread in Crete, and occurred in high incidence in only one field.

Cucumber leafspot virus (CLSV) was isolated from glasshouse cucumbers in Germany with systemic spot symptoms (Weber *et al.*, 1982). The host range was limited to 15 species in five families, and the virus was readily transmitted mechanically. It was determined to be closely serologically related to CFSV (Weber and Stanarius, 1984; Weber *et al.*, 1986) and additional isolates from Lebanon (Vetten, unpublished information).

3. Cucumber Soil-borne Virus

Another virus, cucumber soil-borne virus (CSBV) was isolated from the roots of cucumber seedlings planted in vegetable fields in coastal Lebanon (Koenig *et al.*, 1983). The virus was similar to CNV and CFSV in its ability to cause local lesions on a broad range of host plants, but it did not infect cucurbits systemically, and it was serologically distinct from CNV (Koenig *et al.*, 1983) and also CFSV (Weber and Stannarius, 1984; Koenig, unpublished information). No soil vector was identified, and the virus was considered not to be of economic importance because of its inability to systemically invade plant species commonly grown in the area.

4. Melon Necrotic Spot Virus

Melon necrotic spot virus (MNSV) has been identified in Japan (Kishi, 1966), and California (Gonzalez-Garza et al., 1979) in association with a systemic disease of muskmelons and in the Netherlands in glasshouse cucumbers (Bos et al., 1984). It is serologically unrelated to other cucurbit infecting viruses (CNV, CSBV, CLSV, and CFSV), and has a host range restricted to the Cucurbitaceae. It has been reported to be soil-transmitted by the same vector as CNV (Yoshida et al., 1980; Furuki et al., 1980) and in root leachates of infected plants (Bos et al., 1984). It was also reported to be seed-transmitted in muskmelon (Gonzalez-Garza et al., 1979) as well as by Diabrotica beetles (Coudriet et al., 1979). Seed transmission in cucumber was deemed unlikely because of the limited occurrence of the disease in Holland (Bos et al., 1984), where control was achieved through the use of hygienic measures.

C. Diseases of Legumes

Five serologically distinct, physically similar viruses have been found to occur naturally in leguminous hosts and, except for cowpea mottle virus (CMeV), have host ranges restricted to the Leguminosae. Three of the viruses share the common property of beetle transmission and two appear to be soil-transmitted without the assistance of a vector.

1. Cowpea Mottle Virus

Cowpea mottle virus (CMeV) has been reported to occur only in Nigeria, where it is widely distributed in cowpea (Vigna unguiculata) and bambarra groundnut (Voandzeia subterranea) (Bozarth and Shoyinka, 1979). It causes mottling, bright yellow mosaic, and leaf distortion in infected cowpea. Yield losses of greater than 75% have been reported. The host range has not been thoroughly tested, but systemic infections of many members of the Leguminosae and several members of the Solanaceae were recorded. C. quinoa was reported as a useful local lesion host. The virus was transmitted in up to 10% of seeds collected from diseased cowpea (Shoyinka et al., 1978), but subsequent studies demonstrated that seed transmission was relatively unimportant in the natural spread of the disease. This probably accounts for its limited geographical distribution (Allen et al., 1982). CMeV is principally transmitted by the galerucid beetle Paraluperodes lineata (Bozarth and Shoyinka, 1979). The virus is an important disease threat because of extreme virulence on cowpea, the potential for worldwide distribution in infested seed, and the fact that it has been known to rapidly spread to uninoculated plants in the absence of a known vector. Attempts to control the disease through the use of resistant cultivars have been reported (Allen, 1980).

2. Bean Mild Mosaic Virus

Bean mild mosaic virus (BMMV) was first recognized as a component in a virus disease complex of bean plants (*Phaseolus vulgaris*) sent to the Beltsville Research Center for diagnosis by the Central American Varietal Nursery in El Salvador (Waterworth *et al.*, 1977). The virus produced only mild mosaic symptoms or was latent in bean cultivars tested. It infected six of 18 other legume species, and only two of 42 nonleguminous hosts tested, one of which was *C. quinoa*. No good local lesion host is known. BMMV was readily transmitted by five species of beetles in greenhouse tests, two of which (*Diabrotica balteata* and *Cerotoma raficornis*) were suggested as potentially important natural vectors because of their general distribution in El Salvador (Hobbs, 1981). No information exists on possible seed transmission of the virus, but rapid spread through the soil by an unknown mechanism has been reported (Hampton and Hancock, 1981). The virus is extremely infectious in legumes and like CMeV, it has been reported to spread rapidly among beans in the greenhouse (Waterworth, 1981).

3. Blackgram Mottle Virus

Blackgram mottle virus (BGMV) has been found in naturally infected *Vigna mungo* and in Asia, Australasia, and India (Scott and Hoy, 1981). Naturally infected blackgram plants exhibit mottling and distortion. The experimental host range is reportedly limited to members of the Leguminosae with all cowpea varieties and soybean being susceptible (Phatak, 1974). Different bean cultivars serve as useful propagation and assay species. The virus was seed-transmitted in blackgram, and was also vectored by two different species of beetles (*Cerotoma trifurcata* and *Epilachna varivestis*) tested in greenhouse experiments (Scott and Phatak, 1979).

4. Tephrosia Symptomless Virus

Tephrosia symptomless virus (TSV) is a legume-restricted virus that was isolated from symptomless *Tephrosia villosa* plants in eastern Kenya (Bock *et al.*, 1981; Bock, 1982). It has not been isolated from any economically important legume, nor was it detected in any other wild leguminous plants in the area. TSV caused a systemic mottle in inoculated soybean (*Glycine max*), which is a satisfactory propagation host. It produced only localized infections in *Phaseolus* species. The host range was restricted to the Leguminosae. No vector was identified, and seed transmissibility could not be demonstrated. The virus was isolated from one of five leachates of soil in which infected *T. villosa* were growing, but it did not appear to be transmitted through the soil. The distribution of the virus suggested an efficient insect vector, but potential beetle vectors were not evaluated.

5. Glycine Mottle Virus

Glycine mottle virus (GMeV) is a legume-restricted virus isolated from naturally infected woolly glycine (*Glycine tomentella*) in eastern Australia (Behncken and Dale, 1984). The host range is limited to a few legume species that include the systemic host, *Glycine soya*, and the local lesion host, *Phaseolus vulgaris*. Mechanisms of transmission, other than by mechanical means, have not been evaluated and the virus is not considered to be of economic importance because of its limited host range and restricted distribution.

D. Viruses of Other Hosts

1. Turnip Crinkle Virus

Turnip crinkle virus (TCV) has been identified as the cause of mottling and stunting of *Brassica* crops in Scotland, England, and Yugoslavia, although it is apparently neither widespread nor common (Hollings and Stone, 1972). It can cause severe symptoms in turnip (*Brassica rapa*) and milder mottles in other *Brassica* species. It has a wide experimental host range, infecting species in about 20 plant families. *Chenopodium* species are satisfactory local-lesion hosts, and *Brassica* species are good systemic hosts for purification. TCV is beetle-transmitted by larval and adult flea-beetles of the genera *Phyllotreta* (nine species) and *Psylloides* (two species). Transmission is nonpersistent, so that the vectors rarely remain infectious for more than a day. There are unconfirmed reports of transmission by other insects as well (Hollings and Stone, 1972). No published studies on seed or soil transmission of the virus exist, possibly because it does not appear to be of sufficient importance as a pathogen to have warranted further epidemiological study.

2. Saguaro Cactus Virus

Saguaro cactus virus (SCV) was first identified in symptomless saguaro cactus (*Carnegia gigantea*) in Arizona in connection with a cactus virus survey of the area (Milbrath and Nelson, 1972). A second strain was subsequently identified from a slower growing, colored variety of the cultivated cactus *Chamaecerus sylvestrii* in Germany (Samyn and Welvaert, 1978). It is not associated with any discernible disease in the cactus hosts, and it has been mechanically transmitted to several species in six families. The virus systemically infects *Chenopodium capitatum*, which can serve as a propagation host, and *C. quinoa*, which is a suitable assay host. Field observations suggest a slow rate of spread, and no vectors have been identified. It was not found to be transmitted through seeds of infected cacti and is not considered to be of any economic importance (Nelson and Tremaine, 1975).

3. Galinsoga Mosaic Virus

Galinsoga mosaic virus (GMV) was found when potato weed (*Galinsoga paraviflora*) roots collected in Queensland, Australia were assayed for the presence of virus (Behncken, 1970). Although GMV is of no known economic importance, it has a relatively broad experimental host range, including species in seven families in which it is primarily confined to the inoculated leaves. *Chenopodium* species are satisfactory assay hosts (Behncken *et al.*, 1982). The virus was not transmitted through seed of *G. parviflora*, and attempts failed to demonstrate transmission by two aphid species or by the fungus *Olpidium brassica*. Vector-unassisted soil transmission was the mechanism suggested for distribution of the virus (Shukla *et al.*, 1979).

4. Elderberry Latent Virus

Elderberry latent virus (ELV) was found in naturally infected American Elder (*Sambucus canadensis*) imported into the U.K. from the U.S. (Jones, 1974). The virus is primarily latent in naturally and experimentally infected cultivars, as well as in the majority of the 25 other species in four plant families tested in host range studies. *Chenopodium quinoa* is a usable propagation and assay host (Jones, 1974). There are no known vectors, and distribution of infected propagation materials is likely responsible for dissemination.

5. Plantain Virus 6

Plantain virus 6 (PlV6) was isolated from a single symptomless plant of *Plantago lanceolata* in England (Hammond, 1981). It was mechanically transmissible to two *Nicotiana* species, in which it caused systemic symptoms. It infected beans and celery without producing visible symptoms.

III. PURIFICATION

The viruses are relatively stable and occur in high concentration in infected tissue of the respective propagative hosts. Purification by conventional procedures is generally applicable. A typical protocol includes extraction in neutral phosphate buffer (0.01 to 0.6 M) containing mild reducing agents followed by clarification with organic solvents (n-butanol alone or in combination with chloroform), differential centrifugation, and sucrose density gradient centrifugation. Reported yields range from 30–50 mg/kg for CFSV and NTNV, respectively, to 800 mg/kg for GMeV. In view of the lability of many of these viruses to mild alkali conditions in the presence of EDTA (see next section), extraction buffers such as those reported for TSV, MNSV, and CSBV should be used with caution. Clarification of SCV containing plant extracts by low-speed centrifugation in the absence of organic solvents was achieved by Nelson and Tremaine (1975). They homogenized tissue in 0.2 M sodium acetate buffer at pH 5.0, followed by several cycles of differential centrifugation in lower ionic strength pH 5.0 acetate buffer. This procedure, modified to include a PEG precipitation step, proved most effective for the purification of CarMV (Lommel et al., 1982), as well as TCV, PFBV, GMV, and CSBV (T. J. Morris, unpublished information). Similar acidic extraction protocols have been reported for the purification of GMeV, PlV6, and CNV. Procedures employing an acidification step, however, should be used with some caution because many of the viruses have been reported to be isoelectric at acidic pHs, and the possibility of precipitating the virus during clarification exists.

IV. PROPERTIES OF THE VIRUS PARTICLES

A. Particle Morphology

The morphology of negatively stained preparations of the virus particles listed in Table I are very similar when examined under the electron microscope. The virions appear as rounded to slightly angular particles of 28 to 34 nm in diameter with a granular to somewhat rough surface substructure, the details of which are not easily resolved. Although the appearance of particles in the electron microscope is of limited value in distinguishing between CarMV-like viruses and members of the Tombus- and Dianthovirus groups, the somewhat characteristic appearance is sufficiently distinctive from that of the Sobemo-, Tymo-, and Cucumoviruses to be of some value in determining group affinity (Francki, 1983; Hatta et al., 1983). The CarMV-like viruses are also similar to Tombusviruses with respect to particle integrity when stained with uranyl acetate, but have variable degrees of lability and stain penetration in phosphotungstate (Martelli, 1981). Published electron micrographs of BMMV, BGMV, CFSV, ELV, GMeV, HCRSV, NTNV, PFBV, TSV, and TCV all contained either disrupted or penetrated particles when stained with phosphotungstate, while the particles of CarMV, CSBV, and MNSV did not appear to be affected. Micrographs of uranyl acetate-stained particles only were presented for the remaining viruses listed in Table I. (Note:

The references given in Table I apply to the information given in this section for each of the individual viruses, and they are not reiterated in the text.)

B. Physical and Chemical Properties

1. Sedimentation Properties

The viruses listed in Table I all sediment as single, homogeneous components on rate zonal centrifugation analysis, with sedimentation coefficients that vary from the lowest value of 112 S for ELV, to a high value of 143 S reported for PlV6. The majority fall within the range of 118 to 130 S. This wide degree of variation has been one of the factors responsible for the reluctance to group these viruses. The low sedimentation rate of SCV (118 S), for example, was cited as one of the criteria for excluding it from the Tombusvirus group, while retaining two other viruses (GMV and HCRSV) with the same low rate of sedimentation (Francki et al., 1985a). Some of the variation evident in the viruses might well be due to the different methods and buffer conditions used by different investigators, particularly in view of the tendency of many of the viruses to swell and exhibit lower sedimentation rates between pH 7 and 8 (see next section). Support for this argument comes from our recent comparative study of six of these viruses (CarMV, CSBV, GMV, PFBV, SCV, and TCV) purified by the same procedure from the same host. We were unable to observe any evidence for sedimentation differences in cosedimentation experiments, and all viruses had a significantly slower sedimentation rate (ca. 125 S) than that observed for TBSV at 135 S (T. J. Morris, unpublished information).

Estimations of density in CsCl have been more uniform for those viruses evaluated (CarMV, BGMV, CMeV, CFSV, CSBV, ELV, HCRSV, NTNV, PlV6, SCV, and TSV), with values ranging from 1.34 to 1.36 g/cm^3. The viruses were all stable and sedimented as a single component. The variable sedimentation data would seem incongruous with buoyant density data in view of the observation that the virus with the lowest $S_{20}w$ has the highest reported density (ELV) while the virus with the fastest sedimentation rate has one of the lowest reported density values (PlV6). Hull (1977a) stated that a characteristic feature of CarMV is the presence of two density species in Cs_2SO_4 gradients. Of those viruses analyzed, CFSV, PlV6, and TSV also sedimented as two species while TCV (Hull, 1977b) reportedly sedimented as a single species.

2. Virus Particle Stability

The CarMV-like viruses are generally stable in plant sap with infectivity surviving from 2–6 weeks at room temperature for all but CMeV and ELV, which failed to retain infectivity longer than a few days. Thermal inactivation points were in excess of 80°C for the majority of the viruses, with only HCRSV, SCV, MNSV and CMeV inactivated by 10-min exposures to temperatures of lower than 80°C.

The instability of purified virus preparations under mild alkali conditions (pH 8.25) in the presence of salt (1 M NaCl) and low concentrations of EDTA (10 mM) was one of the first criteria used to define a putative carnation mottle virus group (Hull, 1977a). Subsequent studies have demonstrated that CFSV (Gallitelli *et al.*, 1983), PlV6 (Hammond, 1981), and GMV, SCV, and PFBV (T. J. Morris, unpublished information) are similarly unstable, while TSV (Bock, 1982), TCV, and TBSV are much less affected by the presence of EDTA under conditions of mild alkali (T. J. Morris, unpublished information). Similar studies of the effects of pH in the presence of SDS have further established a variable degree of capsid stability for viruses of this group (Boatman and Kaper, 1976; Ronald and Tremaine, 1976). TCV and SCV were sensitive to low SDS concentrations in the presence of EDTA at pH 7, but not at pH 5.0. CarMV, CNV, and tobacco necrosis virus (TNV) were considered moderately SDS resistant, while TBSV was the most resistant of the viruses of interest examined. These types of studies have been useful in elucidating the structural similarities of TCV and TBSV, which will be discussed in more detail in the next section. It is, however, important to recognize that all of these viruses, with the possible exception of TSV, share the tendency to swell and/or dissociate under mild alkali conditions in the presence of EDTA and high salt, and that the degree of instability is somewhat variable.

3. Virus Particle Composition

The CarMV-like viruses, like the Tombusviruses, have a simple chemical composition with RNA contents estimated from 17–22%, the remainder being composed of protein. The 14.1% RNA estimate for HCRSV seems low in the context of the sedimentation and density estimates (Waterworth, 1980) while the 23% RNA estimate for ELV is relatively high. Where estimates of particle molecular weight have been made, they fall into the same range as that reported for TBSV (8–9×10^6). Particle weight estimates of 7.0×10^6 for CMeV (Bozarth and Shoyinka, 1979) and ELV (Jones, 1974) are on the low side of this group.

The molecular weight estimates for the capsid subunits of all of the viruses (except BMMV) have been reported from electrophoretic mobility measurements made from SDS gels. The original values, which vary from $M_r = 36,400$ to $46,000$, are given in Table I. The accuracy of such estimates should be interpreted with some reservations (Francki, 1983). Our unpublished estimates of coat protein molecular weights are generally in good agreement with the literature for CarMV (38K), CSBV (39K), GMV (37K), TCV (38K), and TBSV (41K), and with calculated molecular weights from nucleotide sequence data for CarMV (Guilley *et al.*, 1985) and TCV (Carrington *et al.*, 1986). We have, however, observed electrophoretic mobilities for the capsid proteins of PFBV (ca. 35K) and SCV (ca. 36K) which are significantly different from published values for these viruses (see Table I).

Minor species of approximately twice the size of the capsid subunit have been reported for TBSV and TCV (Ziegler et al., 1974) as well as for SCV (Nelson and Tremaine, 1975) and HCRSV (Waterworth, 1980). Recent structural studies with TBSV and TCV have established this species to be a stable, covalent dimer of the capsid protein (Stockley et al., 1986) and not a polymerase enzyme as suggested earlier by Butler (1970). Additional minor, lower molecular-weight species have been observed in CarMV, GMeV, NTNV, SCV, and TCV preparations and are presumed to be capsid protein degradation products. A prominent species with a slightly faster mobility than the coat protein is consistently observed in preparations of CarMV (Carrington and Morris, 1985).

The amino acid compositions of CarMV and TCV have been determined from nucleotide sequence data (Carrington, 1986) and for CNV (Tremaine, 1972), GMV (Skotnicki and Gibbs, 1981), HCRSV (S. S. Hearon, personal communication), and SCV (Nelson and Tremaine, 1975) by direct analysis. The compositions are similar, but fail to provide any real clue to possible relatedness.

4. Electrophoretic Properties of the Virus Particles

The electrophoretic properties of only a few of the viruses have been studied. Electrophoretic homogeneity has been established for CNV (isoelectric point, 3.9) (Tremaine, 1972), ELV (isoelectric point, 4.8) (Jones, 1974), NTNV (Mowat et al., 1976) and BGMV (Scott and Phatak, 1979). In contrast, SCV preparations contained three electrophoretic species whose proportion varied on different isolates, and all were infectious (Nelson and Tremaine, 1975). Hearon (1984) demonstrated two infectious electrophoretic components in preparations of HCRSV, and established that the isolates had different amino acid compositions. Similar electrophoretic analysis of CarMV revealed multiple species, which likely represented electrophoretic variants of the virus because stable populations containing a single species could be readily obtained upon single-lesion passage of the virus. All of the viruses used in that comparative study (CarMV, CSBV, GMV, PFBV, SCV, and TCV) contained a single homogeneous electrophoretic component after single-lesion passage (Morris and Carrington, 1984).

C. Virus Particle Structure

Refined structural studies on members of this group, beyond the stage of general particle stability discussed in the previous section, have been performed only on turnip crinkle virus. Several recent studies, including the determination of the particle structure at 3.2 Å resolution (Hogle *et al.*, 1986), the determination of the coat protein sequence (Carrington *et al.*, 1986), and the definition of the mechanism of *in vitro* reassembly (Sorger *et al.*, 1986), have made TCV one of the best-understood viruses from a structural perspective. We shall focus our attention primarily on TCV structure in this section, making appropriate comparisons to the other viruses, for which more limited information is available.

1. Subunit Structure

The complete primary structure of the TCV coat protein has been determined from the nucleotide sequence of a cDNA clone of the 3'-end of the viral genome, and a portion of the amino acid sequence was confirmed directly (Carrington *et al.*, 1986). Detailed structural information is also available for the subunit protein of other structurally similar plant viruses such as TBSV (Harrison *et al.*, 1978) and southern bean mosaic virus (SBMV) (Abad-Zapatero *et al.*, 1980). In addition, the subunit primary structure is also available for CarMV from the nucleotide sequence (Guilley *et al.*, 1985), but refined structural studies have not been performed with this virus. The linear maps of the TCV, TBSV, and SBMV subunit domains, depicted in Fig. 1, show remarkable similarity with respect to the structural organization of the proteins. The TCV subunit is organized into four distinct modules as defined by Harrison (1983) for TBSV. The R domain, present at the N-terminus of the subunit, is connected by an arm to the S domain, which forms the shell of the virion. The S domain is tethered to the P domain, which projects outwardly from the virion surface. Although otherwise structurally similar to TCV and TBSV, the P domain is entirely missing in the smaller SBMV subunit (M_r = 28,500), accounting for its smaller and smoother appearance in the electron microscope. Significant amino acid sequence homology exists among the four subunit proteins in the arm and S domains (Carrington *et al.*, 1986). There is a single gap of one amino acid in the TCV and CarMV sequences, which agree at about 31% of the positions in the arm

FIGURE 1. A comparison of the linear structure maps of the coat proteins of (A) TCV, (B) TBSV, and (C) SBMV. The structural domains of the proteins as described in the text are indicated above the lines. The domain sizes in number of amino acids are represented. The maps for TBSV and SBMV are modified after Harrison (1983) and the TCV map was drawn from the data of Hogle *et al.* (1986).

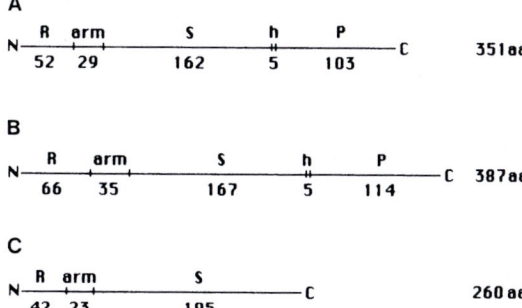

and S domains together and at a level of 37% in the S domain alone (Carrington, 1986). The TCV and TBSV sequence alignments, which agree in about 23% of the positions, have five gaps totaling 15 amino acids (Hogle *et al.,* 1986). Alignment of the TCV and SBMV subunits within the arm and S domains results in six gaps of 41 amino acids and about a 20% homology within the aligned regions. These comparisons suggest that a higher degree of similarity exists between TCV and CarMV than between any other combination, with SBMV being the most distantly related of the four viral proteins. The P domains of TCV and TBSV have nearly identical structures but, surprisingly, little sequence homology is evident. Comparative alignment of the P domain of CarMV is not possible in the absence of high-resolution structural information.

TCV and TBSV are also similar with respect to the presence of a minor structural protein with an apparent molecular weight of 80,000 in the capsid (Ziegler *et al.,* 1974). It has been recently shown that the 80K protein is a stable dimer of two coat protein subunits covalently linked near their N-termini, and that the capsids contain one copy of the covalent dimer plus 178 monomer subunits (Stockley *et al.,* 1986). Similar minor species of about twice the molecular weight of the coat protein subunit have been reported for HCRSV (Hearon, 1984) and SCV (Nelson and Tremaine, 1975), but for no other members of the group.

The structure of TCV has been known for some time to be very similar to that of TBSV (Butler, 1970; Leberman and Finch, 1970; Finch *et al.,* 1970). Proteolytic cleavage of expanded virions suggested a similar domain organization for the viral subunits (Golden and Harrison, 1982), and high-resolution X-ray crystallography showed that the similarity extended to details of polypeptide chain folding and subunit interaction, despite the differences in the amino acid sequence (Harrison, 1983). More recently, Hogle *et al.* (1986) took advantage of this similarity to resolve the structure of TCV at 3.2 Å, using the electron density of TBSV as a starting point for phase refinement by noncrystallographic symmetry. This study allowed for the precise determination of the three-dimensional subunit structure diagrammatically presented in Fig. 2.

The subunit polypeptide backbone of TCV is extremely similar to that of the TBSV subunit. The α-carbon backbone in the arms and S domains of the two viruses are nearly identical except for certain loops between β-strands (see Fig. 2 for S domain folding of the TCV subunit). The S domain is packed in precisely the same way relative to the icosahedral axis. The folded structures of the P domains are almost identical, but are oriented somewhat differently with respect to the S domains. There is a greater tilt of the P domain in TCV, which places the tip at a smaller radius, accounting for a smaller diameter than TBSV.

By analogy to TBSV (see Harrison, 1983, for a detailed review on TBSV structure), there are three distinct packing environments for a subunit in the virus particle, denoted as A, B, and C in Fig. 2. The A and B subunit conformations are very similar in the relative orientation of the

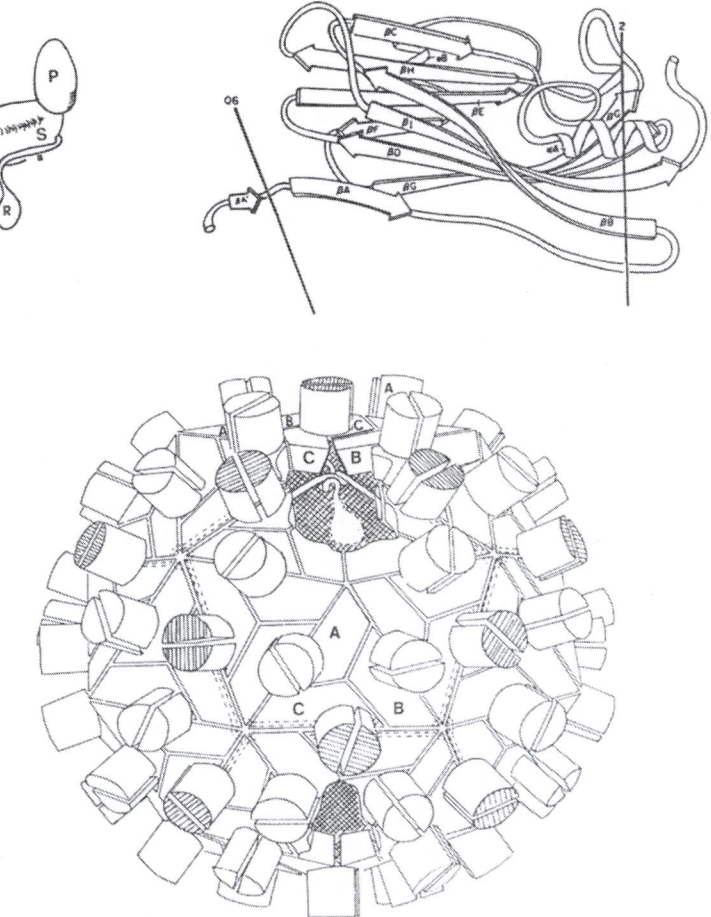

FIGURE 2. The structure of turnip crinkle virus as described by Hogle *et al.* (1986). (A) A representation of the folded subunit of TCV showing the structural domains. (B) Structure of the arm and S domain of the subunit. (C) Packing of the subunits in the virion showing the interdigitating arms of the C-subunits at the cutaway vertex, and the relative positions of the A-, B-, and C-subunits in the shell. From Sorger *et al.* (1986), courtesy of the authors.

hinge angle ("up") between the S and P domains, and the connector arms are disordered. The C subunits have a different orientation of the S and P domains ("hinge angle down") than do the A/B subunits, and the C arms are ordered, extending along the inner edge of the S domain. The arms of the C subunits are important for assembly (see next section) in that they interdigitate in sets of three (termed a β-annulus) forming the framework for a T = 1 cagelike lattice upon which the A/B subunits build to form the final T = 3 structure of the virion. The structure is essentially constructed from two kinds of dimer: the C/C dimers with folded arms that form the cage, and the A/B dimers with the unfolded arms, which fill the shell. The dimeric nature of the subunits is important in the assembly of the virions (Hogle *et al.*, 1986; Sorger *et al.*, 1986).

Expansion of the virus particle at alkaline pH after removal of divalent cations (Ca^{2+}) is another common feature shared by TCV, TBSV, and a number of other plant viruses (Harrison, 1983). The characteristics of the site that regulates the transition in TBSV have been defined, and consist of five aspartic acid side chains and two adjacent sites for divalent cations (Hogle *et al.*, 1983). The analogous site in TCV is different and may involve one cation-mediated salt bridge and one lys-asp salt bridge (Hogle *et al.*, 1986). The analogous site also exists in the CarMV subunit, but again with distinctive features. Thus, the functionally significant site is present for all three viruses, but the number of negative charges and their distribution vary considerably. The variation at this site could account for the previously discussed differences in particle stability.

The TCV structure retains the network of polar interactions around the cation site and at most other interfaces between the S domains that are important in stabilizing the TBSV particle (Harrison, 1980; Hopper *et al.*, 1984). Some of the important contact residues are identical in the two viruses, and there are conservative changes at other positions. The interactions between the subunit interfaces in the TCV particle are, however, rearranged relative to TBSV.

2. Virus Assembly

TCV has proved to be an excellent model for studying assembly mechanisms of spherical viruses because it can be dissociated and reassembled readily in solution (Golden and Harrison, 1982; Sorger *et al.*, 1986). Detailed examination of the *in vitro* assembly revealed that TCV assembly is controlled by specific interaction of a small number of protein subunits with the viral RNA. In an elegant series of experiments, Sorger *et al.* (1986) identified a stable ribonucleoprotein complex (rp-complex) that consisted of six monomer subunits and the 80K dimer attached to the viral RNA after dissociation of the virus at elevated pH and ionic strength. They demonstrated that the rp-complex could be regenerated from naked RNA and coat protein subunits, and that its formation was important in the selective assembly of the virus reconstituted under physiological conditions. A model for assembly (see Fig. 3) was proposed in which three sets of dimers interact with a particular site on the viral RNA (as yet unidentified) to form an initiation complex in which the subunits are connected by a β-annulus and hence take on the C/C conformation. The next set of dimers to associate with the complex adopts the A/B conformation, since additional arms cannot contribute to the β-annulus already present. Intersubunit contact locks the initial dimers into the C/C conformation, and creates new sites for β-annulus formation at the adjacent ends of the C/C dimers already in the complex. Completion of the structure proceeds by alternate addition of dimers into the

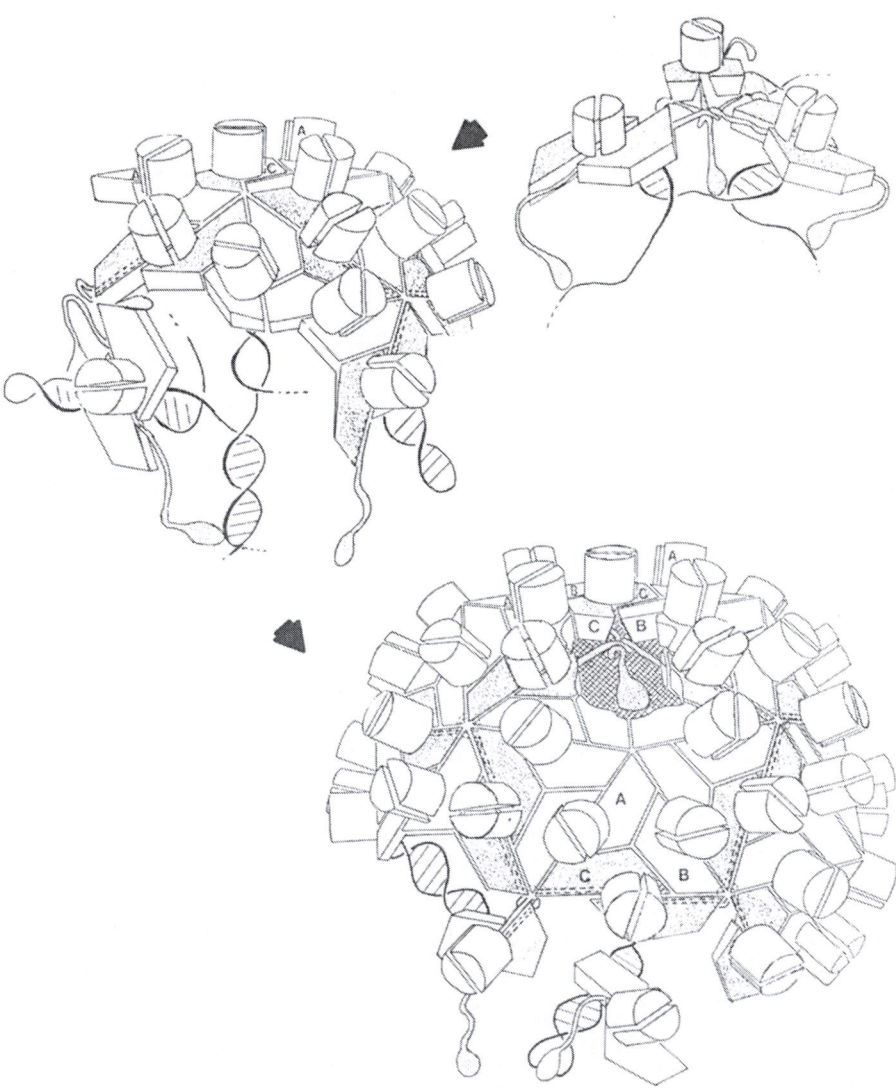

FIGURE 3. Model of TCV self-assembly. The initiation complex forms at a specific site on the viral RNA as three sets of subunit dimers connected with a β-annulus. The R domains of additional subunits that approach the initiation complex interact with the viral RNA and are drawn into the complex. The arms of the first new set of dimers to associate with the complex cannot interact with the existing subunit arms contributing to the β-annulus that is present and they consequently assume the A/B conformation. Intersubunit contact locks the original complex dimer into the extended arm configuration of C-subunits and creates new sites for β-annulus formation at the other end of the C/C dimers. The shell grows by alternate "filling out" steps by A/B dimers and recruitment of C/C dimers into the β-annulus. Courtesy of Sorger et al. (1986).

β-annulus, which assumes the C/C conformation, and "filling-out" steps by dimers that retain the A/B conformation. The R domains assist in the "docking" of the subunits and in gathering the RNA into the shell. Correct closure of the structure is guaranteed by the presence of a unique nucleation site that precludes multiple initiation on the same RNA. Elegant examples of abortive assembly into "monster" particles produced under non-ideal assembly conditions lend graphic support for the model proposed. The authors speculate that the covalent 80K dimer may play a role in the final closure of the shell, but also suggest that it may have a more important function in the initiation of disassembly. The model is an attractive one by virtue of its simplicity: the adoption of an A/B or a C/C conformation is determined completely by interactions that are possible at the time of dimer addition.

A final feature of TCV assembly that should be mentioned is the requirement of an intact arm and R domain to produce a T = 3 particle (Crowther and Amos, 1971; Harrison, 1983). Proteolytic removal of the arm domain from 38K subunits produced 30K derivatives that assembled into T = 1 particles of smaller diameter (20 nm). Such subunits were selectively excluded from particles in competitive reconstitution experiments, further illustrating the importance of the arms in accurate assembly (Sorger *et al.*, 1986). The presence of structurally similar particles serologically related to the virus particle were previously described in association with TCV (Tremaine and Chidlow, 1972) and were also recognized in preparations of CNV (Tremaine, 1972). Their presence may suggest a similar mechanism of assembly for that virus.

V. SEROLOGICAL PROPERTIES

The CarMV-like viruses are generally good immunogens, and antisera suitable for virus identification using such simple techniques as immunodiffusion have been reported for all the viruses listed with the exception of P1V6. The majority of the viruses have been extensively evaluated for possible serological relatedness to structurally similar small RNA viruses in immunodiffusion tests that have generally included antisera to most of the viruses that were available at the time the virus was described. In no case has serological cross-reactivity been reported among any of the viruses listed in Table I except the closely related CFSV and CLSV (Weber and Stanarius, 1984; Weber *et al.*, 1986). Not all of the viruses, however, have been tested reciprocally. Host range differences generally rule out the likelihood of close relationships between most tentative members of the group. The use of more sensitive tests such as indirect ELISA for the detection of distant serological relationships (Koenig, 1981; Lommel *et al.*, 1982) has not been entirely satisfactory when applied to the study of members of this group (T. J. Morris, unpublished information). Burgermeister and Koenig (1984) used an electro-blot immunoassay (Western blot) to detect group-specific cross-reactivity between a large number of small RNA viruses. Several of the Tombusvirus and CarMV-like viruses cross-reacted in these tests. Curiously, several of the Tombusvirus antisera also reacted to Potexviruses while others seemed to react to all antisera tested. In similar analyses, we were unable to detect any signal on Western blots of the coat proteins of TCV, GMV, CSBV, SCV, PFBV, TBSV, and SBMV probed with antisera prepared to CarMV, while a significant level of cross-reactivity was detected to GMV protein using an antiserum prepared to SDS-disrupted TCV (T. J. Morris, unpublished information). Distant serological relationships between Tombus- and CarMV-like viruses may not be unexpected in view of the structural similarity of coat proteins of TBSV, TCV, and CarMV, but it will likely require the diligent application of monoclonal antibody techniques to clearly resolve any questions of distant serological relatedness.

VI. GENOME PROPERTIES AND REPLICATION

A. Genome Structure

1. Size and Properties of the Viral RNA

It is generally recognized that a single species of ssRNA comprises the genome of members of this group. RNA molecular weight estimations from electrophoretic studies must be interpreted with caution, however, as many of the determinations were made in the absence of effective RNA denaturants. The values reported in Table I vary from 1.27×10^6 for BMMV to 1.6×10^6 for NTNV and PlV6. The low value reported for BMMV and the absence of an estimate for the coat protein size suggests that it is probably premature to seriously consider its inclusion in the group. Similarly, the revised estimate for ELV at 1.32×10^6 in glyoxal gels and the presence of additional species of $M_r = 1.2$ and 0.54×10^6 suggest that the status of this virus should also be considered tenuous (Jones and Badenoch, 1981). Although many of the reported molecular weight estimates of the genomic RNAs of these viruses overlap the values reported in the literature for true Tombusviruses, two recent comparative studies have indicated that this may be due to inaccuracies inherent in the methods used. A direct comparison by Gallitelli et al. (1985) of several Tombusvirus RNAs with identical estimated lengths of 4700 nt (ca. $M_r = 1.6 \times 10^6$) clearly showed that the genome of TCV and GMeV was smaller (c. 3900 nt or $M_r = 1.3 \times 10^6$) and that SCV and GMV RNAs had even slightly faster electrophoretic mobilities (c. 3500 nt or $M_r = 1.3 \times 10^6$). In a complementary study, T. J. Morris (unpublished information) estimated the TBSV RNA length to be 4800 nt (c. $M_r = 1.63 \times 10^6$) relative to CarMV, TCV, CSBV, PFBV, and SCV at 4000 nt (c. $M_r = 1.36 \times 10^6$)

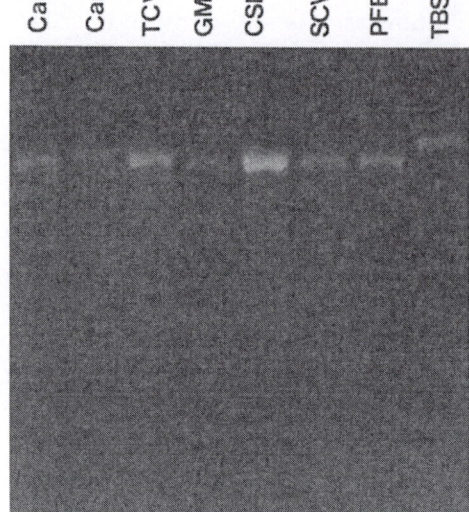

FIGURE 4. Electrophoretic analysis of Carnation mottle viruslike genomic RNAs in a 1.5% agarose gel containing 6% formaldehyde. The viral RNAs were denatured at 65°C in the presence of 6% formaldehyde and 50% formamide prior to electrophoresis. The RNA was visualized by staining with ethidium bromide. The lanes contain viral RNAs of carnation mottle virus, strain P (CarP) and strain B (CarB); turnip crinkle virus (TCV); galinsoga mosaic virus (GMV); cucumber soil-borne virus (CSBV); saguaro cactus virus (SCV); pelargonium flower break virus (PFBV); and tomato bushy stunt virus (TBSV).

and GMV at 3900 nt (*ca.* $M_r = 1.33 \times 10^6$). As can be seen from Fig. 4, all the viral RNAs (except for GMV RNA) have the same mobility as CarMV RNA in denaturing gels, and the values are in close agreement with the actual genome length of 4003 nt determined for CarMV RNA (Guilley *et al.*, 1985). These comparative studies provide the first evidence that viruses such as CarMV, CSBV, GMV, GMeV, PFBV, SCV, and TCV are genomically smaller than true Tombusviruses.

The base compositions for BMMV, BGMV, CNV, CSBV, GMV, HCRSV, SCV, and TCV have been reported. The values provide few clues to possible relationships, and the reader is referred to the original papers listed in Table I for the results.

The genomes of CarMV and TCV are the only ones to have been characterized beyond estimation of electrophoretic mobilities in gels. Labeling studies have suggested that the genomic RNA of CarMV possesses a cap structure at the 5'-end (Guilley *et al.*, 1985), and direct evidence suggests the absence of poly (A) tails on the 3'-ends of both CarMV RNA (Carrington, 1986) and TCV RNA (Carrington *et al.*, 1986).

2. RNA Sequence Homologies

The synthesis of DNA complementary to CarMV RNA and its use in hybridization analysis was reported by Kummert (1980), but no comparative hybridization to viral RNAs other than TYMV was reported in this study. Gallitelli *et al.* (1985) used cDNA in a solid phase hybridization assay to estimate relatedness by nucleic acid homology among seven Tombusviruses and four members of the CarMV-like group (TCV, GMeV, SCV, and GMV). The results showed that all the definitive Tombusviruses tested had sequence relationships, while none of the CarMV-like viruses had detectable sequence homology to the Tombusviruses or to each other. T. J. Morris (unpublished information) compared the sequence relationships of TBSV and six CarMV-like viruses (TCV, GMV, CSBV, SCV, and PFBV) by the liquid hybridization procedure of Gould and Symons (1983), and could not detect any significant level of homology among any of the viral genomes. Together, these studies suggest that each of the CarMV-like viruses examined are genetically distinct and have no relationship to definitive members of the Tombusvirus group.

3. RNA Primary Structure

The complete nucleotide sequence has only been determined to date for CarMV (Guilley *et al.*, 1985). The nucleotide sequence at the 3'-end of TCV, including the entire coat protein cistron, has been determined from a cDNA clone of the 3'-proximal portion of the viral genome (Carrington *et al.*, 1986). Nucleotide sequences covering substantial portions of the remainder of the genome have also been determined (P. G. Stockley and S. C. Harrison, personal communication). Primary-structure information has not been published for any of the other viruses listed in Table I.

For CarMV, the nucleotide sequence was determined from cDNA clones (Carrington and Morris, 1984) of the genomic RNA which collectively represented over 99% of the sequence. The remainder was determined by direct-sequence analysis of RNA and cDNA transcripts (Guilley *et al.*, 1985). The entire sequence consisted of 4003 nucleotides, and contained two long open reading frames which together accounted for the major polypeptide products observed in cell-free translation experiments (see Fig. 5 for the sequence map, and the next section for details on the polypeptides). Starting from the 5'-end, a leader sequence of 69 nt precedes the first AUG codon at nt 70, which starts an open reading frame that terminates with UAA at nt 2677. This long open reading frame is punctuated by two amber termination codons at nt 805 and nt 2359.

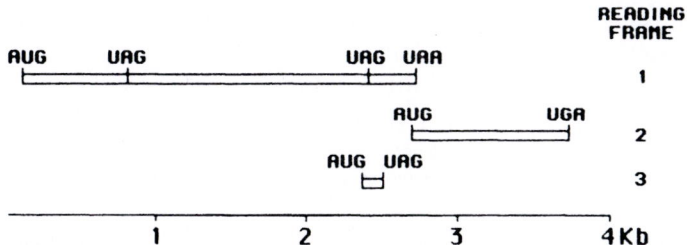

FIGURE 5. Map of the major open reading frames in carnation mottle virus genomic RNA. The hollow boxes represent the coding regions with initiation codons and termination codons at the indicated positions. Refer to text (page 97) and Guilley *et al.* (1985) for details.

Translation of the coding sequence would produce a polypeptide of M_r = 26,841 (p27) up to the first UAG codon, and two readthrough polypeptides of M_r = 85,831 (p86) and M_r = 97,738 (p98) by termination at the second UAG and UAA codons, respectively. The second long coding region, which begins with AUG at nt 2669 and ends with UGA at nt 3713, would encode a protein of M_r = 37,787 (p38). This coding region overlaps, although it is out of phase with, the first long open reading frame and is expressed through the synthesis of a subgenomic RNA, which is translated into the viral coat protein (see next section). A third, smaller open reading frame, out of phase with the nt 70-2677 coding region but entirely within it, extends from nt 2334 to nt 2517 and has the potential to encode a polypeptide of M_r = 6746 (p7). Evidence for the expression of this gene from a second subgenomic RNA will be discussed below. A 3'-noncoding region of 290 nt completes the CarMV sequence. CarMV has less than one half of the nonstructural protein coding capacity of TMV. However, they share a number of gene organizational features, such as the production of readthrough proteins from 5'-proximal open reading frames and at least two 3'-derived subgenomic RNAs of which the smallest encodes the coat protein.

B. Regulation and Expression of the Genome

1. Carnation Mottle Virus

Prior to 1980, there was very little known about the nonstructural proteins of these viruses and their mechanisms of gene expression. Salomon *et al.* (1978) studied the production of translation products directed by CarMV RNA in a cell-free wheat germ system, in which they reported the production of three unrelated polypeptides using genomic-sized RNA as template. A translational map based on this study was proposed by Davies and Hull (1982), which suggested a polycistronic translational strategy. Evidence for the presence of subgenomic species in RNA extracts of CarMV-infected plants was not consistent with the proposed polycistronic model (Morris, 1983). The production of a cDNA clone representing almost the entire genome of CarMV permitted Carrington and Morris (1984) to identify two subgenomic species of 1472 and 1689 bases in length, which mapped toward the 3'-end of the genome. The subgenomic RNAs were encapsidated in virions and synthesized in infected plants (Fig. 6). The genomic coordinates from which the two subgenomic RNAs originate were precisely defined on the viral RNA sequence by high-resolution S_1 nuclease protection and primer-extension mapping techniques (Carrington and Morris, 1986). The two 3'-proximal subgenomic RNAs initiate at nt 2532 and nt 2315 on the sequence map and extend to the 3'-end of the genome. Cell-free translation of authentic viral mRNAs and synthetic transcripts suggest that the open reading frames expressed

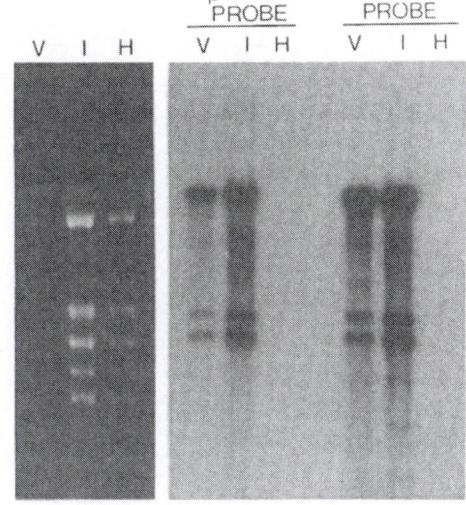

FIGURE 6. Northern blot analysis of Carnation mottle virus-related single-stranded RNAs. Virion (V) RNA, and total single-stranded RNA from infected (I) and healthy (H) *Chenopodium quinoa* were electrophoresed through a 1.5% agarose/6% formaldehyde gel. The set on the left was stained with ethidium bromide and two additional sets were transferred to nitrocellulose and hybridized with ^{32}P-labeled probes derived from a clone representing the 3'-end of the CarMV genome (pCarMV), or random CarMV cDNA. Note the presence of 1.5- and 1.7-kb subgenomic RNAs in both the virion and tissue-derived RNA samples (From Carrington and Morris, 1984).

from the 1.5-kb and 1.7-kb subgenomic RNAs code for p38 and p7, respectively, as identified on the genome map in Fig. 7. These coding regions contain the first AUG codons encountered from the 5'-ends of the 1.5- and 1.7-kb RNAs.

The authenticity of the polypeptides identified from the sequence is supported by a number of cell-free translation studies. Harbison *et al.* (1984) and Carrington and Morris (1985) independently identified the synthesis of three major polypeptides in cell-free, rabbit reticulocyte lysate translation experiments. These correspond to the polypeptides p86, p38, and p27 shown in Fig. 8. To avoid confusion, we have used the accurate polypeptide sizes determined from the genome sequence as opposed to the less precise electrophoretic estimates reported in the original papers. Both studies identified p38 as the coat protein, and demonstrated that it was synthesized primarily when a subgenomic fraction of viral RNA was used to program the reticulocyte lysate. In addition, p27 and p86 were recognized as the primary products synthesized from genome-sized template. Analysis of α-chromotryptic cleavage peptides of these two proteins demonstrated that they contained overlapping sequences, suggesting that p86 was a readthrough product of p27 (Carrington and Morris, 1985). Harbison *et al.* (1985) identified the synthesis of the double-readthrough polypeptide, p98, in reticulocyte lysates programmed with CarMV RNA under conditions that optimized for stop-codon readthrough. Confirmation of the p27-p86 gene order at the 5'-end was achieved by cell-free translation of defined transcripts synthesized *in vitro* from SP6 promotor/CarMV cDNA recombinant plasmids (Carrington and Morris, 1986). Using similar methods, this study also provided solid evidence for synthesis of p38 from the 1.5-kb subgenomic RNA and p7 from the 1.7-

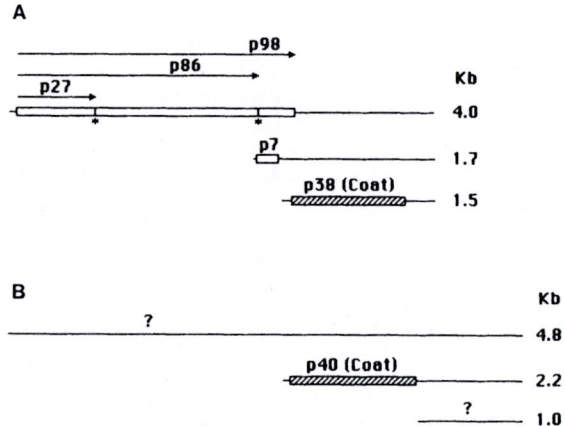

FIGURE 7. Models of the organization and expression of (A) the carnation mottle virus (CarMV) and turnip crinkle virus (TCV) genomes and (B) the tomato bushy stunt virus (TBSV) genome. Lines indicate the scaled positions of the mapped viral RNAs. Open boxes indicate open reading frames potentially expressed from the RNAs on which they are shown. The sizes of encoded gene products are indicated above the boxes and the sizes of the genomic and subgenomic RNAs are given at the end of each line. The positions of the UAG amber termination codons are indicated by asterisks. The CarMV map reflects the precise size and location of the polypeptides and subgenomic RNAs as determined by Guilley *et al.* (1985) and Carrington and Morris (1986). Evidence for a similar genome arrangement for TCV was provided by Dougherty and Kaesberg (1981), Altenbach and Howell (1982), and Carrington *et al.* (1986). The positions of p27 and p86 on the map correspond to the p25 and p80 polypeptides identified in those studies. Neither an open reading frame corresponding to p7 on the 1.7-kb subgenomic RNA nor a polypeptide corresponding to p98 have been shown for TCV. The partial TBSV map was constructed from analysis of virus-associated RNAs (Hillman *et al.*, 1985) and preliminary sequence data from our laboratory (Hillman, 1986). The TBSV map is presented with the reservation that there have been no published cell-free translation studies with this virus.

kb subgenomic RNA. Together, these studies have allowed us to construct the refined CarMV genetic map presented in Fig. 7. As yet, evidence for the existence of only p38 and p98 in CarMV-infected plant cells has appeared (Harbison *et al.*, 1985).

2. Turnip Crinkle Virus

The absence of complete sequence information for the 5'-proximal region of the TCV genome precludes the presentation of a precise genome map for this virus. Two cell-free translation studies (Dougherty and Kaesberg, 1981; Altenbach and Howell, 1982) do support the hypothesis that the TCV genome has a very similar coding properties to that of CarMV (see Fig. 7). Evidence for synthesis of a p25 from a 5'-proximal gene, a readthrough p80 polypeptide, and the synthesis of a p38 coat protein encoded by subgenomic RNA could be collectively interpreted from both

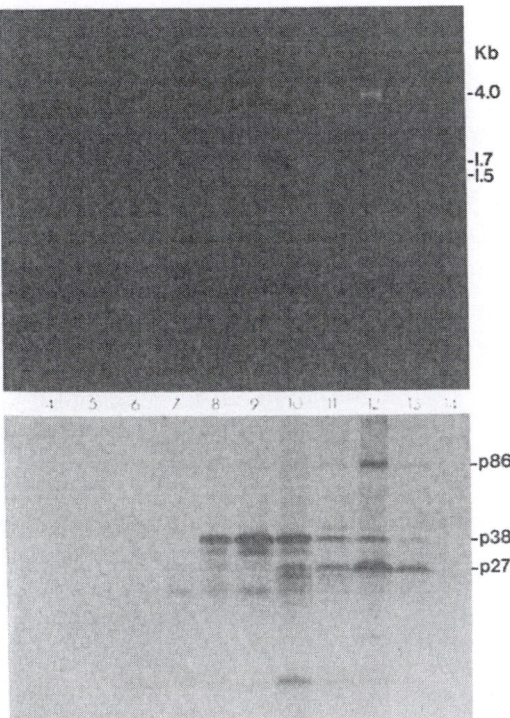

FIGURE 8. Gel electrophoresis and cell-free translation products of CarMV RNA which was fractionated on a sucrose density gradient. RNA from the gradient fractions was electrophoresed in a denaturing, 1.5% agarose gel and stained with ethidium bromide (top). RNA samples from the same fractions were translated in reticulocyte lysate reactions and the ³⁵S-labeled products were analyzed by SDS–PAGE and fluorography (bottom). Note the predominant synthesis of polypeptides of 27- and 86K from genomic-length templates, and the higher level of coat protein (p38) synthesis from subgenomic RNA template (From Carrington and Morris, 1985).

studies. A number of other minor species were observed in these studies, including p56 and p36 polypeptides that were shown to have peptide sequences in common with p80 (Dougherty and Kaesberg, 1981). These studies allowed Davies and Hull (1982) to present a genome map that predicted the synthesis of p36 from a non-3'-proximal subgenomic RNA. It now seems more likely that this product was an artifact of the cell-free translation system, and may have been generated from fragmented RNA. The recent identification of two subgenomic RNAs (about 1700 nt and 1446 nt) which map at the 3'-end of the genomic RNA, and the demonstration that the smaller subgenomic RNA codes for the TCV coat protein (Carrington et al., 1986), provide strong support for the map depicted in Fig. 8.

3. Other CarMV-like Viruses

Little information is available on the genetic organization of any of the other viruses listed in Table I except for the identification of two subgenomic species in RNA extracts of infected plants of CSBV, GMV, PFBV, and SCV (see Fig. 9). These were approximately 1.7 and 1.5 kb for each virus examined (T. J. Morris, unpublished information), the same size as those identified for both CarMV and TCV. In contrast, subgenomic

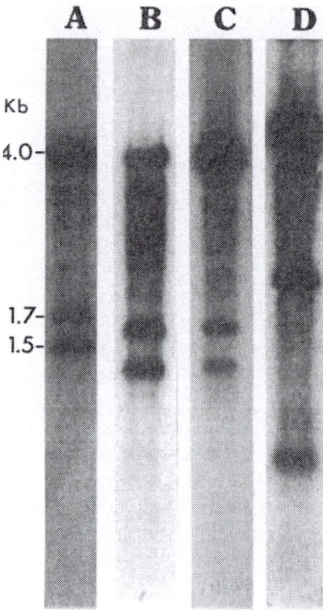

FIGURE 9. Northern blot analysis of carnation mottlelike, virus-related, single-stranded RNAs. The analysis was performed essentially as described in Fig. 6. Total single-stranded RNA from infected *Chenopodium quinoa* was electrophoresed, transferred to nitrocellulose, and probed with homologous, ^{32}P-cDNA probes prepared to each of the respective viral RNAs. The tracts depict the viral-related RNAs in infected tissue extracts of (A) carnation mottle virus, (B) galinsoga mosaic virus, (C) turnip crinkle virus, and (D) tomato bushy stunt virus. The calculated sizes of the two major subgenomic RNAs were approximately 1.5 and 1.7 kb for each of the CarMV-like viruses and 1.0 and 2.2 kb for TBSV.

species of 2.2 and 1.0 kb were detected in TBSV-infected plants (Hillman *et al.*, 1985; Hillman, 1986), while others have reported from three to four subgenomic species associated with TBSV infection (Hayes *et al.*, 1984; Gallitelli *et al.*, 1985). Although a resolution to the question of the TBSV genome organization must await more definitive studies, our partial sequence data supports a somewhat surprising feature of the genome map depicted in Fig. 7 (Hillman, 1986). The TBSV coat protein appears to map at the 5'-end of the 2.2-kb subgenomic RNA and not at the 3'-end of the genome. This feature alone is very different from the genetic organization now established for both CarMV and TCV.

C. Viral Replication

Information pertaining to kinetics, enzymology, and other features of the replication of CarMV and similar viruses is relatively sparse. Although the time course of CarMV infection in *Dianthus* protoplasts has been reported (Kluge *et al.*, 1983), no studies have directly addressed the kinetics of genomic and subgenomic RNA synthesis or the role of RNA-dependent RNA polymerase in the infection process. A putative polymerase encoded by CarMV has been identified by Guilley *et al.* (1985) through identification of amino acid sequences conserved among RNA-dependent RNA polymerases of a number of different RNA viruses (Kamer and Argos, 1984). This homology was located within the p86 readthrough domain, suggesting p86 and/or p98 function in viral RNA replication. Henriques and Morris (1979) identified a dsRNA species in TCV-infected plants which they suggested was the replicative form of the viral genomic RNA. Subsequent studies (Morris, 1983; Morris, unpublished information) have identified three major dsRNA species in lithium chloride-soluble RNA extracts of plant tissue infected with CarMV, TCV, CSBV, GMV, PFBV, and SCV (see Fig. 10). All of the viruses produced a dsRNA species of the size expected for the genomic RNA replicative form (about 4.0 kb) in addition to dsRNA species reflective of the sizes of the subgenomic RNAs identified by Northern blot hybridization (1.5 and 1.7 kb). Similar dsRNA species have also been reported for HCRSV (Hearon, 1984) and tobacco necrosis virus (Condit and Fraenkel-Conrat, 1979). A greater complexity of dsRNA species has been detected in TBSV-infected plants (Morris, 1983; Hillman et al., 1985; Hayes et al., 1985); the distinct sizes of these species relative to the CarMV-like viral RNAs (see Fig. 10) reflect the different sizes of the genomic and subgenomic species identified for this virus.

VII. CYTOPATHOLOGY

Cytopathological alterations in cells infected with Tombusviruses have received considerable attention. It is generally accepted that such features as the induction of membraneous, "multivesicular bodies" with peripheral vesicles in the cytoplasm of infected cells, infection and/or alteration of the nucleus, and the accumulation of virus in "bubblelike" extensions into the vacuole are features sufficiently characteristic to be of value in grouping these viruses (Martelli, 1981; Francki et al., 1985a; and Martelli et al., Chapter 2 in this volume). The CarMV-like viruses,

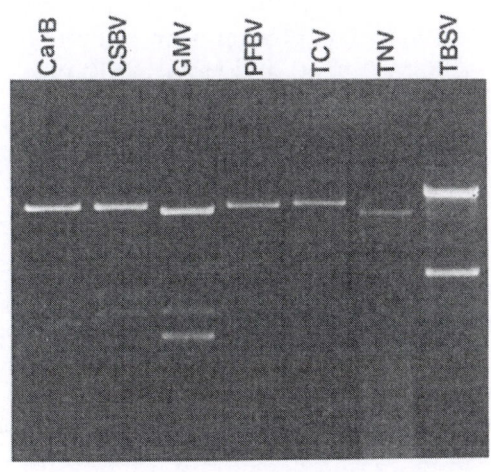

FIGURE 10. Double-stranded RNAs isolated from infected *Chenopodium quinoa* leaf tissue analyzed by electrophoresis in 1.5% agarose gels and stained with ethidium bromide. The double-stranded species correspond in size to genomic and subgenomic RNAs from tissue infected with carnation mottle virus, strain B (CarB); cucumber soilborne virus (CSBV), galinsoga mosaic virus (GMV); pelargonium flower break virus (PFBV); turnip crinkle virus (TCV); tobacco necrosis virus (TNV); and tomato bushy stunt virus (TBSV).

on the other hand, have not been so extensively and comparatively studied in the context of abnormal cytopathology, and few generalizations can be made in this regard for the group.

Ultrastructural observations of this group have shown that virus particles accumulate to sufficiently high concentrations in the cytoplasm of infected cells to be easily recognized, although crystalline arrays are not usually formed. Crystals of viruslike particles have been described only in HCRSV-infected cells and in the roots of TCV-infected plants (S. S. Hearon, personal communication). Reports of inclusions are uncommon, with only PLPV reported to induce cytoplasmic inclusion bodies containing virus particles embedded in a matrix (Stefanac et al., 1982). Also, fibrillar cytoplasmic inclusions were described in CSBV-infected tissue (Koenig et al., 1983). A feature of ultrastructural observations on the majority of the viruses (BGMV, CFSV, CSBV, GMV, HCRSV, SCV, and TCV) is a lack of virus particles associated with the nucleus. However, both CarMV (Robleda, 1973) and GMeV (Behncken and Dale, 1984) have been observed in close proximity to the nucleolus. Vacuolar involvement is a feature associated with only a few of the viruses. TCV was reported to occur in the vacuole (Russo and Martelli, 1982) as was HCRSV (Waterworth, 1980), while NTNV appeared to be embedded in a vacuolar matrix (Mowat et al., 1976). Only GMeV produced the cytoplasmic extrusions into the vacuole that are more typical of the Tombusviruses (Martelli, 1981). Multivesicular-like bodies similar to those associated with Tombusvirus infection have been reported for several of the CarMV-like viruses. The vesiculated proxisomes reported in close association with chloroplasts of GMeV-infected plants (Behncken and Dale, 1984) closely resemble the multivesicular bodies of Tombusviruses. Similar vesiculated and swollen peroxisomes, less typical of multivesicular bodies, have been observed in HCRSV-infected cells (S. S. Hearon, personal communication). In addition, similar structures of mitochondrial origin with conspicuous peripheral vesiculation have been observed in plants infected with TCV (Russo and Martelli, 1982), and also GMV (Hatta et al., 1983). The proliferation of cytoplasmic vesicles is a feature reported for BGMV (Scott and Hoy, 1981), NTNV (Mowat et al., 1976), and CSBV (Koenig et al., 1983).

VIII. SATELLITES

The occurrence of satellites and satellite RNAs in association with CarMV-like viruses is limited to reports that TCV is capable of supporting the replication of a small RNA (RNA C, also designated S-TCV) of about 350 nucleotides in turnip plants. Altenbach and Howell (1981) demonstrated that RNA C had many of the properties commonly associated with other plant viral satellite RNAs. It was encapsidated, lacked extensive homology to genomic TCV RNA, and was dependent on TCV RNA for replication. In addition, RNA C appeared to intensify the systemic symptoms produced by the helper virus, and failed to show messenger activity *in vitro* (Altenbach and Howell, 1982). The status of additional satellite-associated RNA species, also identified in infected plants, was addressed using a cDNA clone of RNA C (Altenbach and Howell, 1984). A multimeric set of RNA C-related species was detected and was suggested to represent intermediates of satellite RNA replication as described for other viral satellites (see Francki, 1985 for a review). In addition to the multimeric forms, several smaller satellite-related species were identified. Of these, RNA E appeared to be a truncated form of the satellite, while RNA F was suggested to be a different satellite because it did not reappear in plants inoculated with extensively purified RNA C. The most curious RNA C-related species identified was RNA D, which was produced in plants inoculated with viral genomic RNA alone. It was speculated that this species may have been derived from the host plant genome, because satellite-related sequences were detected in the DNA of uninfected plants. The status of the satellite species associated with TCV was further clarified with more extensive cDNA cloning of RNA C and the sequence determinations of several independent clones (Simon and Howell, 1986). RNA C was shown to be a linear molecule whose 3'-terminal sequence (about 150 nt) was extensively homologous to the 3'-end sequence of TCV genomic RNA as determined by Carrington *et al.* (1986). The 5'-proximal portion of the satellite RNA sequence was not identified in any region of the available TCV genomic sequence, but was present in RNAs D and F. These two RNAs appeared to lack the 3'-proximal, TCV-related sequences. It is interesting to speculate on the possibility that the TCV satellite RNA may have originated from viral and host sequences through a recombination event.

Koenig *et al.* (1983) noted the presence of an additional minor RNA species ($M_r = 0.16$–0.19×10^6) in some preparations of CSBV RNA, which was not required for infection. Further evidence of the possible satellite nature of this RNA has not been reported.

IX. CONCLUSIONS

In view of the extensive information that now exists about some of the CarMV-like viruses, it would seem appropriate to suggest that serious consideration be given to the official classification of some of the members. Clearly, two possibilities could be entertained. A new plant virus group could be defined with the suggested name, the **Carmoviruses**, in keeping with existing nomenclature. Alternatively, a subgroup (analogous to a subfamily) of the Tombusviruses could be defined to include these viruses. This latter possibility is attractive in view of close structural and physicochemical similarity of the two virus "groups." It also accommodates the generally accepted criteria for defining virus families (and analogous plant virus groups) most recently reiterated by Matthews (1985). It provides the additional advantage of allowing a more complete provisional classification of the viruses based on particle structure and RNA and protein composition without requiring the more thorough characterization necessary for assignment to either subgroup. In addition, such criteria would set a precedent for defining additional logical subgroupings that might include the beetle-transmitted, legume-infecting viruses, for example. Either of the two possible actions would have a beneficial impact on the existing classification system by reducing the list of unclassified plant viruses, and by eliminating the current state of confusion that exists with respect to the affinity status of this large number of plant viruses.

Regardless of the ultimate taxonomic designation, it is now quite apparent that there are some unifying aspects of the CarMV-like viruses that clearly distinguish them from the Tombusviruses. The criteria that seem of less importance in distinguishing between the groups include subtle differences in the physicochemical properties, such as a lower sedimentation value, a slightly smaller coat protein, and greater pH lability of the virion, and biological differences including transmission mechanisms and host cytopathology. The most striking distinction, and one rarely considered in defining group status because of a general lack of precise information for most viruses, is the genetic organization of the viral genome. Compelling evidence now exists that viruses such as CarMV and TCV are genetically different from the Tombusviruses in the sense that they have a smaller genomic RNA (4.0 versus 4.8 kb), from which two 3'-proximal subgenomic RNAs of a characteristic size (1.5 and 1.7 kb) are expressed. Unlike TBSV, the coat protein gene is located at the 3'-end of the genome, and the coat protein is translated from the smaller of the two subgenomic RNAs. Conclusive studies on the genome organization of the majority of the CarMV-like and Tombusviruses do not exist. Such detailed knowledge may be difficult to obtain because of an understandable lack of enthusiasm to perform the exhaustive studies that would be necessary. It would seem reasonable to accept the determination of subgenomic RNA size and number by the relatively simple approaches of Northern hybridization and replicative RNA analysis as one of the criteria for assignment of group affinity. This type of analysis has been accomplished for CarMV, TCV, CSBV, GMV, PFBV, and SCV (Morris and Carrington, 1984) as well as for HCRSV (Hearon, 1984). If such a criterion is adopted for this and other groups of plant viruses, it will be interesting to see how the classification of tobacco necrosis virus (TNV) is eventually dealt with. TNV has two subgenomic RNAs of similar size as CarMV; it shares fungus transmissibility with another CarMV-like virus (CNV); and yet it defines a monotypic group. It might be speculated that the only significant difference between the TNV and CarMV-like genomes is a truncated coat protein gene that resulted from loss of a dispensable portion of the P domain of the subunit at some point in the evolution of the virus. If this speculation proves valid, then it should not be unreasonable to demote this group to subgroup status within the CarMV-like virus group.

X. RECENT DEVELOPMENTS

The unsettled taxonomic status of the many different cucurbit-infecting CarMV-like viruses has been recently evaluated with respect to the size and number of subgenomic RNA and dsRNA species produced in infected hosts. On the basis of these unpublished observations, melon necrotic spot viruses can be considered as a CarMV-like virus while cucumber necrosis virus produces viral-specific RNA species more typical of members of the Tombusvirus group. CNV differs from CarMV-like viruses and resembles Tombusviruses with respect to the location of its coat protein gene, which maps in a non-3'-proximal position on the viral genome (D'Ann Rochon, personal communication). This arrangement is in agreement with the physical location and subgenomic coding assignments which we have also determined from sequence data of another Tombusvirus genome (Hillman, 1986). Similar analyses of viral-specific RNAs associated with the closely related cucumber fruit streak and cucumber leafspot viruses suggest that these viruses might be considered as possible members of the Tombusvirus group as well (H. J. Vetten, personal communication).

ACKNOWLEDGMENTS. The authors express particular appreciation to Dr. Steven Harrison for providing manuscripts and illustrations prior to publication and for his cooperation in our studies on TCV structure. We also wish to thank Drs. Suzanne Hearon, D'Ann Rochon, and Brad Hillman for providing unpublished results and for helpful discussions. A. O. Jackson and D. C. Stenger are also thanked for their assistance and encouragement during the preparation of the manuscript. The research on CarMV and TCV was supported by a USDA Competitive Grant (# 8500068).

REFERENCES

Abad-Zapatero, C., Abdel-Meguid, S. S., Johnson, J. E., Leslie, A. G. W., Rayment, I., Rossman, M. G., Suck, D., and Tsukihara, T., 1980, Structure of southern bean mosaic virus at 2.8Å resolution, *Nature* **286**:33.

Allen, D. J., 1980, Identification of resistance to cowpea mottle virus, *Trop. Agric., Trinidad* **57**:325.

Allen, D. J., Anno Nyako, F. O., Ochieng, R. S., and Ratinam, M., 1981, Beetle transmission of cowpea mottle and southern bean mosaic viruses in West Africa, *Trop. Agric., Trinidad* **58**:171.

Allen, D. J., Thottappilly, G., and Rossel, H. W., 1982, Cowpea mottle virus: field resistance and seed transmission in virus-tolerant cowpea, *Ann. Appl. Biol.* **100**:331.

Altenbach, S. B., and Howell, S. H., 1981, Identification of a satellite RNA associated with turnip crinkle virus, *Virology*, **112**:25.

Altenbach, S. B., and Howell, S. H., 1982, *In vitro* translation products of turnip crinkle virus RNA, *Virology* **118**:128.

Altenbach, S. B., and Howell, S. H., 1984, Nucleic acid species related to the satellite RNA of turnip crinkle virus in turnip plants and virus particles, *Virology* **134**:72.

Behncken, G. M., 1970, Some properties of a virus from *Galinsoga parviflora*, *Aust. J. Biol. Sci.* **23**:497.

Behncken, G. M., Francki, R. I. B., and Gibbs, A. J., 1982, Galinsoga mosaic virus, *CMI/AAB Descriptions of Plant Viruses* No. 252.

Behncken, G. M., and Dale, J. L., 1984, Glycine mottle virus: A possible member of the tombusvirus group, *Intervirology* **21**:159.

Boatman, S., and Kaper, J. M., 1976, Molecular organization and stabilizing forces of simple RNA viruses. IV. Selective interference with protein-RNA interactions using sodium dodecyl sulfate, *Virology* **70**:1.

Bock, K. R., 1982, Tephrosia symptomless virus, *CMI/AAB Descriptions of Plant Viruses* No. 256.

Bock, K. R., Gutherie, E. J., and Figueiredo, G., 1981, Purification and some properties of tephrosia symptomless virus, *Ann. Appl. Biol.* **97**:277.

Bos, L., Van Dorst, H. J. M., Huttinga, H., and Maat, D. Z., 1984, Further characterization of melon necrotic virus causing severe disease in glasshouse cucumbers in the Netherlands and its control, *Neth. J. Pl. Pathol.* **90**:55.

Bozarth, R. F., and Shoyinka, S. A., 1979, Cowpea mottle virus, *CMI/AAB Descriptions of Plant Viruses* No. 212.

Brierley, P., 1964, Effects of four viruses on yield and quality of King Cardinal carnations, *Plant Dis. Rep.* **48**:5.

Brierley, P., and Smith, E. F., 1957, Carnation viruses in the United States, *Phytopathology* **47**:714.

Burgermeister, W., and Koenig, R., 1984, Electro-blot immunoassay—A means for studying serological relationships among plant viruses? *Phytopath. Z.* **111**:15.

Butler, P. J. G., 1970, Structures of turnip crinkle and tomato bushy stunt viruses. III. The chemical subunits: Molecular weights and number of molecules per particle, *J. Mol. Biol.* **52**:589.

Carrington, J. C., 1986, Structure and function of the carnation mottle virus genome, Ph.D Thesis, Department of Plant Pathology, University of California, Berkeley.

Carrington, J. C., and Morris, T. J., 1984, Complementary DNA cloning and analysis of carnation mottle virus RNA, *Virology* **139**:22.

Carrington, J. C., and Morris, T. J., 1985, Characterization of the cell-free translation products of carnation mottle genomic and subgenomic RNAs, *Virology* **144**:1.

Carrington, J. C., and Morris, T. J., 1986, High resolution mapping of carnation mottle virus-associated RNAs, *Virology* **150**:196.

Carrington, J. C., Morris, T. J., Stockley, P. G., and Harrison, S. C., 1987, Structure and assembly of turnip crinkle virus. IV: Analysis of the coat protein gene and implications of the subunit primary structure. *J. Mol. Biol.*, **194**:265.

Condit, C., and Fraenkel-Conrat, H., 1979, Isolation of replicative forms of 3'-terminal subgenomic RNAs of tobacco necrosis virus, *Virology* **97**:122.

Courdriet, D. L., Kishaba, A. N., and Carroll, J. E., 1979, Transmission of muskmelon necrotic spot virus in muskmelons by cucumber beetles, *J. Econ. Ent.* **72**:560.

Crowther, R. A., and Amos, L. A., 1971, Three-dimensional image reconstruction of some small spherical viruses, *Cold Spring Harbor Symp. Quant. Biol.* **36**:489.

Davies, J. W., and Hull, R., 1982, Genome expression of plant positive-strand RNA viruses, *J. Gen. Virol.* **61**:1.

Dias, H. F., 1970, The relationship between cucumber necrosis virus and its vector, *Olpidium cucurbitacearum*, *Virology* **42**:204.

Dias, H. F., and McKeen, C. D., 1972, Cucumber necrosis virus. *CMI/AAB Descriptions of Plant Viruses* No. 82.

Dougherty, W. G., and Kaesberg, P., 1981, Turnip crinkle virus RNA and its translation in rabbit reticulocyte and wheat embryo extracts, *Virology* **115**:45.

Finch, J. T., Klug, A., and Leberman, R., 1970, The structures of turnip crinkle and tomato bushy stunt viruses. II. The surface structure: Dimer clustering patterns, *J. Mol. Biol.* **50:**215.

Francki, R. I. B., 1983, Current problems in plant virus taxonomy, in: *A Critical Appraisal of Viral Taxonomy* (R. E. F. Matthews, ed.), pp. 63–104, CRC Press, Boca Raton, Fla.

Francki, R. I. B., 1985, Plant virus satellites, *Annu. Rev. Microbiol.* **39:**151.

Francki, R. I. B., Milne, R. G., and Hatta, T. (eds.), 1985a, *Atlas of Plant Viruses,* Volume I, CRC Press, Boca Raton, Fla.

Francki, R. I. B., Milne, R. G., and Hatta, T. (eds.), 1985b, *Atlas of Plant Viruses,* Volume II, CRC Press, Boca Raton, Fla.

Furuki, I., Hibi, T., Honda, Y., Saito, Y., and Komuro, Y., 1980, Relationship between cucumber necrosis virus and melon necrotic spot virus, *Ann. Phytopath. Soc. Japan* **46:**419.

Gallitelli, D., Vovlas, C., and Avegelis, A., 1983, Some properties of cucumber fruit streak virus, *Phytopath. Z.* **106:**149.

Gallitelli, D., Hull, R., and Koenig, R., 1985, Relationships among viruses in the tombusvirus group: Nucleic acid hybridization studies, *J. Gen. Virol.* **66:**1523.

Golden, J. S., and Harrison, S. C., 1982, Proteolytic dissection of turnip crinkle virus subunit in solution, *Biochemistry* **16:**3862.

Gonzalez-Garza, R., Gumpf, D. J., Kishaba, A. N., and Bohn, G. W., 1979, Identification, seed transmission, and host range pathogenicity of a California isolate of melon necrotic spot virus, *Phytopathology* **69:**340.

Gould, A. R., and Symons, R. H., 1983, A molecular biological approach to relationships among viruses, *Annu. Rev. Phytopath.* **21:**179.

Guilley, H., Carrington, J. C., Balazs, E., Jonard, G., Richards, K., and Morris, T. J., 1985, Nucleotide sequence and genome organization of carnation mottle virus RNA, *Nucleic Acids Res.* **13:**6663.

Hakkaart, F. A., 1964, Description of symptoms and assessment of loss caused by some viruses in the carnation cultivar "William Sim," *Neth. J. Plant Path.* **70:**53.

Hammond, J., 1981, Viruses occurring in *Plantago* species in England, *Pl. Path.* **30:**237.

Hampton, R. O., and Hancock, C. L., 1981, Soil-related greenhouse spread of bean mild mosaic virus. *Phytopathology* **71:**223.

Harbison, S.-A., Wilson, T. M. A., and Davies, J. W., 1984, An encapsidated, subgenomic messenger RNA encodes the coat protein of carnation mottle virus, *Biosci. Rep.* **4:**949.

Harbison, S.-A., Davies, J. W., and Wilson, T. M. A., 1985, Expression of high molecular weight polypeptides by carnation mottle virus RNA, *J. Gen. Virol.* **66:**2597.

Harrison, S. C., 1980, Protein interfaces and intersubunit bonding. The case of tomato bushy stunt virus, *Biophys. J.* **32:**139.

Harrison, S. C., 1983, Virus structure: High-resolution perspectives, *Adv. Virus Res.* **28:**175.

Harrison, S. C., Olsen, A. J., Schutt, C. E., Winkler, F. K., and Bricogne, G., 1978, Tomato bushy stunt virus at 2.9Å resolution, *Nature* **276:**368.

Hatta, T., Francki, R. I. B., and Grivell, C. J., 1983, Particle morphology and cytopathology of galinsoga mosaic virus, *J. Gen. Virol.* **64:**687.

Hayes, R. J., Buck, K. W., and Brunt, A. A., 1984, Double-stranded and single-stranded subgenomic RNAs from plant tissue infected with tomato bushy stunt virus, *J. Gen. Virol.* **65:**1293.

Hearon, S. S., 1984, A virus from *Hibiscus rosa-sinensis* with properties of a tombusvirus, *Phytopathology* **74:**862.

Henriques, M. C., and Morris, T. J., 1979, Evidence for different replicative strategies in the plant tombusviruses, *Virology* **99:**66.

Hillman, B. I., Morris, T. J., and Schlegel, D. E., 1985, Effects of low-molecular-weight RNA and temperature on tomato bushy stunt virus symptom expression, *Phytopathology* **75:**361.

Hillman, B. I., 1986, Genome organization, replication and defective RNAs of tomato bushy stunt virus, Ph.D Thesis, Department of Plant Pathology, University of California, Berkeley.

Hobbs, H. A., 1981, Transmission of bean curly dwarf mosaic virus and bean mild mosaic virus by beetles in Costa Rica, *Plant Disease* **65**:491.

Hogle, J., Kirschhausen, T., and Harrison, S. C., 1983, Divalent cation sites in tomato bushy stunt virus. Difference maps at 2.9Å resolution, *J. Mol. Biol.* **171**:95.

Hogle, J. M., Maeda, A., and Harrison, S. C., 1986, Structure and assembly of turnip crinkle virus, I: X-ray crystallographic structure analysis at 3.2Å resolution, *J. Mol. Biol.* **191**:625.

Hollings, M., and Stone, O. M., 1970, Carnation mottle virus, *CMI/AAB Descriptions of Plant Viruses* No. 7.

Hollings, M., and Stone, O. M., 1972, Turnip crinkle virus. *CMI/AAB Descriptions of Plant Viruses* No. 109.

Hollings, M., and Stone, O. M., 1974, Pelargonium flower break virus, *CMI/AAB Descriptions of Plant Viruses* No. 130.

Hollings, M., Stone, O. M., and Smith, D. R., 1977, Productivity of virus-tested carnation clones and the rate of re-infection with virus, *J. Hort. Sci.* **47**:141.

Hopper, P., Harrison, S. C., and Sauer, R. T., 1984, Structure of tomato bushy stunt virus V. Coat protein sequence determination and its structural implications, *J. Mol. Biol.* **177**:701.

Hull, T., 1977a, The grouping of small spherical plant viruses with single RNA components, *J. Gen. Virol.* **36**:289.

Hull, R., 1977b, The stabilization of particles of turnip rosette and of other members of the southern bean mosaic virus group, *Virology* **79**:58.

Jones, A. T., 1974, Elderberry latent virus, *CMI/AAB Descriptions of Plant Viruses* No. 127.

Jones, A. T., and Badenoch, S., 1981, Elderberry latent virus, *Rep. Scott. Hort. Res. Inst.* **1981**:119.

Jones, D. R., and Behncken, G. M., 1980, Hibiscus chlorotic ringspot, a widespread virus disease in the ornamental *Hibiscus rosa-sinensis*, *Australas. J. Pl. Path.* **9**:4.

Kamer, C., and Argos, P., 1984, Primary structural comparison of RNA-dependent polymerases from plant, animal and bacterial viruses, *Nucleic Acids Res.* **12**:7269.

Kishi, K., 1966, Necrotic spot of melon, a new virus disease, *Ann. Phytopath. Soc. Japan* **32**:138.

Kluge, S., Kirsten, U., and Ortel, C., 1983, Infection of *Dianthus* protoplasts with carnation mottle virus, *J. Gen. Virol.* **64**:2485.

Koenig, R., 1981, Indirect ELISA methods for broad specificity detection of plant viruses, *J. Gen. Virol.* **55**:53.

Koenig, R., Lesemann, D. E., Huth, W., and Makkouk, K. M., 1983, Comparison of a new soilborne virus from cucumber with tombus-, diantho-, and other similar viruses, *Phytopathology* **73**:515.

Koenig, R., and Lesemann, D. E., 1985, Plant viruses in German rivers and lakes. I. Tombusviruses, a potexvirus and carnation mottle virus, *Phytopath. Z.* **112**:105.

Kummert, J., 1980, Synthesis and characterization of DNA complementary to carnation mottle virus RNA, *Virology* **105**:35.

Leberman, R., and Finch, J. T., 1970, The structures of turnip crinkle and tomato bushy stunt viruses. I. A small protein particle derived from turnip crinkle virus, *J. Mol. Biol.* **50**:209.

Lommel, S. A., McCain, A. H., and Morris, T. J., 1982, Evaluation of indirect enzyme-linked immunosorbent assay for the detection of plant viruses, *Phytopathology* **72**:1018.

Lommel, S. A., McCain, A. H., Mayhew, D. E., and Morris, T. J., 1983a, Survey of commercial carnation cultivars for four viruses in California by indirect enzyme-linked immunosorbent assay, *Plant Disease* **67**:53.

Lommel, S. A., Stenger, D. C., and Morris, T. J., 1983b, Evaluation of virus diseases of commercial carnations in California, *Acta Hortica* **141**:79.

Makkouk, K. M., and Shahab, S., 1980, Identification, properties and incidence of carnation mottle virus on carnations in Lebanon, *Z. Pflanzenkrankheit Pflanzenschutz* **87**:557.

Martelli, G. P., 1981, Tombusviruses, in: *Handbook of Plant Virus Infections and Comparative Diagnosis* (E. Kurstak, ed.), pp. 61–90, Elsevier, Amsterdam.

Martelli, G. P., Russo, M., and Quacquarelli, A., 1977, Tombusvirus (tomato bushy stunt virus) group, in: *The Atlas of Insect and Plant Viruses* (K. Maramorosch, ed.), pp. 257–279, Academic Press, New York.

Matthews, R. E. F., 1982, Classification and nomenclature of viruses. Fourth report of the International Committee on the Taxonomy of Viruses, *Intervirology* 17:1.

Matthews, R. E. F., 1985, Viral taxonomy for the nonvirologist, *Ann. Rev. Microbiol.* 39:451.

McKeen, C. D., 1959, Cucumber necrosis virus, *Can. J. Bot.* 37:913.

Milbrath, G. M., and Nelson, M. R., 1972, Isolation and characterization of a virus from saguaro cactus, *Phytopathology* 62:739.

Morris, T. J., 1983, Virus-specific double-stranded RNA: Functional role in RNA virus infection, in: *Plant Infectious Agents: Viruses, Viroids, Virusoids, and Satellites* (H. D. Robertson, S. H. Howell, M. Zaitlin, R. L. Malmberg, eds.), pp. 80–83, Cold Spring Harbor Laboratory, Cold Spring Harbor, N.Y.

Morris, T. J., and Carrington, J. C., 1984, A carnation mottle virus group: A comparison of some tentative members, *Phytopathology* 74:807.

Morris-Krsinich, B. A. M., and Milne, K. S., 1977, Natural infection of daphne by carnation mottle virus, *Plant Dis. Reptr.* 61:675.

Mowat, W. P., Asjes, C. J., and Brunt, A. A., 1976, Narcissus tip necrosis virus, *CMI/AAB Descriptions of Plant Viruses* No. 166.

Nelson, M. R., and Tremaine, J. H., 1975, Physicochemical and serological properties of a virus from saguaro cactus, *Virology* 65:309.

Nelson, M. R., Yoshimura, M. A., and Tremaine, J. H., 1975, Saguaro cactus virus, *CMI/AAB Descriptions of Plant Viruses* No. 148.

Phatak, H. C., 1974, Seed-borne plant viruses—Identification and diagnosis in seed health testing, *Seed Sci. Technol.* 2:3.

Plese, N., and Stefanac, Z., 1980, Some properties of a distinctive isometric virus from pelargonium, *Acta Hortica.* 110:183.

Robleda, S. C., 1973, Ultrastructure of cells infected with carnation mottle virus, *Phytopathol. Z.* 78:134.

Ronald, W. P., and Tremaine, J. H., 1976, Comparison of the effects of sodium dodecyl sulfate on some isometric viruses, *Phytopathology* 66:1302.

Russo, M., and Martelli, G. P., 1982, Ultrastructure of turnip crinkle and saguaro cactus virus-infected tissues, *Virology* 118:109.

Salomon, R., Bar-Joseph, M., Soreq, H., Gozes, I., and Littauer, U. A., 1978, Translation *in vitro* of carnation mottle virus RNA. Regulatory function of the 3'-region, *Virology* 90:288.

Samyn, G., and Welvaert, W., 1978, About an isometric virus on a cultivated cactus: *Chamaecereus sylvestrii 'aureus,'* *Phytopath. Z.* 91:276.

Scott, H. A., and Hoy, J. W., 1981, Blackgram mottle virus, *CMI/AAB Descriptions of Plant Viruses* No. 237.

Scott, H. A., and Phatak, H. C., 1979, Properties of blackgram mottle virus, *Phytopathology* 69:346.

Shoyinka, S. A., Bozarth, R. F., Reese, J., and Rossel, H. W., 1978, Cowpea mottle virus: A seed-borne virus with distinctive properties infecting cowpeas in Nigeria, *Phytopathology* 68:693.

Shukla, D. D., Shanks, G. J., Teakle, D. S., and Behncken, G. M., 1979, Mechanical transmission of galinsoga mosaic virus in soil, *Aust. J. Biol. Sci.* 32:267.

Simon, A. E., and Howell, S. H., 1986, The virulent satellite of turnip crinkle virus is a chimeric RNA molecule with one domain homologous to the 3'-end of the helper virus genome, *EMBO J.* 5:3423.

Skotnicki, A., and Gibbs, A., 1981, Some properties of the virions of galinsoga mosaic virus, *Australas. Pl. Path.* 10:27.

Smookler, M., and Lobenstein, G., 1975, Viruses of carnation identified in Israel, *Phytopath. Medit.* 14:1.

Sorger, P. K., Stockley, P. G., and Harrison, S. C., 1986, Structure of turnip crinkle virus II. Mechanism of *in vitro* assembly, *J. Mol. Biol.* **191**:639.

Stefanac, Z., Plese, N., and Wrischer, M., 1982, Intracellular changes provoked by pelargonium line pattern virus, *Phytopathol. Z.* **105**:288.

Stobbs, L. W., Cross, G. W., and Manocha, M. S., 1982, Specificity and methods of transmission of cucumber necrosis virus by *Olpidium radicale* zoospores, *Can. J. Pl. Path.* **4**:134.

Stockley, P. G., Kirsch, A. L., Chow, E. P., Smart, J. E., and Harrison, S. C., 1986, Structure of turnip crinkle virus III. Identification of a unique coat protein dimer, *J. Mol. Biol.* **191**:721.

Stone, O. M., 1980, Nine viruses isolated from pelargonium in the United Kingdom, *Acta Hortic.* **110**:177.

Stone, O. M., and Hollings, M., 1973, Some properties of pelargonium flower-break virus, *Ann. Appl. Biol.* **75**:15.

Stone, O. M., and Hollings, M., 1977, Pelargonium, *Annu. Rep. Glasshouse Crops Res. Inst.* **1976**:121.

Tremaine, J. H., 1972, Purification and properties of cucumber necrosis virus and a smaller top component, *Virology* **48**:582.

Tremaine, J. H., and Chidlow, J., 1972, Serological relationship of viruses and their reassembly products, *Virology* **50**:247.

Waterworth, H., 1980, Hibiscus chlorotic ringspot virus, *CMI/AAB Descriptions of Plant Viruses* No. 227.

Waterworth, H., 1981, Bean mild mosaic virus, *CMI/AAB Descriptions of Plant Viruses* No. 231.

Waterworth, H. E., and Kaper, J. M., 1972, Purification and properties of carnation mottle virus and its ribonucleic acid, *Phytopathology* **62**:959.

Waterworth, H. E., Meiners, J. P., Lawson, R. H., and Smith, F. F., 1977, Purification and properties of a virus from El Salvador that causes mild mosaic in bean cultivars, *Phytopathology* **67**:169.

Weber, I., Proll, E., Ostermann, W. D., Leiser, R. M., Stanarius, A., and Kegler, H., 1982, Characterization of cucumber leaf spot virus, a virus unknown so far in glasshouse cucumber, *Arch. Phytopathol. PflSch.* **18**:137.

Weber, I., and Stanarius, A., 1984, Comparison of cucumber leafspot virus with four newly isolated viruses from cucumber by immuno-electronmicroscopy, *Arch. Phytopathol. Pflanzenschutz* **20**:447.

Weber, I., Stanarius, A., and Kalinina, I., 1986, Weitere Untersuchungen zur serologischen verwandtschaft zwischen dem Gurkenblattfleckenvirus (cucumber leafspot virus) und dem Gurkenfruchtstreifenvirus (cucumber fruit streak virus), *Arch. Phytopathol. Pflanzenschutz* **22**:169.

Yoshida, K., Goto, T., Nemato, M., and Tsuchizaki, T., 1980, Five viruses isolated from melon in Hokkaido. *Ann. Phytopath. Soc. Japan* **46**:339.

Zandvoort, R., 1973, The spread of carnation mottle virus in carnations in glasshouses, *Neth. J. Pl. Path.* **79**:81.

Ziegler, A., Harrison, S. C., and Leberman, R., 1974, The minor proteins in tomato bushy stunt virus and turnip crinkle viruses, *Virology* **59**:509.

CHAPTER 4

The Sobemovirus Group

R. HULL

I. INTRODUCTION

Hull (1977a) proposed a new group of plant viruses based upon southern bean mosaic virus (SBMV). This group was accepted by the International Committee on Taxonomy of Viruses (ICTV) (Matthews, 1982) under the name Sobemovirus (sigla for *so*uthern *be*an *mo*saic *virus*) group. Although the ICTV only accepted two viruses as full members and four as possible members, for the purposes of this chapter I will consider ten viruses as being full members and six as being possible members (Table I). LTSV, SCMV, and VTMoV have been suggested as comprising another possible group (the velvet tobacco mottle virus group) (Matthews, 1982), but it is now generally accepted (see Francki *et al.*, 1985a) that these viruses have close affinities to Sobemoviruses. The possible members show some similarities to Sobemoviruses but as more information on them becomes available, at least some of them may be grouped elsewhere.

Sobemoviruses are characterized by being mechanically transmissible, having relatively narrow host ranges, and having isometric particles that sediment between 110 and 120 S and form one band in CsCl (some form several bands in Cs_2SO_4), and having a single viral coat protein species of mol. wt. about 30,000 (30K). The genome is a single species of single-stranded RNA of mol. wt. about 1.4×10^6 (about 4500 nucleotides) and has a plus- or messenger-sense polarity. For those viruses that it is known, the 5'-terminus of the RNA has a genome-linked protein (VPg). The particles of some members (those in the originally proposed velvet tobacco mottle virus group) also contain a viroidlike RNA, which is a satellite in that it differs in sequence from the genomic RNA and is

R. HULL • Department of Virus Research, John Innes Institute, Colney Lane, Norwich NR4 7UH, United Kingdom.

TABLE I. Some Biological Properties of Subemoviruses

Virus	Abbreviation	Vector	Seed transmission	Particles in		Reference
				Cytoplasm	Nuclei	
Southern bean mosaic	SBMV	Beetles	+	+C[b]	+	Tremaine and Hamilton (1983)
Blueberry shoestring	BBSSV	Aphids	-	+C	-	Ramsdell (1979a)
Cocksfoot mottle	CfMV	Beetles	-	+C	+	Catherall (1970), Mohamed and Mossop (1981)
Lucerne transient streak	LTSV	ND[a]	-	+	+	Forster and Jones (1980)
Rice yellow mottle	RYMV	Beetles	-	+C	-	Bakker (1975)
Solanum nodiflorum mottle	SNMV	Beetle, mirid, flea-beetles	-	+	+	Greber (1981)
Sowbane mosaic	SoMV	Leafminer, leafhopper	+	+C	+	Kado (1971)
Subterranean clover mottle	SCMV	ND	ND	+C	+	Francki et al. (1983)
Turnip rosette	TRosV	Beetles	ND	+C	+	Hollings (1973), Hull (unpublished observations)
Velvet tobacco mottle	VTMoV	Mirid, beetles	-	+	+	Randles et al. (1981), Francki et al. (1985a)
Possible members						
Cocksfoot mild mosaic	CMMV	Aphid, beetle	-	+	ND	Huth and Paul (1972), W. Huth (personal communication)
Cynosurus mottle	CyMV	Aphid	ND	+	(+)[b]	Mohamed and Mossop (1981)
Ginger chlorotic fleck	GCFV	ND	ND	ND	ND	Thomas (1986)
Maize chlorotic mottle	MCMV	Beetle	-	+	-	Gordon et al. (1984), D. E. Lesemann (unpublished observations)
Olive latent virus-1	OLV-1	ND	ND	ND	ND	Gallitelli and Savino (1986)
Panicum mosaic	PMV	Beetle	+	+C	ND	Niblett and Toler (1977)

[a] ND, not determined.
[b] C, crystalline arrays; +, rare occurrence.

completely dependent upon the genomic RNA for its replication (see Francki et al., 1985b for review). Panicum mosaic virus has a satellite particle (Buzen et al., 1984).

The group has been reviewed by Sehgal (1981) and Francki et al. (1985a). Individual viruses have also been reviewed; where possible I will refer to these reviews.

II. THE VIRUSES AND DISEASES

A. Southern Bean Mosaic Virus

The type member of the group, southern bean mosaic virus (SBMV) (Tremaine and Hamilton, 1983), was first reported by Zaumeyer and Harter (1942), who soon after provided a fuller description (Zaumeyer and Harter, 1943). It was first named both bean mosaic virus 4 and southern bean mosaic virus (after being isolated from *Phaseolus* beans in Louisiana), but the name SBMV has taken preference. It has also been named *Marmor laesiofaciens* (Zaumeyer and Harter, 1944) and *Phaseolusvirus laedens* (Roland, 1959). The virus has been reported from the United States, Central and South America, Africa, France, and India.

There are four major strains of SBMV, which are differentiated by the legume hosts they infect and by the symptoms they induce. All the strains have narrow host ranges restricted primarily to the Leguminosae. The type strain (the bean strain) (SBMV-B) infects most common cultivars of bean (*Phaseolus vulgaris*), many of them systemically. In this host it causes a mosaic which, depending upon cultivar, varies from severe to mild; in very susceptible cultivars it causes leaf distortions. In some cultivars, e.g., Pinto, it causes local lesions. It does not infect cowpea. The cowpea strain (SBMV-CP) (Shepherd and Fulton, 1962) infects most cowpea cultivars systemically, causing vein clearing, mosaic, and leaf distortion; in some cultivars it gives local lesions. The cowpea strain does not infect beans except for cv. Pinto in which it causes symptomless infection of the inoculated leaves (Tremaine and Hamilton, 1983). The severe bean or Mexican strain (SBMV-M) (Yerkes and Patino, 1960; Grogan and Kimble, 1964) infects many cultivars of bean and gives systemic symptoms in cv. Pinto; some bean cultivars, e.g., Sutter Pink, give local lesions with both the bean and Mexican strains (Grogan and Kimble, 1964). The Mexican strain also infects cowpeas (Tremaine and Hamilton, 1983). The Ghana strain of SBMV (SBMV-GH) (Lamptey and Hamilton, 1974) infects many cultivars of cowpea systemically and induces local lesions or systemic symptomless infection in some cultivars of bean. Table II lists differential host reactions of these four strains.

It is possible that there may be other strains of SBMV. For instance, Singh and Singh (1975) described an isolate from *Phaseolus mungo* that

TABLE II. Differential Host Reactions of Four Strains of SBMV

Species and cultivar	Strain			
	SBMV-B	SBMV-M	SBMV-CP	SVMV-GH
Bean cv. Prince	S[a]	S	NS	S
Bean cv. Pinto	LL	LL,S	NS	LL
Cowpea cv. Blackeye	NS	SUS	S	S
Cowpea cv. Clay	NS	SUS	LL	LL,S

[a] Abbreviations: S, systemic infection; NS, no symptoms; LL, necrotic local lesions; SUS, cowpea susceptible but no details of reaction in cultivars given.

infected cowpea, and Jayasinghe (1982) reported an isolate that infected beans, cowpea, and peas. Strains of SBMV-B and SBMV-CP that break the hypersensitive reactions of some bean and cowpea cultivars have been reported by Valverde and Fulton (1982) and McGovern and Kuhn (1982).

Within one strain of SBMV, the bean strain, there are various electrophoretic and buoyant density variants (Magdoff-Fairchild, 1967).

B. Blueberry Shoestring Virus

Blueberry shoestring virus (BBSSV) (Ramsdell, 1979a) is restricted to *Vaccinium* spp. where it causes the leaves to become narrow and strap-shaped with occasional reddish vein-banding. Reddish lesions are found on the current year and one-year-old stems, and there is occasionally flower color breaking.

The virus is limited to the New Jersey, Michigan, and Washington states of the United States and to Nova Scotia, Canada.

C. Cocksfoot Mottle Virus

Cocksfoot mottle virus (CfMV) (Catherall, 1970) has a narrow host range restricted to a few members of the Graminae. In cocksfoot (*Dactylis glomerata*) and wheat it causes a conspicuous yellow streaking and mottling of the leaves. In oats and barley the symptoms are much milder but include some necrotic spotting.

This virus has been found in England (Sergeant, 1964), France (Hariri and Lapierre, 1978), East Germany (Rabenstein and Schmidt, 1979), New Zealand (Mohamed, 1980), and Japan (Toriyama, 1982).

D. Lucerne Transient Streak Virus

Lucerne transient streak virus (LTSV) (Forster and Jones, 1979) has been found naturally only in lucerne (*Medicago sativa*) in Australia, New

Zealand, and Canada (Forster and Jones, 1979; Paliwal, 1982), but has a moderately wide experimental host range including members of the *Chenopodiaceae, Leguminosae,* and *Solanaceae.*

In lucerne, the symptoms of chlorotic streaking along the lateral veins and sometimes distortion of leaves are often transient and disappear. LTSV gives local lesions in *Chenopodium amaranticolor* and *C. quinoa.* When the viroidlike satellite RNA 2 is present the local lesions are necrotic; when it is absent they are chlorotic (Jones *et al.,* 1983). No other differences in host range or symptomatology were found for satellite-containing and satellite-free isolates of the virus.

E. Rice Yellow Mottle Virus

Rice yellow mottle virus (RYMV) (Bakker, 1975) is restricted to a few species of Graminae. In rice it causes stunting of the plants and mottling of the leaves. There are no known local-lesion hosts.

This virus occurs in Kenya, Ivory Coast, Sierra Leone, Nigeria, Upper Volta, Niger, and Malawi (Bakker, 1975; Fauquet and Thouvenel, 1977; Anon., 1979; John *et al.,* 1984).

F. Solanum Nodiflorum Mottle Virus

Solanum nodiflorum mottle virus (SNMV) (Greber, 1981) was first reported causing a prominent mottle of *Solanum nodiflorum* by Greber (1973). This virus infects a number of Solanaceae species, where the symptoms vary from moderate systemic mottles in various *Solanum* spp. to symptomless infection of inoculated leaves in species including *Petunia hybrida* and *Lycopersicon esculentum.* SNMV gives necrotic local lesions or necrotic rings (depending upon environmental conditions) in *Nicotiana debnyi* (Jones and Mayo, 1984). SNMV, freed from the satellite RNA 2 by passage through single lesions in *N. debnyi* gave the same symptoms as satellite-containing isolates (Jones and Mayo, 1984).

SNMV is found only in north-eastern Australia.

G. Sowbane Mosaic Virus

Sowbane mosaic virus (SoMV) (Kado, 1971) infects mainly members of the Chenopodiaceae. In *Chenopodium amaranticolor* and *C. quinoa* it may cause chlorotic local lesions and systemic yellow flecking. However, infection is often latent. As this virus is transmitted efficiently by seed, it often infects the standard *Chenopodium* spp. test plants. Some reported viruses, e.g., apple latent virus 2, are really SoMV.

SoMV has been reported from Australia, Europe, Japan, North and South America, and South Africa.

H. Subterranean Clover Mottle Virus

Subterranean clover mottle virus (SCMoV) (Francki *et al.*, 1983) was isolated from subterranean clover (*Trifolium subterraneum*) in western Australia. Infected plants were stunted and had severe mottle or mosaic. The host range appears limited, the virus also causing necrotic local lesions in *Pisum sativum* and small gray local lesions in *Chenopodium murale*.

I. Turnip Rosette Virus

Turnip rosette virus (TRosV) (Hollings and Stone, 1973) was first described by Blencowe and Broadbent (1957) and Broadbent and Heathcote (1958) from turnips (*Brassica rapa*) in Scotland. Its host range is mainly limited to various *Brassica* spp. in which it causes severe stunting and rosetting and a vein-banding mottle of the leaves. In *B. pekinensis* it induces necrotic and sometimes chlorotic local lesions, and it causes symptomless local infection of *Nicotiana clevelandii* and *N. bigelovii*.

J. Velvet Tobacco Mottle Virus

Velvet tobacco mottle virus (VTMoV) (Randles *et al.*, 1981) causes a yellow mosaic and blistering of leaves of velvet tobacco (*Nicotiana velutina*) in south Australia. The virus will infect other *Nicotiana* species, giving systemic symptoms in *N. clevelandii* (together with local lesions) and *N. glutinosa*, and local lesions in *N. glauca*. Symptomless infection is found in *petunia hybrida* (systemic) and in the inoculated leaves of *N. tabacum*, *Datura stramonium*, and *Gomphrena globosa*.

K. Cocksfoot Mild Mosaic Virus

The name cocksfoot mild mosaic virus (CMMV) (Huth and Paul, 1972) covers a range of virus isolates (strains) from cocksfoot (*Dactylis glomerata*) and various other grasses, which have been named phleum mottle, brome stem leaf mottle, festuca mottle, and holcus transitory mottle viruses (Torrance and Harrison, 1981). Molinia streak virus (Huth, 1974) is also very distantly related to CMMV (Querfurth and Bercks, 1976). The symptoms of CMMV and its various strains are mild to severe mottles in various members of the Graminae. Some isolates may cause severe necrotic streaking. There are some symptom and host range differences between various isolates and strains of CMMV (see Torrance and Harrison, 1981).

CMMV and its various strains are found in the United Kingdom, France, and West Germany.

L. Cynosurus Mottle Virus

Cynosurus mottle virus (CyMV) (Catherall *et al.*, 1977) was originally described under the name lolium mottle virus (A'Brook, 1972), but the name was changed by Catherall *et al.* (1977) as they were unable to infect *Lolium* spp. with it. CyMV has a host range very similar to CfMV (Catherall *et al.*, 1977), but CyMV did not infect cocksfoot and CfMV did not infect *Cynosurus cristatus*. In most of its hosts CyMV causes a severe mottle and in *C. cristatus* a lethal yellow or necrotic mottle.

CyMV has been reported from the United Kingdom, West Germany, and New Zealand.

M. Ginger Chlorotic Fleck Virus

The recently reported ginger chlorotic fleck virus (GCFV) (Thomas, 1986) appears to be limited in host range to ginger (*Zingiber officinale*), where it causes prominent chlorotic flecks parallel to leaf veins.

The virus occurs in South East Asia.

N. Maize Chlorotic Mottle Virus

Maize chlorotic mottle virus (MCMV) (Gordon *et al.*, 1984) is found naturally in maize in Peru, Argentina, Mexico, and the United States. In maize it induces a chlorotic mottle, leaf necrosis, and stunting or even premature death of the plant. Some maize genotypes are tolerant and give mild symptoms or latent infections. The host range of MCMV is restricted to the Graminae.

O. Olive Latent Virus-1

Olive latent virus-1 (OLV-1) (Gallitelli and Savino, 1985) was isolated from symptomless olive trees (*Olea europea*) from southern Italy. The virus causes systemic infection in *Nicotiana benthamiana*, showing as a mosaic and leaf crinkling and necrotic local lesions in various species.

P. Panicum Mosaic Virus

There are two major strains of Panicum mosaic virus (PMV) (Niblett and Toler, 1977), the type strain and the St. Augustin decline (SAD) strain. PMV is closely serologically related to Molinia streak virus (Paul *et al.*, 1980), which perhaps should be considered as a third strain. The host range of all strains is restricted to the Graminae. The type strain causes mild mosaic symptoms in many grass species but does not infect St.

Augustin grass (*Stenotaphrum secundatum*). The SAD strain infects St. Augustin grass, where it causes general chlorosis, internodal shortening, and leaf and stolon necrosis. Most isolates of this virus contain a satellite virus that is separately encapsidated. There is no report of any effect of this satellite on host range or symptom expression.

PMV is found in the United States and Mexico, the type strain being reported from Kansas and the SAD strain from Texas, Louisiana, and Mexico.

III. TRANSMISSION

All members of the group except BBSSV are mechanically transmissible, most with ease although GCFV was transmitted with difficulty (Thomas, 1986).

The natural vectors of most Sobemoviruses are insects (Table I). SBMV, CfMV, CMMV, PMV, RYMV, TRosV, and MCMV are transmitted efficiently by chrysomelid beetles (Walters, 1969; A'Brook and Benigno, 1972; Bakker, 1974; Gordon *et al.*, 1984; Nault *et al.*, 1978; W. Huth, personal communication). The acquisition time varies between a few minutes to one day and in most cases, the virus is retained by the vector for a few (2–14) days. The interaction between SBMV and its vectors *Cerotoma trifurcata* (SBMV-C) and *Epilachna varivestis* (SBMV-B) have been studied in the most detail. Virus can be recovered from the insects' intestinal tracts (Walters, 1969), and regurgitant fluids (Walters, 1969; Scott and Fulton, 1978; Kopek and Scott, 1983). Virus particles accumulate in the chitinized linings of the mouthparts, pharynx, membranes of the proventriculus, and multicellular pockets of the midgut (M. R. McLaughlin, quoted in Sehgal, 1981). This is suggestive of a noncirculative interaction with the vector with transmission through regurgitation. However, virus particles are also found in the hemolymph, and insects can be made viruliferous by injection of purified virus into the hemolymph (Slack and Scott, 1971; Scott and Fulton, 1978). Furthermore, hemocytes of beetles injected with SBMV-CP contained virus particles (Kim *et al.*, 1977). This is suggestive of a circulative interaction with the vectors (Walters, 1969; Fulton *et al.*, 1975). Gergerich *et al.* (1983) suggested that the vector specificity is controlled by effects of factors present in the regurgitant on the virus particles.

SNMV and VTMoV are transmitted by coccinellid beetles, and SNMV (less efficiently) by flea beetles (Greber, 1981). In the field, coccinellid beetles are the major vector of SNMV. Both viruses are acquired by the beetle vectors in less than 2 hr; VTMoV persists in the vector longer than does SNMV (Greber, 1981).

Several of the Sobemoviruses (BBSSV, CMMV, and CyMV) are transmitted by aphids. BBSSV has a circulative (and most probably nonpropagative) relationship with its vector *Illinoia pepperi* (Morimoto *et al.*, 1985).

Virus particles are found in the insects' salivary glands within 2 hr of the start of feeding (Petersen, 1984). As BBSSV can be transmitted by feeding aphids on purified virus through membranes (Morimoto *et al.*, 1985), it would seem unlikely that a noncapsid helper component is involved in the virus–vector interaction. CyMV is transmitted in a noncirculative manner by *Rhopalosiphum padi* (Mohamed, 1978a). CMMV is transmitted inefficiently by *Myzus persicae* (Huth and Paul, 1972).

SoMV is transmitted by a variety of insects: the leafminer fly *Liriomyza langei*, the leafhopper *Circulifer tenellus*, and the fleahopper *Halticus citri* (Bennett and Costa, 1961). It is likely that this transmission is purely mechanical with no real virus–vector interaction.

VTMoV and SNMV are both transmitted by the mirid bug *Cyrtopeltis nicotianae* (Greber, 1981; Francki *et al.*, 1985b); in the field it is the major vector of VTMoV (Randles *et al.*, 1981). These are the only confirmed reports of plant virus transmission by a mirid bug. The interaction between VTMoV and its vector has many of the features of the semipersistent mode (Francki *et al.*, 1985b).

Three of the Sobemoviruses are transmitted through seed. SBMV strains are transmitted in the seed of beans (SBMV-B, 1–5%; Zaumeyer and Harter, 1943) and in cowpea (SBMV-CP, SBMV-GH, 5–40%) (Shepherd and Fulton, 1962; Lamptey and Hamilton, 1974; Givord, 1981). There is some controversy as to whether the virus is in the embryo, but there is evidence for infection of germinating seedlings from contaminated seed coats (McDonald and Hamilton, 1972; Uyemoto and Grogan, 1977). There is no evidence for pollen transmission to the seed or pollinated plant (Hamilton *et al.*, 1977). SoMV is seed-transmitted at a very high rate (up to 83%) in *Chenopodium* spp. (Bennett and Costa, 1961; Dias and Waterworth, 1967; Kado, 1967). It has been suggested that the virus is carried internally (Bennett and Costa, 1961) but in which tissue was not ascertained. The SAD strain of PMV is seed-transmitted in *Setaria italica* (Niblett and Toler, 1977).

IV. PURIFICATION

Most Sobemoviruses occur in relatively high concentrations and can be grown easily in common herbaceous hosts. However, as was pointed out by Hull (1985), in designing good and reliable purification procedures consideration must be taken of the effects of buffers and other chemicals on the structure and stability of the virus particles. As discussed in more detail below (Section V.F.1), the particles of most Sobemoviruses are stabilized by pH and divalent cation-dependent bonds. Thus buffers below pH 7.0 should be used and chelating agents avoided if possible. A method that works satisfactorily for several Sobemoviruses is that described by Denloye *et al.* (1978) and modified by Brisco *et al.* (1985a). Leaves are blended at 4°C in 0.1 M sodium acetate pH 5.0 (with a reducing agent if

TABLE III. Some Physical Properties of Sobemoviruses

Virus[a]	Particle diameter (nm)	S_{20w}	Banding density in CsCl (g/cc)	Multiple banding in Cs_2SO_4	Extinction coefficient $A260^{0.1\%}_{mm}$ 1cm	EDTA-NaCl stability[b]	Virion RNA MW genomic ($\times 10^{-6}$)	Viroidlike RNA	Virion coat protein MW $\times 10^{-3}$	Reference
SBMV	28	115	1.36	Yes	5.85	L	1.4	No	29	Tremaine and Hamilton (1983)
BBSSV	27	120	1.39	Yes	5.2	L	1.45	ND	30	Ramsdell (1979a,b)
CfMV	30	112–118	1.39	Yes	ND	L	1.45	ND	29.5–32	Catherall (1970), Mohamed and Mossop (1981), Hull (1977a)
LTSV	27–28	113	1.37	Yes	5.2	L	1.4	Yes	32(29)[f]	Forster and Jones (1980)
RYMV	25	109	1.37	Yes	6.5	L	1.4	ND	31	Hull (1977a), Bakker (1975)
SNMV	28–30	115	1.37	ND[c]	ND	ND	1.5	Yes	31	Greber (1981), Francki et al. (1985a)
SoMV	26	113	1.35	Yes	4.9	L	1.3	No	31	Hull (1977a), Kado (1971)
SCMV	30	115	1.37	ND	ND	ND	1.5	Yes	29(26)	Francki et al. (1983)
TRosV	28	112	1.34	Yes	ND	L	1.4	No	30	Hull (1977a), Hollings and Stone (1973)
VTMoV	30	115	1.37	ND	ND	ND	1.5	Yes	37(33,31.5)	Randles et al. (1981)
CMMV	25–30	108	1.39,1.38	ND	ND	(L)	1.55 – 1.6,0.5[d]	No	25–37	Huth and Paul (1972), Torrence and Harrison (1981)
CyMV	28	113	1.40	No	7.3	L	1.45	ND	30.5	Mohamed (1978b)
GCFV	30	111	1.36	No	ND	L	1.5	ND	29	Thomas (1986)
MCMV	30	109	1.37	ND	ND	ND	1.4–1.6	No	23.9–27.0	Gordon et al. (1984)
OLV-1	30	111	1.36	ND	ND	L	1.35	ND	32	Gallitelli and Savino (1986)
PMV virus	29–30	109	ND	ND	ND	ND	28S[e]	ND	30.0	Niblett and Toler (1977),
satellite	15–16	42	ND	ND	ND	ND	14S	ND	16.0	Buzen et al. (1984)

[a] For abbreviations see Table I. [b]Effect on the virus particles of 1M NaCl at pH 7.0 in the presence of EDTA; L, labile (particle dissociated). [c]ND, not determined. [d]Both species of RNA found in virus particles but not determined if both needed for infection. [e]Values quoted as sedimentation coefficients. [f]Values in parentheses are minor polypeptide species.

necessary), and the homogenate strained through cheesecloth and centrifuged at 10,000g for 10 min. To the supernatant is added: polyethylene glycol (6000 mol. wt.) to give 10% w/v, NaCl to give 1% w/v, and Triton X-100 to give 1% v/v. After stirring for 1 hr at 4°C, the solution is centrifuged at 10,000g for 10 min and the pellet resuspended in 0.1 M sodium acetate pH 5.0. This is followed by a clarifying low speed centrifugation (10,000g, 10 min) and then high speed centrifugation (65,000g, 3.5 hr) through a cushion of 25% w/v sucrose in acetate buffer. Further purification can be effected by isopycnic centrifugation in CsCl at pH 5.0 (acetate buffer). The CsCl banding densities for Sobemoviruses are given in Table III.

This method has not been tested for all the Sobemoviruses, so not all the individual problems may have been recognized. It is likely that for some viruses variations on this method will be needed. For instance SoMV precipitates at pH 5 (Hull, 1977b) and so should be purified in acetate buffer at pH 6.0.

V. PARTICLE PROPERTIES

A. Size and Shape of Particles

All the Sobemoviruses have isometric particles of 25–30 nm diameter (Table III) (Fig. 1). Horne *et al.* (1977) noted that slight variation in both

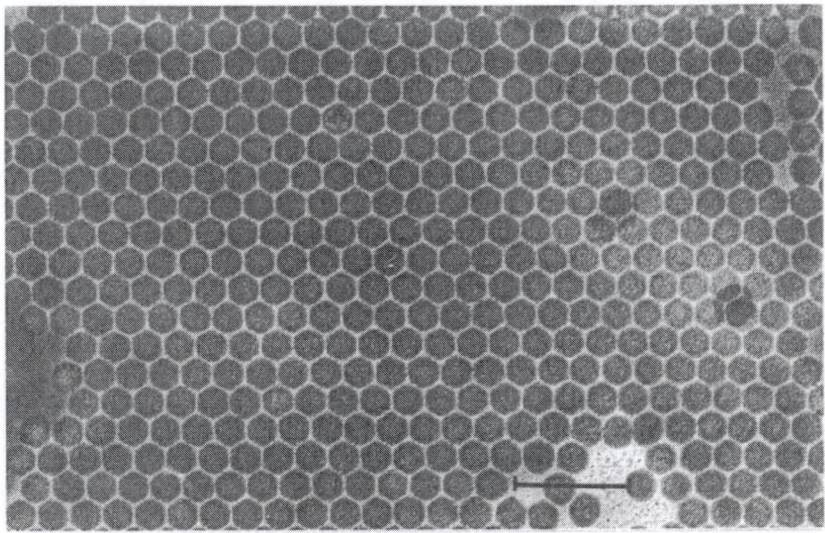

FIGURE 1. Electron micrograph of particles of SBMV, prepared by the method of Horne *et al.* (1977) and stained with ammonium molybdate and uranyl acetate. Scale bar, 100 nm.

size and shape of TRosV capsids was revealed when the virus was examined in crystalline arrays. They also noted that the central region was
either electron dense or transparent (depending upon treatment).

No detailed surface structure can be seen on Sobemovirus capsids in
the electron microscope. Even when crystalline arrays of TRosV were
examined by linear photographic techniques no structure could be resolved (Horne *et al.*, 1977). However, much is known of the detailed
structure of SBMV particles from neutron small-angle and X-ray diffraction studies, and the particles have been shown to have T = 3 symmetry.
It is most likely from size considerations that the particles of all the other
Sobemoviruses have a similar symmetry. The structure of SBMV is discussed in Section V.F.4.

B. Sedimentation

Purified preparations of all the Sobemoviruses, with the exception of
PMV, sediment as single components with S_{20w} values in the range of
108–120S (Table III). In some cases, e.g., SBMV, RYMV, SoMV, TRosV,
LTSV, SCMoV, SNMV, and VTMoV, measurements have been made in
the same laboratory. In other cases comparative measurements have not
been made. PMV sediments as two components, one being the virus and
the other the satellite (see Section II.P).

C. Isopycnic Banding

All the Sobemoviruses that have been tested are stable in CsCl, and
all except CMMV form a single band in CsCl gradients. As shown in
Table III, the banding density varies from 1.34–1.39 g/cc. This is likely
not to be a reflection of different nucleic acid contents, but to be due to
differences in Cs^+ binding.

Preparations of CMMV form two bands in CsCl gradients. It is considered likely that the particles in the more dense band contain the larger
RNA, and those in the less dense band contain several molecules (presumably three) of the smaller RNA (Torrance and Harrison, 1981). It is
interesting to note in this context that LTSV, SCMoV, SNMV, and VTMoV
each form one band in CsCl. Since the particles of each of these viruses
contain both genomic and viroidlike satellite RNAs, either each particle
must encapsidate one molecule of the genomic and an equal number of
satellite RNAs, or some particles contain genomic and others contain
several copies of the satellite RNA. If it is the latter case, the complement
of satellite RNA must equal in molecular weight that of the genomic
RNA.

Hull (1977a) suggested that the formation of multiple bands of Cs_2SO_4
gradients was a characteristic feature of Sobemoviruses. Table III shows

that many of the members of the group show this feature but the possible members CyMV, GCMV, and OLV-1 only form one band in Cs_2SO_4. The phenomenon of multiple banding in Cs_2SO_4, first reported for SBMV by Magdoff-Fairchild (1967), was examined in detail by Hull (1977b). No differences in properties (nucleic acid content, protein or nucleic acid composition, specific infectivity) were found between SBMV particles banding in the heavy and light zones; the banding patterns were also not due to the preparation being mixtures of strains. The distribution of SBMV between heavy and light zones was affected by time of exposure to Cs_2SO_4, pH, and the presence or absence of divalent cations. Hull (1977b) concluded that the multiple banding in Cs_2SO_4 was due to variations in particle structure, which resulted in variations in the penetration of the SO_4^{2-} ion.

D. Composition of Particles

Sobemovirus particles comprise just coat protein and RNA, although there is the suggestion that those of SBMV may associate with a host compound in seed coats (McDonald and Hamilton, 1973). The extinction coefficient (E_{260nm}) for those viruses for which it has been determined varies from 4.9–6.5 (Table III). There is also variation in the $E_{260nm/280nm}$ ratio (data not shown) but, as both these values depend upon the coat protein composition as well as the nucleic acid content, little of significance can be deduced.

E. Coat Proteins

As can be seen from Table III, most Sobemoviruses have a single species of coat protein. The main exceptions would appear to be VTMoV, in which several coat protein species are found, and PMV, in which there are two. However, Chu and Francki (1983) showed that for VTMoV it is likely that the smaller proteins are derived from the larger one by proteolysis. Kiberstis and Zimmern (1984) reported that the coat protein of SNMV, which is closely related antigenically to VTMoV, is a product of 38,000 mol. wt. (38K) when translated *in vitro* from a subgenomic mRNA. They also showed, by immune precipitation, that the 38K polypeptide was present in protoplasts. They suggested that this product may be rapidly proteolyzed to the 31K product found in SNMV capsids. The smaller of the PMV coat proteins is from particles that contain the satellite RNA (Niblett and Toler, 1977).

The sizes of the coat proteins of Sobemoviruses range from 24–37K, but most are around 30K. The three exceptions are: MCMV, for which there are estimates ranging from 23.9–27.0K (Gordon *et al.*, 1984), CMMV at 25–27K (Huth and Paul, 1972; Torrance and Harrison, 1981), and VTMoV at 37K (Chu and Francki, 1983; Francki *et al.*, 1985b). In VTMoV (and SNMV), as discussed above, the coat proteins are processed. The possibility has been raised (Francki *et al.*, 1985b) that other Sobemovirus coat proteins may be stable products from proteolysis. However, *in vitro* translation of SBMV (Salerno-Rife, *et al.*, 1980), TRosV (Morris-Krsinich and Hull, 1981), and PMV (Buzen *et al.*, 1984) RNAs indicate that the 29–30K coat protein polypeptide is the primary translation product.

On gel electrophoresis of the coat proteins of many of the Sobemoviruses, a minor protein band of twice the major coat protein molecular weight is found. This is a dimer of the coat protein (Sehgal and Hsu, 1977), which is not disrupted to monomer size by sodium dodecyl sulfate, reducing agents, or various other chemical treatments. How the two monomer subunits are linked together and what the significance of the stable dimers is are questions as yet unanswered.

The amino acid sequence of SBMV-CP coat protein has been determined (Dietrich *et al.*, 1981; Hermodson *et al.*, 1982). The protein comprises 260 amino acids and has an acetylated N-terminal residue. The N-terminal 41 residues are highly basic and also contain many glutamine and proline residues.

F. Capsid Stabilization and Structure

One of the commonalities throughout the Sobemovirus group appears to be the factors that stabilize the capsids. Most of the information on this has been derived from experiments on SBMV and TRosV, but other members of the group show features that are in accord with these findings (Table III).

1. Stabilization

Wells and Sisler (1969) first showed that divalent cations were involved in the stabilization of SBMV capsids; this finding was supported by the experiments of Sehgal and Sinha (1974) and Hsu *et al.* (1976). Hull (1977c), in a detailed study on TRosV, recognized three types of bond that stabilized the virus. As well as the divalent cation protein–protein bond, there is a pH-dependent protein–protein link and protein–RNA interactions. SBMV, CfMV, RYMV, and SoMV behaved in a similar manner to TRosV (Hull, 1977c).

The pH-dependent bond resembled the carboxyl–carboxylate bond of bromoviruses in that it ionized at pHs above 7.0 (Hull, 1977c).

The divalent cation-mediated bond was examined in more detail by Hull (1978). He showed that for TRosV it was Ca^{2+} rather than Mg^{2+}, which had been suggested by previous workers (Wells and Sisler, 1969; Hsu *et al.*, 1976), that played the major role. He estimated that approximately 180 Ca^{2+} ions are bound per virus particle with a pKd of 4.15 at pH 6, whereas at pH 7 about 300 Ca^{2+} ions were bound with an average pKd of 3.91. SBMV appears to bind Ca^{2+} in a similar manner (for further details of Ca^{2+} binding sites see Section V.F.4).

When both the divalent cation and pH-dependent bonds are removed, the virus particles swell. They sediment more slowly and become labile to high salt concentrations (Hsu et al., 1976; Hull, 1977c). This salt lability is indicative of protein–RNA salt links. At moderately high salt concentrations (0.4 M KCl) approximately two-thirds of the protein subunits are removed from SBMV particles to give a "subviral entity" (Sehgal and Sinha, 1974). Swollen SBMV progressively loses more protein subunits when the salt concentration is raised from 0.25 to 0.55 M; above 0.55 M most, if not all, the protein subunits are removed from the RNA (Tremaine et al., 1982).

Addition of purified RNA to swollen particles of SBMV and TRosV also results in the production of a slower sedimenting component (Hull and Aldous, 1980). Using radioactively labeled protein and RNA, they showed that about two-thirds of the coat protein molecules were removed from the virus particles and bound to the added RNA. RNAs from alfalfa mosaic and brome mosaic viruses, yeast RNA, and electronegative polymers such as sodium dextran sulfate acted in a manner similar to SBMV and TRosV RNA (Hull and Aldous, 1980; Tremaine et al., 1983). The full implications of this observation on the overall protein–RNA interactions in these viruses have not yet been elucidated.

Brisco et al. (1986b) studied the kinetics of SBMV swelling using photon correlation spectroscopy. The hydrodynamic diameter of virions at pH 8.25 increased from 29.9 nm to 44.0 nm within 3 min of the addition of the chelator EGTA; over the next 15–20 min it increased to about 56 nm. It was suggested that extrusion of the N-terminal domain of the coat protein contributed to this large hydrodynamic diameter.

2. Other Factors

Freezing and thawing preparations of SBMV in distilled water apparently changes the conformation of the capsid (Sehgal and Das, 1975), and the virus particles swell. Urea has various effects on SBMV virions depending upon pH and temperature (Sehgal et al., 1979a). These range from irreversible precipitation to partial dissociation.

Unswollen SBMV, CfMV, and SoMV are apparently unaffected by treatment with sodium dodecyl sulfate (SDS) (Ronald and Tremaine, 1976; Sehgal, 1980); however, on swelling, particles of these viruses are dissociated by 0.5% SDS.

3. Reassembly

SBMV (bean and Mexican strains) and SoMV can be reassembled from partial-dissociation products (Hsu et al., 1977) or from completely separated protein and RNA in high salt concentration (Tremaine and Ronald, 1977; Tremaine *et al.*, 1982) when the ionic strength is lowered. The number of subunits per RNA molecule increased from 30 to 145 when the salt concentration was reduced from 0.5 to 0.1 M (Tremaine *et al.*, 1982). Savithri and Erickson (1983) reassembled SBMV-CP from protein and RNA at low ionic strength. T = 3 particles were found in the presence of divalent cations when high-molecular-weight RNA was used at pH 7 and 9 and low-molecular-weight RNA at pH 9; if low-molecular-weight RNA was used at pH 7 or 5, T = 1 particles were made. In the absence of divalent cations, T = 3 particles were made at pH 7 with high-molecular-weight RNA and T = 1 particles at pH 5 with low-molecular-weight RNA.

T = 1 particles also result from limited digestion of swollen SBMV by trypsin (Tremaine and Ronald, 1978; Sehgal *et al.*, 1979b). The N-terminal 61 amino acids are removed by trypsin (Erickson and Rossmann, 1982; Mackenzie and Tremaine, 1986). In the low ionic strength reassembly by Savithri and Erickson (1983) there was some proteolysis of coat protein in the absence of RNA, in spite of the use of a protease inhibitor.

The reassembly and protease observations led Savithri and Erickson (1983) to suggest that the RNA nucleates reassemble, probably by electrostatic interaction with the very basic N-terminal region of the protein. The pH and Ca^{2+} ion binding then affects the bonding arrangement of the subunits through the charge configuration of the carboxyl group clusters. This determines whether the resulting particle is T = 3 or T = 1.

4. Particle Structure

Various physical techniques have been used to elucidate the structure of SBMV particles. As a result, many details of the structure are known at the molecular level.

Neutron small-angle scattering reveals the structure of unswollen SBMV as four concentric shells (Kruse *et al.*, 1982). In the outer compact shell (110–143 Å radius) reside about 88% of the protein. The three other shells comprise RNA and protein at different packing densities, with that of the innermost shell being the greatest. There is no major reorganization of the structure upon swelling of the virus particles.

Nuclear magnetic resonance (NMR) studies have been used to analyze the motional state of components of SBMV particles. Using [^{13}C]-NMR, McCain *et al.* (1982a) showed that in unswollen particles SBMV coat protein had only a few highly mobile amino acid side chains. The number increased upon swelling and increased even further on partial dissociation of the particles. Thus there was a progressive loosening of the coat protein structure associated with the conformational changes of the particle. From [^{31}P]-NMR, McCain *et al.* (1982b) found that in unswollen SBMV the RNA was in a relatively rigid, compact form but on swelling of the particles the RNA became much more mobile.

From circular dichroism (CD) studies the RNA in unswollen virions of both TRosV and SBMV appeared to have considerable base pairing (about 70%) (Denloye *et al.*, 1978; Odumosu *et al.*, 1981). Swelling of the virions of both viruses results in a small reduction in base stacking. The CD spectra of both viruses showed that the coat protein contained a relatively high proportion of β-structure and low proportion of α-helix. There was a slight increase in the proportion of β-structure on swelling the particles.

SBMV was one of the first viruses to be crystallized (Price, 1946; Miller and Price, 1946). Electron microscopy indicated that the particles in these crystals had a cubic packing with cell edge length of 345 Å (Labaw and Wyckoff, 1957). X-ray diffraction studies showed that SBMV could crystallize in various forms: the orthorhombic system (Magdoff, 1960); rhombohedral and hexagonal crystal forms (Johnson *et al.*, 1974; Akimoto *et al.*, 1975). From a series of X-ray diffraction analysis to greater and greater resolution—11 Å (Rayment *et al.*, 1978) 5 Å (Suck *et al.*, 1978), and 2.8 Å (Abad-Zapatero *et al.*, 1980; Rossmann *et al.*, 1983a; Silva and Rossman, 1985)—a detailed picture of the structure of the coat protein subunit and of the distribution and interactions between coat protein subunits in the capsid has been obtained. There are considerable similarities between the basic molecular structure of SBMV and that of particles of other viruses with isometric particles, such as the tomato bushy stunt and satellite tobacco necrosis viruses (Rossman *et al.*, 1983b), human rhinovirus type 14 (Rossman *et al.*, 1985), and poliovirus type 1 (Hogle *et al.*, 1985). The following is a simplified description of the structure of SBMV. For more detail, the reader is referred to the papers listed above and to a review by Rossman (1984).

The basic structure of the particle is T = 3 icosahedral symmetry with 180 protein subunits. Each protein subunit comprises two domains, the R (random) and the S (shell or surface) domain. The N-terminal part of the polypeptide chain makes up the R domain which, in the three types of subunit (see below), is for the A type 62 residues, B 64 residues, and C 38 residues long. As noted above (Section V.E), this domain is rich in arginines, lysines, prolines, and glutamines. The S domain consists of a core made up of a bundle of eight antiparallel β-sheets (termed the β-barrel) together with two α-helical regions. This domain is narrower at one end, giving it a wedge shape, which allows close subunit packing. The proportions of beta and alpha structures are close to those derived from CD analyses mentioned above. The X-ray analysis data, coupled with the primary sequence of the polypeptide, give information for the positioning of each amino acid.

Because of the quasi-equivalent symmetry of icosahedral particles (Caspar and Klug, 1962), each icosahedral unit comprises one each of three configurations of subunit, termed A, B, and C. As noted above, it is the presence of the amino-terminal sequence, the R domain, which determines that a T = 3 particle is assembled; removal of this domain results in T = 1 particles [however, in certain circumstances subunits

with R domains can form T = 1 particles (Savithri and Erickson, 1983)].
The key to the formation of the T = 3 structure is the interaction of part
of the N-terminal region of the polypeptide chain (residues 38–52, which
are next to the R domain) of adjacent type C subunits to form what is
termed a β-annulus. The interacting C subunits can be considered to form
a network structure into which the A and B subunits fit and about which
the rest of the particle is assembled.

Two types of binding site for Ca^{2+} have been located. One is on the
quasi-threefold axis and is made up of the three carboxylate groups of
glutamic acid at position 194 of the A, B, and C subunits; this accounts
for 60 Ca^{2+} ions per virion. The second is generated by aspartic acids at
position 138 of the C and A subunits and position 141 of the B subunit,
together with the carboxyl group of the phenylalanine at position 199 of
the A and B subunits and of the asparagine at position 259 of the C subunit.
This site binds 180 Ca^{2+} ions per virion. It is not known if these two
types of Ca^{2+} binding sites reflect the two types which differ in binding
affinity reported for TRosV by Hull (1978).

The interaction between the protein subunits and the RNA involves
the R domain and also basic amino acids on the inner surface of the S
domain. There are 23 positive charges per subunit, which over the whole
virion almost balances the total negative charge of the RNA. The spacing
of the basic groups on the inner side of the coat protein predicted from
the structure corresponds well to the separation of phosphates on the
nucleic acid backbone. It is suggested that the basic R domains have a
histone-like function and fold the RNA during viral assembly.

Although many features of the structure of SBMV are known in great
detail, there are other features about which little is known. These include
the positioning and significance of the stable dimers of coat protein described in Section V.E, and the positioning of the 5' VPg in relation to
the coat protein subunits. Knowledge of these may be important in understanding early events in particle disassembly and initial translation
(see Section VII.C.3.).

VI. SEROLOGY

Most of the Sobemoviruses are not serologically related to each other.
However, there are some exceptions. SNMV is closely related to VTMoV
but not to SCMoV (Randles et al., 1981; Greber, 1981; Chu and Francki,
1983; Francki et al., 1983). SCMov is distantly related to LTSV but not
to SNMV or VTMoV (Francki et al., 1983). A New Zealand isolate of
CfMV was reported to be related to CyMV (Mohamed, 1978a; Mohamed
and Mossop, 1981), in contrast to the lack of serological relationships
between European isolates of CfMV and CyMV (Paul et al., 1980).

Serological interactions have also been reported between the type
strain of PMV and the phleum mottle, but not the type strain of CMMV,

and also between the type strain of PMV and molinia streak virus (Niblett and Toler, 1977).

The various strains of SBMV are serologically related to each other, but give spurs on gel double diffusion and immunoelectrophoresis tests (Grogan and Kimble, 1964; Lamptey and Hamilton, 1974). Most of the strains of CMMV reacted serologically with each other; however, in some combinations, e.g., CMMV Netherlands isolate and festuca mottle English isolate, there was no reaction (Torrance and Harrison, 1981). There are three serotypes of MCMV that can be differentiated in gel double diffusion tests by the formation of spurs (Niblett and Claflin, 1978; Uyemoto, 1980). The type and SAD strains of PMV are distinguishable serologically (Holcomb, 1974; Berger and Toler, 1983).

Monoclonal antibodies have been used to probe the structure of SBMV particles. Tremaine *et al.* (1985a) found two major types of monoclonal antibody to a panel raised against SBMV-B particles. Antibodies B4, B7, and B11 did not react with native particles but did against swollen virus, indicating that their antigenic sites were on the subunit interfaces. Antibodies B5, B6, and B10 reacted well with unswollen virus and poorly or not at all with swollen virus. B5, B6, and B10 reacted only with SBMV-B and not SBMV-CP, whereas B4, B7, B9, and C1 and C3 (raised against SBMV-CP) reacted with both SBMV-B and SBMV-CP (Tremaine *et al.*, 1985b). Tremaine *et al.* (1986) showed that B6 and B10 had 180 and 90 binding sites, respectively, on SBMV-B, and two antibodies raised against SBMV-CP had, for 1D6, 180 binding sites on SBMV-B and SBMV-CP and, for 1C4, 180 binding sites on SBMV-CP and 60 sites on SBMV-B. Using the above observations and competitive binding assays, Tremaine *et al.* (1986) suggested the positions of the binding sites relative to the known structure of the virus particles.

A monoclonal antibody raised against the N-terminal cyanogen bromide peptide of the coat protein of SBMV-CP bound to swollen virus but not to native (MacKenzie and Tremaine, 1986). Its binding was abolished by trypsin treatment of the particles, indicating that the recognition site was in the N-terminal 30 amino acids of the coat protein. This antibody bound to virus particles that had been recompacted by pH adjustment or by the addition of Ca^{2+} after swelling. This suggested that in the recompaction the N-terminus of the coat protein was not withdrawn inside the particles.

VII. GENOME PROPERTIES, EXPRESSION, AND REPLICATION

A. Virion RNA

The genomic nucleic acid of all Sobemoviruses (except possibly CMMV) is a single molecule of plus-strand RNA. The base ratio compositions of the RNAs of those definitive Sobemoviruses for which it has been reported (SBMV, RYMV, SoMV, and TRosV) are all reasonably similar (G 25–29%; A 21–26%; C 22–27%; U 23–27%) (Tremaine, 1966; Ghabrial *et al.*, 1967; Kado, 1971; Hollings and Stone, 1973; Bakker, 1975). The RNAs of CMMV and PMV have similar base ratios to those of the definitive Sobemoviruses (Huth and Paul, 1972; Buzen *et al.*, 1984) but that of MCMV (G 22–25%; A 24–26%; C 15–16%; U 34–37%) is different (Gordon *et al.*, 1984). However, Huth (personal communication) has recently determined the base ratio of MCMV RNA to be G 23.4%; A 28.3%; C 25.8%; and U 22.5%, values that are much closer to those of Sobemovirus RNAs.

Table III shows that the molecular weights of Sobemovirus genomic RNAs are in the range 1.3–1.6 × 10⁶ (about 4000–4800 nucleotides). Particles of CMMV also contain a significant amount of an RNA of mol. wt. 0.5 × 10⁶; it has not yet been determined whether this is required for infection or is an encapsidated subgenomic RNA. Subgenomic or possible subgenomic RNAs are encapsidated in relatively small amounts by SBMV, TRosV, LTSV, SNMV, SCMoV, and VTMoV (Rutgers *et al.*, 1980; Ghosh *et al.*, 1981; Morris-Krsinich and Hull, 1981; Gould, 1981; Francki *et al.*, 1983; Kiberstis and Zimmern, 1984); these subgenomic RNAs will be discussed in Section V.II.C.2.

Particles of LTSV, SNMV, SCMoV, and VTMoV also contain a viroid-like satellite RNA species. This is the subject of a review by Francki *et al.* (1985b) and will not be discussed further here. As noted previously, purified preparations of PMV comprise two particle types, one containing the genomic RNA and the other a satellite RNA.

In the absence of direct proof from cDNA clones that the genomic RNA comprises just one species, there is only circumstantial evidence that there are not two species of the same size. Hull and Aldous (1979) examined the hybridization kinetics of the RNAs of TRosV and two strains of SBMV with their homologous cDNAs (Fig. 2), and compared these kinetics with those of tobacco mosaic virus (TMV) RNA to its cDNA. The Crt (Crt is concentration of ribonucleotides in mole/liter × sec) at 50% hybridization was 7.7 × 10⁻³ for SBMV-B and 7.4 × 10⁻³ for SBMV-CP, values that were all less than that for TMV (10.5 × 10⁻³). The complexities of TRosV, SBMV-B, and SBMV-CP were calculated from the Crt values (taking TMV to be 2 × 10⁶) as being 1.47 × 10⁶, 1.37 × 10⁶, and 1.41 × 10⁶, respectively. These are close to the molecular weights measured by gel electrophoresis, indicating that the full genome is just one RNA species.

The 5′-termini of SBMV and TRosV RNAs have been shown to have a covalently linked protein (VPg) (Ghosh *et al.*, 1979; Hull and Morris-Krsinich, 1980) of apparent mol. wt. 10–12,000 (Mang *et al.*, 1982). If these VPgs are like other VPgs in being very basic, these molecular weights are probably very overestimated (Daubert and Bruening, 1984). The VPg

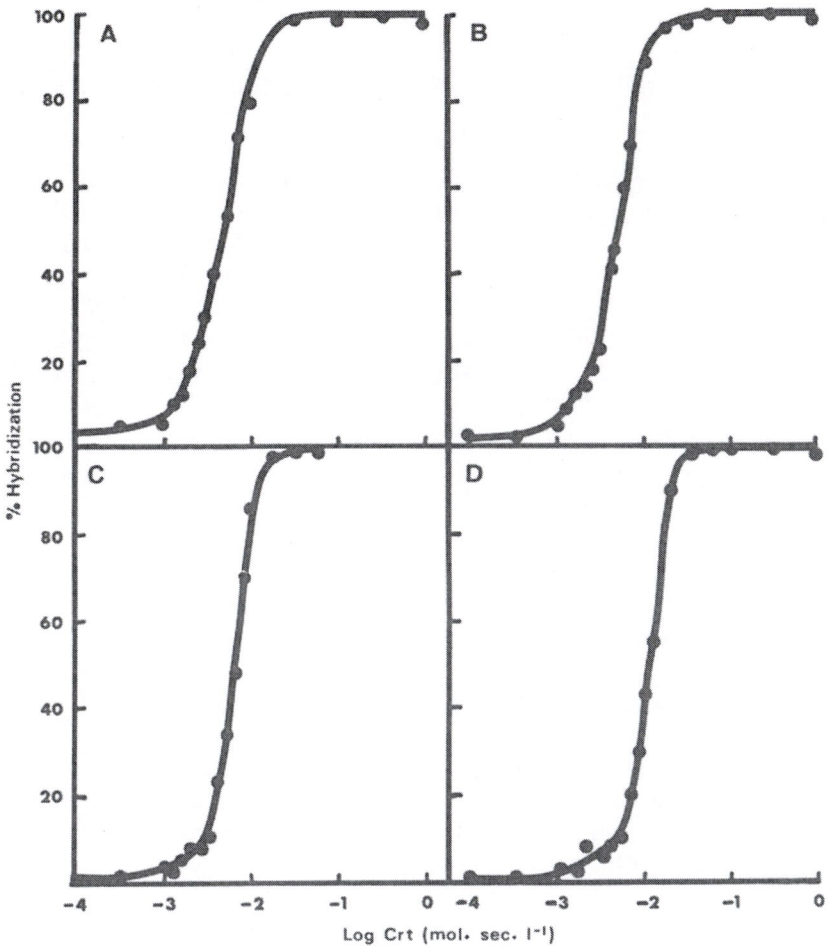

FIGURE 2. Hybridization kinetics of the RNAs of (A) SBMV-B; (B) SBMV-CP; (C) TRosV; (D) TMV, with their respective cDNAs. From Hull and Aldous (1979).

is essential for the infectivity of SBMV and TRosV RNAs (Veerisetty and Sehgal, 1980; Hull and Morris-Krsinich, 1980). The RNA of LTSV loses its infectivity on treatment with proteinase K (Forster and Jones, 1980), a property that suggests that it also might have a VPg.

The 3'-termini of SBMV and TRosV RNAs are not polyadenylated (Ghosh *et al.*, 1979; Hull and Aldous, 1979); those of SBMV-B and SBMV-CP end in . . . G-OH (Mang *et al.*, 1982). Sequencing of the RNAs of SBMV-B and SBMV-CP (Mang *et al.*, 1982) revealed a 3'-terminal non-coding region (129 bases in SBMV-B and 136 bases in SBMV-CP), which had no obvious sequence homology between the two viruses.

B. Nucleic Acid Homologies

There have been few studies on the homologies between the genomic nucleic acids of Sobemoviruses.

Mohamed and Mossop (1981) reported that there was about 5–8% sequence homology between the genomes of CfMV and CyMV as determined in liquid–liquid hybridization tests. In similar tests the RNAs of SBMV-B and SBMV-CP each had about 12–15% homology with that of TRosV, while there was about 17–20% homology between the genomes of SBMV-B, SBMV-CP, and SBMV-GH (Hull and Aldous, 1979). Mang *et al.* (1982) showed that although there was no sequence homology between the 3' noncoding regions of SBMV-B and SBMV-CP, there was considerable homology between the sequences coding for the 90 or so C-terminal amino acids of the viral coat protein cistron.

C. Replication and Expression

In contrast to our detailed knowledge of the structure and composition of the particles of at least one Sobemovirus, there is relatively little detailed information on their replication and expression. What is available is fragmentary, and is derived from studies using protoplasts and tissue culture, translation of Sobemovirus RNAs, and the isolation of RNA species intermediate in replication and expression. At the time of writing no full nucleic acid sequence of a Sobemovirus genome is available, although no doubt such data will be obtained soon. Thus we do not know fully what the coding potential of the virus is.

1. Protoplasts and Tissue Culture

As has been noted many times previously, replication that approaches synchrony is usually only obtained in protoplasts or cell suspensions. There are methods of quasi-synchronizing virus replication in plants (Dawson and Schlegel, 1973) but these have not been applied to Sobemoviruses.

The conditions for infecting soybean protoplasts with SBMV (Jarvis and Murakishi, 1980) and turnip protoplasts with TRosV (Morris-Krsinich *et al.*, 1979) have been determined. Poly-L-ornithine is required for infection in both systems. About 30–35% soybean protoplasts and up to 70% of the turnip protoplasts became infected under optimal conditions. TRosV showed a single step multiplication curve, with the virus first being detectable about 24 hr post-infection and reaching a maximum of 3×10^6 particles per protoplast after about 60 hr (Morris-Krsinich *et al.*, 1979).

SBMV has also been studied in soybean callus culture (Wu and Murakishi, 1978, 1979). In this system the cells were held at 6°C for four days and then moved to 25°C in an attempt to improve the synchronization of virus multiplication (White *et al.*, 1977). Synthesis of SBMV, as measured by the incorporation of [³H]uridine into virions and by infectivity assays, rose rapidly for the first 35–40 hr and then more slowly up to 80 hr after transfer to 25°C. An RNA species with properties suggestive of being SBMV dsRNA was synthesized mainly from 16–24 hr after the temperature transfer. This RNA species also accumulated in cells at 6°C.

2. Translation and Subgenomic RNAs

The RNAs of SBMV, TRosV, VTMoV, SNMV, and PaMV have been translated in *in vitro* systems (Salerno-Rife *et al.*, 1980; Rutgers *et al.*, 1980; Mang *et al.*, 1982; Brisco *et al.*, 1985b; Morris-Krsinich and Hull, 1981; Francki *et al.*, 1985b; Kiberstis and Zimmern, 1984; Buzen *et al.*, 1984) (Table IV).

Both SBMV-B and SBMV-CP RNAs translate to give four proteins (Table IV). The two largest proteins are translated from the genomic RNA and are related; Mang *et al.* (1982) suggested that the larger is a readthrough product of the smaller. The third product has been identified as coat protein and is translated from a subgenomic RNA of 0.3–0.4×10^6 (Rutgers *et al.*, 1980; Ghosh *et al.*, 1981). The sequence data of Mang *et al.* (1982) positions the coat protein cistron at the 3'-end of the genomic RNA. Thus the subgenomic mRNA for coat protein must be 3'-coterminal (or nearly so) with the genomic RNA. It is not clear what is the mRNA for the smallest product.

TRosV RNA also translates to give four proteins (Table IV), although differing in size from those of SBMV. The three largest ones are translated from the genomic RNA and have been shown by peptide analysis to be related. Morris-Krsinich and Hull (1981) suggested that at least the 67K product is derived from the 105K product by proteolysis. The smallest of the TRosV products has been shown to be coat protein and is encoded by a subgenomic RNA of 0.5×10^6.

Four polypeptides can also be translated from SNMV RNA (Kiberstis and Zimmern, 1984). The 100K and 67K products are related, although it is not known if this is by readthrough or processing. The 38K product is the precursor to the coat protein and is translated from a subgenomic mRNA of about 0.4×10^6. The origin of the 28K product, which does not contain any methionines, was not determined.

There is less information on the products of VTMoV RNA translation. Francki *et al.* (1985b) tentatively conclude that the primary 115K product is a readthrough protein comprising the 60K, 30K, and 19K products. The 37K product is thought to be the undegraded coat protein (see Section III. F.1) and is possibly encoded by a subgenomic RNA of 0.63×10^6 described by Gould (1981).

The products of translation of PaMV RNA (Table IV) have not been

TABLE IV. *In Vitro* Translation Products and RNA Sizes of Sobemoviruses

Virus[a]	Translation[b] system	Major polypeptides (mol. wt. × 10^-3)	RNA size (mol. wt. × 10^-6)	Reference
SEMV-B	WE and RL	105, 75, 29 (coat protein), 14	}1.4, 0.3–0.4(SG)[c], ? 1.4	Salerno-Rife et al. (1980), Rutgers et al. (1980), Ghosh et al. (1981), Mang et al. (1982)
SEMV-CP	WE and RL	100, 70, 30 (coat protein), 20	}1.4, (SG), ? 1.4	Mang et al. (1982)
SNMV	WE	100, 67, 38 (coat protein), 28	}1.5, 1.5(SG), ? 1.5	Kiberstis and Zimmern (1984)
TRosV	RL	105, 67, 35, 30 (coat protein)	}1.4, 0.5(SG)	Morris-Krsinich and Hull (1981)
VTMoV	WG	115, 60, 37 (coat protein), 19, 15, 14	}1.5, 0.63(SG)	Francki et al. (1985b)
PaMV	WG	50, 30 (coat protein), 20	ND, ND	Buzen et al. (1984)

[a] Abbreviations in Table I and text.
[b] RL, rabbit reticulocyte lysate; WE, wheat embryo; WG, wheat germ.
[c] SG, subgenomic.

analyzed in any detail. The assignment of the 30K product as coat protein is based on size (Buzen *et al.*, 1984).

Thus there are many similarities between the translation of, at least, SBMV, TRosV, SBMV, and VTMoV. The picture that seems to be emerging is of a large product being translated from the genomic RNA and the coat protein from a subgenomic RNA. The large product is either a readthrough of one or two others, or is processed to give smaller ones. In view of the constraints of eukaryotic ribosome translating the 5' cistron, the large product must come from the 5' end of the genomic RNA. This (and the sequence data of Mang *et al.*, 1982) places the coat protein (and the subgenomic RNA) to the 3' end of the virion RNA, which accords with other plant RNA viruses that use subgenomic RNAs for their expression (Davies and Hull, 1982; Goldbach, 1986).

The function of the products other than coat protein is at present unknown. Zaitlin and Hull (1987) point out that many of the RNA plant viruses encode proteins involved in replication and a protein that facilitates cell-to-cell spread. By analogy with other viruses, the proteins encoded at the 5' end of the genomic RNA are likely candidates for involvement in replication. Whether Sobemoviruses encode a cell-to-cell spread factor is unknown. One observation that suggests that they do, by analogy with the experiments of Taliansky *et al.* (1982), is that in *Dactylis glomerata* and *Cynosurus cristatus* jointly infected with CfMV and CyMV both viruses are found systemically, whereas in single infections CfMV only infected *D. glomerata* and CyMV only infected *C. cristatus* (Catherall and Andrews, 1979).

3. Uncoating of Particles

Brisco *et al.* (1985b, 1986a) showed that the RNA in SBMV particles could be translated in *in vitro* systems. For successful translation the particles had to be swollen (i.e., their structure relaxed). This phenomenon resembles the cotranslational disassembly of tobacco mosaic virus (see Wilson, 1985), which had been shown to occur *in vivo* (Shaw *et al.*, 1986). The picture presented for SBMV forms a rational story of the particles entering the cells, swelling, and the virion RNA being translated with concomitant disassembly. If the products of the 5'-cistrons are involved with replication of the viral RNA it would be an advantage to have them translated as early as possible in the virus replication cycle. Whether there are any structural mechanisms for locating the 5' end of the RNA is at present unknown.

4. Replication of RNA

Little is known of the detailed molecular biology of the replication of Sobemoviruses. There are three observations that throw some light on this topic.

Only one species of double-stranded (ds) RNA is found associated with SBMV and VTMoV infections (Morris and Dodds, 1979; Wu and Murakishi, 1979; Chu et al., 1983). This is of the size expected for the genomic RNA, and no ds form of the subgenomic RNA has been reported. This suggests, as might be expected, that the replication of the genomic RNA is plus-stranded RNA → minus-strand RNA → plus-strand RNA. It also suggests that, like several other plant RNA viruses, e.g., Bromoviruses and alfalfa mosaic virus (see Hull and Maule, 1985) the subgenomic mRNA is derived from internal initiation in the genomic minus-strand RNA.

In *Nicotiana clevelandii* leaves infected with VTMoV there is a rapid rise in the RNA-dependent RNA polymerase activity (Francki et al., 1985b). RNA-dependent RNA polymerase activity capable of synthesizing VTMoV dsRNA was found in crude extracts from infected leaves. The template RNA is tightly bound to the polymerase complex (Francki et al., 1985b). The dsRNA and the polymerase activity were found almost exclusively in the soluble cytoplasmic fraction. Francki et al. (1985b) suggest that viral RNA replication takes place in the cytoplasm and does not involve membranes.

As noted below (Section VIII), cells infected with many of the Sobemoviruses contain cytoplasmic fibrils that might be ds viral RNA. In some viruses these fibrils are enveloped in endoplasmic reticulum-derived vesicles and in others not. Thus it would appear that the replication of most, if not all, Sobemoviruses occurs in the cytoplasm although there are questions still to be answered about the involvement of membranes.

VIII. CYTOPATHOLOGY

Most of the Sobemoviruses occur at relatively high concentrations in infected plants and, for all members that have been examined (Table I), particles are found in the cytoplasm of most cells. In at least some of the cases the particles are found in crystalline arrays in the cytoplasm (Bakker, 1974; Edwardson et al., 1966; Hartmann et al., 1973; Milne, 1967; Weintraub and Ragetli, 1970a,b; Mohamed and Mossop, 1981). Particles of several of the members are also found within nuclei (Table I) (see Francki et al., 1985a). Structures resembling virus particles were found occasionally in nuclei of plants infected with CyMV (Mohamed and Mossop, 1981). The significance of the presence of particles in the nuclei to the replication of these viruses is unclear. No particles have been detected in chloroplasts or mitochondria.

Soybean tissue culture cells infected with SBMV-B show the distribution of virus described above (Wu et al., 1982).

Cells infected with many of the Sobemoviruses contain cytoplasmic fibrils, some of which are enveloped in endoplasmic reticulum-derived vesicles: There is some suggestion that these fibrils are double-stranded

RNA (see Francki *et al.*, 1985a). Characteristic tubules, often aggregated into bundles, are found in cells of plants infected with BBSSV, RYMV, and SNMV (Greber, 1981; Bakker, 1974; Francki *et al.*, 1985b). The nature of these structures is unknown.

IX. CONCLUSIONS

The Sobemoviruses present a reasonably coherent group with many properties in common. It is obvious from Tables I and III that more information about the possible members needs to be obtained before it can be determined if they are true members or not.

Sobemoviruses are among the simplest of viruses. It is therefore somewhat surprising that, although the physical structure of the particles (for at least SBMV) is known at the molecular level, there is not much information on their biology at that level. Because of their apparent simplicity it would seem that members of this group are ideal subjects for the study of molecular biology of plant RNA virus–host interactions.

Note in Proof

Recent nucleic acid sequence analyses indicate that MCMV has closer affinities to carnation mottle virus than to SBMV (S. Lommel, personal communication). Thus there are considerable doubts as to whether MCMV should be included in the Sobemovirus group.

REFERENCES

Abad-Zapatero, C., Abdel-Mequid, S. S., Johnson, J. E., Leslie, A. G. W., Rayment, I., Rossmann, M. G., Suck, D., and Tsukihara, T., 1980, Structure of southern bean mosaic virus at 2.8Å resolution, *Nature (Lond.)* **286**:33.

A'Brook, J., 1972, Lolium mottle virus, *Plant Pathology* **32**:118.

A'Brook, J., and Benigno, D. A., 1972, The transmission of cocksfoot mottle and phleum mottle viruses by *Oulema melanopa* and *O. lichenis*, *Ann. Appl. Biol.* **72**:169.

Akimoto, T., Wagner, M. A., Johnson, J. E., and Rossmann, M. G., 1975, The packing of southern bean mosaic virus in various crystal cells, *J. Ultrastr. Res.* **53**:306.

Anon., 1979, Rice diseases, *Ann. Rept. Int. Inst. Trop. Ag.* **1978**:20.

Bakker, W., 1974, Characterisation and ecological aspects of rice yellow mottle virus in Kenya, *Agric. Res. Rep.* No. 829, Wageningen 152 pp.

Bakker, W., 1975, Rice yellow mottle virus, *CMI/AAB Descriptions of Plant Viruses* No. 149.

Bennett, C. W., and Costa, A. S., 1961, Sowbane mosaic virus caused by a seed-transmitted virus, *Phytopathology* **51**:546.

Berger, P. H., and Toler, R. W., 1983, Quantitative immunoelectrophoresis of panicum mosaic virus and strains of St. Augustine decline, *Phytopathology* **73**:185.

Blencowe, J. W., and Broadbent, L., 1957, Viruses of cruciferous plants, *Annual Report Rothamsted Experimental Station*, **1956**:105.

Brisco, M. J., Hull, R., and Wilson, T. M. A., 1985a, The effect of extraction protocol on the yield, purity and translation products of RNA from an isometric plant virus, *J. Virol. Methods* **10**:195.

Brisco, M. J., Hull, R., and Wilson, T. M. A., 1985b, Southern bean mosaic virus-specific proteins are synthesized in an *in vitro* system supplemented with intact, treated virions, *Virology* **143**:392.

Brisco, M. J., Hull, R., and Wilson, T. M. A., 1986a, Swelling of isometric and of bacilliform plant virus nucleocapsids is required for virus-specific protein synthesis *in vitro*, *Virology* **148**:210.

Brisco, M. J., Haniff, C., Hull, R., Wilson, T. M. A., and Sattelle, D. B., 1986b, The kinetics of swelling of southern bean mosaic virus: A study using photon correlation spectroscopy, *Virology* **148**:218.

Broadbent, L., and Heathcote, G. D., 1958, Properties and host range of turnip crinkle, rosette and yellow mosaic viruses, *Ann. Appl. Biol.* **46**:585.

Buzen, F. G., Niblett, C. L., Hooper, G. R., Hubbard, J., and Newman, M. A., 1984, Further characterization of panicum mosaic virus and its associated satellite virus, *Phytopathology* **74**:313.

Caspar, D. L. D., and Klug, A., 1962, Physical principles in the construction of regular viruses, *Cold Spring Harbor Symp. Quant. Biol.* **27**:1.

Catherall, P., 1970, Cocksfoot mottle virus, *CMI/AAB Descriptions of Plant Viruses* No. 23.

Catherall, P. L., and Andrews, P. A., 1979, Cocksfoot mottle (CfMV) and cynosurus mottle (CyMV) viruses, *Welsh Plant Breeding Station Ann. Rept.* **1978**:199.

Catherall, P. L., Andrews, P. A., and Chamberlain, J. A., 1977, Host ranges of cocksfoot mottle and cynosurus mottle viruses, *Ann. Appl. Biol.* **87**:233.

Chu, P. W. G., and Francki, R. I. B., 1983, Chemical and serological comparison of the coat proteins of velvet tobacco mottle and *Solanum nodiflorum* mottle viruses, *Virology* **129**:350.

Chu, P. W. G., Francki, R. I. B., and Randles, J. W., 1983, Detection, isolation and characterisation of high molecular weight double-stranded RNAs in plants infected with velvet tobacco mottle virus, *Virology* **126**:480.

Daubert, S. D., and Bruening, G., 1984, Detection of genome-linked proteins of plant and animal viruses, *Methods Virol.* **8**:347.

Davies, J. W., and Hull, R., 1982, Genome expression of plant positive-strand RNA viruses, *J. Gen. Virol.* **61**:1.

Dawson, W. O., and Schlegel, D. E., 1973, Differential temperature treatment of plant greatly enhances multiplication rates, *Virology* **53**:476.

Denloye, A. O., Homer, R. B., and Hull, R., 1978, Circular dichroism studies on turnip rosette virus, *J. Gen. Virol.* **41**:77.

Dias, H. F., and Waterworth, H. E., 1967, The identity of a seed-borne mosaic virus of *Chenopodium amaranticolor* and *C. quinoa*, *Can. J. Bot.* **45**:1285.

Dietrich, J. B., Peter, R., Das, B. C., Peter, C., and Collot, D., 1981, Primary structure of the southern bean mosaic virus coat protein: first results, *Arch. Biochem. Biophys.* **210**:794.

Edwardson, J. R., Purcifull, D. E., and Christie, R. G., 1966, Electron microscopy of two small spherical plant viruses in thin sections, *Can. J. Bot.* **44**:821.

Erickson, J. W., and Rossman, M. G., 1982, Assembly and crystallization of a T = 1 icosahedral particle from trypsinized southern bean mosaic virus coat protein, *Virology* **116**:128.

Fauquet, C., and Thouvenel, J.-C., 1977, Isolation of the rice yellow mottle virus in Ivory Coast, *Pl. Dis. Reptr.* **61**:443.

Forster, R. L. S., and Jones, A. T., 1979, Properties of lucerne transient streak virus, and evidence of its affinity to southern bean mosaic virus, *Ann. Appl. Biol.* **93**:181.

Forster, R. L. S., and Jones, A. T., 1980, Lucerne transient streak virus, *CMI/AAB Descriptions of Plant Viruses* No. 224.

Francki, R. I. B., Randles, J. W., Hatta, T., Davies, C., Chu, P. W. G., and McLean, G. D., 1983, Subterranean clover mottle virus: another virus from Australia with encapsidated viroid-like RNA, *Plant Pathology* **32**:47.

Francki, R. I. B., Milne, R. G., and Hatta, T. (eds.), 1985a, *Atlas of Plant Viruses*, Volume 1, pp. 153–169, CRC Press, Boca Raton, Fl.

Francki, R. I. B., Randles, J. W., Chu, P. W. G., Rohozinski, J., and Hatta, T., 1985b, Viroid-like RNAs incorporated in conventional virus capsids, in: *Subviral Pathogens of Plants and Animals: Viroids and Prions* (K. Maramorosch and J. J. McKelvey, eds.), pp. 265–297, Academic Press, New York.

Fulton, H. P., Scott, H. A., and Gamez, R., 1975, Beetle transmission of legume viruses, in: *Tropical Diseases of Legumes* (J. Bird and K. Maramorosch, eds.), p. 123, Academic Press, New York.

Gallitelli, D., and Savino, V., 1985, Olive latent virus-1, an isometric virus with a single RNA species isolated from olive in Apulia, Southern Italy, *Ann. Appl. Biol.* **106**:295.

Gergerich, R. C., Scott, H. A., and Fulton, J. P., 1983, Regurgitant as a determinant of specificity in the transmission of plant viruses by beetles, *Phytopathology* **73**:936.

Ghabrial, S. A., Shepherd, R. J., and Grogan, R. G., 1967, Chemical properties of three strains of southern bean mosaic virus, *Virology* **33**:17.

Ghosh, A., Dasgupta, R., Salerno-Rife, T., Rutgers, T., and Kaesberg, P., 1979, Southern bean mosaic viral RNA has a 5'-linked protein but lacks 3' terminal poly(A), *Nucleic Acids Res.* **7**:2137.

Ghosh, A., Rutgers, T., Mang, K.-Q., and Kaesberg, P., 1981, Characterization of the coat protein mRNA of southern bean mosaic virus and its relationship to the genomic RNA, *J. Virol.* **39**:87.

Givord, L., 1981, Southern bean mosaic virus isolated from cowpea (*Vigna unguiculata*) in the Ivory Coast, *Plant Disease* **65**:755.

Goldbach, R. W., 1986, Molecular evolution of plant RNA viruses, *Annu. Rev. Phytopath.* **24**:289.

Gordon, D. T., Bradfute, O. E., Gingery, R. E., Nault, L. R., and Uyemoto, J. K., 1984, Maize chlorotic mottle virus, *CMI/AAB Descriptions of Plant Viruses* No. 234.

Gould, A. R., 1981, Studies on encapsidated viroid-like RNA. II. Purification and characterization of a viroid-like RNA-associated with velvet tobacco mottle virus (VTMoV), *Virology* **108**:123.

Greber, R. S., 1973, A beetle-transmitted isometric virus from *Solanum nodiflorum*, *Aust. Plant Pathol. Soc. Newsl.* **2**:3.

Greber, R. S., 1981, Some characteristics of solanum nodiflorum virus—a beetle-transmitted isometric virus from Australia, *Aust. J. Biol. Sci.* **34**:369.

Grogan, R. G., and Kimble, K. A., 1964, The relationship of severe bean mosaic from Mexico to southern bean mosaic virus and its related strain from cowpea, *Phytopathology* **54**:75.

Hamilton, R. I., Leung, E., and Nichols, C., 1977, Surface contamination of pollen by plant viruses, *Phytopathology* **67**:395.

Hariri, D., and Lapierre, H., 1978, Le virus de la nécrose et Mosaïque du dactyle (*Dactylis glamerata* L.), 281, Centre National de Recherches Agronomique INRA, Versailles, France.

Hartmann, J. X., Bath, J. E., and Hooper, G. R., 1973, Electron microscopy of viruslike particles from shoestring-diseased high bush blueberry, *Vaccinium corymbosum* L., *Phytopathology* **63**:432.

Hermodson, M. A., Abad-Zapatero, C., Abdel-Mequid, S. S., Pundak, S., Rossmann, M. G., and Tremaine, J. H., 1982, Amino acid sequence of southern bean mosaic virus coat protein and its relation to the three-dimensional structure of the virus, *Virology* **119**:133.

Hogle, J. M., Chow, M., and Filman, D. J., 1985, Three-dimensional structure of poliovirus at 2.9Å resolution, *Science* **229**:1358.

Holcomb, G. E., 1974, Serological strains of panicum mosaic virus, *Proc. Am. Phytopathol. Soc.* **1**:21.

Hollings, M., and Stone, O. M., 1973, Turnip rosette virus, *CMI/AAB Descriptions of Plant Viruses* No. 125.

Horne, R. W., Harnden, J. M., and Hull, R., 1977, The *in vitro* crystalline formations of turnip rosette virus. I. Electron microscopy of two- and three-dimensional arrays, *Virology* **82**:150.

Hsu, C. H., Sehgal, O. P., and Pickett, E. E., 1976, Stabilising effect of divalent metal ions on virions of southern bean mosaic virus, *Virology* **69**:587.

Hsu, C. H., White, J. A., and Sehgal, O. P., 1977, Assembly of southern bean mosaic virus for its two subviral intermediates, *Virology* **81**:471.

Hull, R., 1977a, The grouping of small spherical plant viruses with single RNA components, *J. Gen. Virol.* **36**:289.

Hull, R., 1977b, The banding behaviour of the viruses of the southern bean mosaic virus group in gradients of cesium sulphate, *Virology* **79**:50.

Hull, R., 1977c, The stabilization of the particles of turnip rosette virus and of other members of the southern bean mosaic virus group, *Virology* **79**:58.

Hull, R., 1978, The stabilization of the particles of turnip rosette virus. III. Divalent cations, *Virology* **89**:418.

Hull, R., 1985, Purification, biophysical and biochemical characterization of viruses with special reference to plant viruses, in: *Virology, A Practical Approach* (B. W. J. Mahy, ed.), p. 1, IRL Press, Oxford.

Hull, R., and Aldous, D., 1979, The nucleic acid of southern bean mosaic and turnip rosette viruses, *John Innes Institute Annu. Rep.* **1978**:98.

Hull R. and Aldous, D., 1980, Stabilization of turnip rosette and southern bean mosaic virus particles, *John Innes Institute Annu. Rep.* **1979**:98.

Hull, R., and Maule, A. J., 1985, Virus multiplication, in: *The Plant Viruses, Volume I, Polyhedral Viruses with Tripartite Genomes* (R. I. B. Francki, ed.), p. 83, Plenum Press, New York.

Hull, R., and Morris-Krsinich, B. A. M., 1980, Protein covalently linked to the RNA of turnip rosette virus, *John Innes Institute Annu. Rep.* **1979**:98.

Huth, W., 1974, *Molinia* streak virus—a new virus on grasses, *Acta Biol. Jugoslavica B* **11**:195.

Huth, W., and Paul, H. L., 1972, Cocksfoot mosaic virus, *CMI/AAB Descriptions of Plant Viruses* No. 107.

Jarvis, N. P., and Murakishi, H. H., 1980, Infection of protoplasts from soybean cell culture with southern bean mosaic and cowpea mosaic viruses, *J. Gen. Virol.* **48**:365.

Jayasinghe, W. U., 1982, Chlorotic mottle of bean (*Phaseolus vulgaris* L.), *Centro Internacional de Agricultura Tropical Monograph*, Series 09E(2)82.

John, V. T., Thottappilly, G., and Awoderu, V. A., 1984, Occurrence of rice yellow mottle virus in some Sahelian countries in West Africa, *FAO Plant Protection Bull.* **32**:86.

Johnson, J. E., Rossmann, M. G., Smiley, I. E., and Wagner, M. A., 1974, Single crystal X-ray diffraction studies of southern bean mosaic virus, *J. Ultrastr. Res.* **46**:441.

Jones, A. T., Mayo, M. A., and Duncan, G. H., 1983, Satellite-like properties of small circular RNA molecules in particles of lucerne transient streak virus, *J. Gen. Virol.* **64**:1167.

Jones, A. T., and Mayo, M. A., 1984, Satellite nature of the viroid-like RNA-2 of solanum nodiflorum mottle virus and the ability of other plant viruses to support the replication of viroid-like RNA molecules, *J. Gen. Virol.* **65**:1713.

Kado, C. I., 1967, Biological and biochemical characterization of sowbane mosaic virus, *Virology* **31**:217.

Kado, C. I., 1971, Sowbane mosaic virus, *CMI/AAB Descriptions of Plant Viruses* No. 64.

Kiberstis, P. A., and Zimmern, D., 1984, Translation strategy of *Solanum nodiflorum* mottle virus RNA: synthesis of a coat protein precursor *in vitro* and *in vivo*, *Nucleic Acids Res.* **12**:933.

Kim, K. S., Scott, H. A., and Robison, M. D., 1977, Ultrastructural responses of bean leaf beetle hemocytes to beetle-transmitted and non-transmitted plant viruses, *Proc. Am. Phytopath. Soc.* **4**:130.

Kopek, J. A., and Scott, H. A., 1983, Southern bean mosaic virus in Mexican bean beetle and bean leaf beetle regurgitants, *J. Gen. Virol.* **64**:1601.

Kruse, J., Timmins, P. A., and Witz, J., 1982, A neutron scattering study of the structure of compact and swollen form of southern bean mosaic virus, *Virology* **119**:42.

Labaw, L. W., and Wyckoff, R. W. G., 1957, The structure of southern bean mosaic virus protein cystals, *Arch. Biochem. Biophys.* **67**:225.

Lamptey, P. N. L., and Hamilton, R. I., 1974, A new cowpea strain of southern bean mosaic virus from Ghana, *Phytopathology* **64**:1100.

MacKenzie, D. J., and Tremaine, J. H., 1986, The use of a monoclonal antibody specific for the N-terminal region of southern bean mosaic virus as a probe of virus structure, *J. Gen. Virol.* **67**:727.

Magdoff, B. S., 1960, Subunits in southern bean mosaic virus, *Nature* **185**:673.

Magdoff-Fairchild, B. S., 1967, Electrophoretic and buoyant density variants of southern bean mosaic virus, *Virology* **31**:142.

Mang, K.-Q., Ghosh, A., and Kaesberg, P., 1982, A comparative study of the cowpea and bean strains of southern bean mosaic virus, *Virology* **116**:264.

Matthews, R. E. F., 1982, Classification and nomenclature of viruses, Fourth Report of the International Committee on Taxonomy of Viruses, *Intervirology* **17**:1.

McCain, D. C., Virudachalam, R., Markley, J. L., Abdel-Mequid, S. S., and Rossmann, M. G., 1982a, Carbon-13 NMR study of southern bean mosaic virus, *Virology* **117**:501.

McCain, D. C., Virudachalam, R., Santini, R. E., Abdel-Mequid, S. S., and Markley, J. L., 1982b, Phosphorus-31 nuclear magnetic resonance study of internal motion in ribonucleic acid of southern bean mosaic virus, *Biochemistry* **21**:5390.

McDonald, J. G., and Hamilton, R. I., 1972, Distribution of southern bean mosaic virus in the seed of *Phaseolus vulgaris*, *Phytopathology* **62**:387.

McDonald, J. G., and Hamilton, R. I., 1973, Analytical density-gradient centrifugation of southern bean mosaic virus from seed coats of *Phaseolus vulgaris*, *Virology* **56**:181.

McGovern, M. H., and Kuhn, C. W., 1982, Southern bean mosaic virus strain derived by preferential systemic movement, *Phytopathology* **72**:937.

Miller, G. L., and Price, W. C., 1946, Physical and chemical studies on southern bean mosaic virus. II. Crystallization by dialysis, *Arch. Biochem.* **11**:329.

Milne, R. G., 1967, Electron microscopy of leaves infected with sowbane mosaic virus and other small polyhedral viruses, *Virology* **32**:589.

Mohamed, N. A., 1978a, Cynosurus mottle virus, a virus affecting grasses in New Zealand, *N.Z. J. Agric. Res.* **21**:709.

Mohamed, N. A., 1978b, Physical and chemical properties of cynosurus mottle virus, *J. Gen. Virol.* **40**:379.

Mohamed, N. A., 1980, Cocksfoot mottle virus in New Zealand, *N.Z. J. Agric. Res.* **23**:273.

Mohamed, N. A., and Mossop, D. W., 1981, Cynosurus and cocksfoot mottle viruses: A comparison, *J. Gen. Virol.* **55**:63.

Morimoto, K. M., Ramsdell, D. C., Gillett, J. M., and Chaney, W. G., 1985, Acquisition and transmission of blueberry shoestring virus by its aphid vector *Illinoia pepperi*, *Phytopathology* **75**:709.

Morris, T. J., and Dodds, J. A., 1979, Isolation and analysis of double-stranded RNA from virus-infected plant and fungal tissue, *Phytopathology* **69**:854.

Morris-Krsinich, B. A. M., and Hull, R., 1981, Translation of turnip rosette virus RNA in rabbit reticulocyte lysates, *Virology* **114**:98.

Morris-Krsinich, B. A. M., Hull, R., and Russo, M., 1979, Infection of turnip leaf protoplasts with turnip rosette virus, *J. Gen. Virol.* **43**:339.

Nault, L. R., Styer, W. E., Coffey, M. E., Gordon, D. T., Negi, L. S., and Niblett, C. L., 1978, Transmission of maize chlorotic mottle virus by chrysomelid beetles, *Phytopathology* **68**:1071.

Niblett, C. L., and Claflin, L. E., 1978, Corn lethal necrosis—a new virus disease of corn in Kansas, *Plant Dis. Reptr.* **62**:15.

Niblett, C. L., and Toler, R. W., 1977, Panicum mosaic virus, *CMI/AAB Descriptions of Plant Viruses No. 177*.

Odumosu, A. O., Homer, R. B., and Hull, R., 1981, Circular dichroism studies on southern bean mosaic virus, *J. Gen. Virol.* **53**:193.

Paliwal, Y. C., 1982, Lucerne transient streak—a virus of alfalfa newly recognised in North America, *Phytopathology* **72**:989.

Paul, H. L., Querfurth, G., and Huth, W., 1980, Serological studies on the relationships of some isometric viruses of Graminae, *J. Gen. Virol.* **47**:67.

Petersen, M. A., 1984, Localization of blueberry shoestring virus (BBSSV) in the blueberry aphid, *Illinoia pepperi* (MacGillivray), Masters Thesis, Michigan State University, East Lansing.

Price, W. C., 1946, Purification and crystallization of southern bean mosaic virus, *Am. J. Bot.* **33**:45.

Querfurth, G., and Bercks, R., 1976, Relative importance of IgG- and IgM-antibodies in distant serological cross-reactivities of isometric *Molinia* streak and cocksfoot mild mosaic virus and of tobacco mosaic virus, *Phytopath. Zeitschrift* **85**:193.

Rabenstein, F., and Schmidt, H. B., 1979, Nachweis des Knaulgrasscheckungs-Virus (cocksfoot mottle virus) in der DDR, *Arch. Phytopathol. Pflanzenschutz* **15**:351.

Ramsdell, D. C., 1979a, Blueberry shoestring virus, *CMI/AAB Descriptions of Plant Viruses* No. 204.

Ramsdell, D. C., 1979b, Physical and chemical properties of blueberry shoestring virus, *Phytopathology* **69**:1087.

Randles, J. W., Davies, C., Hatta, T., Gould, A. R., and Francki, R. I. B., 1981, Studies on encapsidated viroid-like RNA. 1. Characterization of velvet tobacco mottle virus, *Virology* **108**:111.

Rayment, I., Johnson, J. F., Suck, D., Akimoto, T., and Rossmann, M. G., 1978, An 11 Å-resolution electron density map of southern bean mosaic virus, *Acta Cryst.* **B34**:567.

Roland, G., 1959, Rapport sur l'activité de la Commission de Nomenclature des Virus 1954–1959, *Taxon* **8**:126.

Ronald, W. P., and Tremaine, J. H., 1976, Comparison of the effects of sodium dodecyl sulfate on some isometric viruses, *Phytopathology* **66**:1302.

Rossman, M. G., 1984, The structure of southern bean mosaic virus, in: *Biological Macromolecules and Assemblies.* Vol. 1 (A. McPherson, ed.), p. 46, John Wiley & Sons Inc., New York.

Rossman, M. G., Abad-Zapatero, C., Hermodson, M. A., and Erickson, J. W., 1983a, Subunit interactions in southern bean mosaic virus, *J. Mol. Biol.* **166**:37.

Rossmann, M. G., Abad-Zapatero, C., Murthy, M. R. N., Liljas, L., Jones, T. A., and Strandberg, B., 1983b, Structural comparison of some small spherical plant viruses, *J. Mol. Biol.* **165**:711.

Rossmann, M. G., Arnold, E., Erickson, J. W., Frankenberger, E. H., Griffith, J. P., Hecht, H.-J., Johnson, J. E., Kamer, G., Luo, M., Mosser, A. G., Rueckert, R. R., Sherry, B., and Vriend, G., 1985, Structure of a human common cold virus and functional relationship to other picornaviruses, *Nature* **317**:145.

Rutgers, T., Salerno-Rife, T., and Kaesberg, P., 1980, Messenger RNA for the coat protein of southern bean mosaic virus, *Virology* **104**:506.

Salerno-Rife, T., Rutgers, T., and Kaesberg, P., 1980, Translation of southern bean mosaic virus RNA in wheat embryo and rabbit reticulocyte extracts, *J. Virol.* **34**:51.

Savithri, H. A., and Erickson, J. W., 1983, The self-assembly of the cowpea strain of southern bean mosaic virus: Formation of T = 1 and T = 3 nucleoprotein particles, *Virology* **126**:328.

Scott, H. A., and Fulton, J. P., 1978, Comparison of the relationships of southern bean mosaic virus and the cowpea strain of tobacco mosaic virus with the bean leaf beetle, *Virology* **84**:207.

Sehgal, O. P., 1980, Effect of protein cross-linking reagents and sodium dodecyl sulfate on southern bean mosaic virus, *Phytopathology* **70**:342.

Sehgal, O. P., 1981, Southern bean mosaic virus group, in: *Handbook of Plant Virus Infections and Comparative Diagnosis* (E. Kurstak, ed.), pp. 91–121, Elsevier/North Holland Biomedical Press, Amsterdam.

Sehgal, O. P., and Das, P. D., 1975, Effect of freezing on conformation and stability of the virions of southern bean mosaic virus, *Virology* **64**:180.

Sehgal, O. P., and Hsu, C. H., 1977, Identity and location of a minor protein component in virions of southern bean mosaic virus, *Virology* **77**:1.

Sehgal, O. P., and Sinha, R. C., 1974, Characteristics of a nucleoproteinaceous subviral entity resulting from partial degradation of southern bean mosaic virus, *Virology* **59**:499.

Sehgal, O. P., Van, M., and White, J. A., 1979a, pH-dependent urea sensitivity of southern bean mosaic virus, *Virology* **94**:479.

Sehgal, O. P., Hsu, C. H., White, J. A., and Van, M., 1979b, Enzymic sensitivity of conformationally altered virions of southern bean mosaic virus, *Phytopath. Z.* **95**:167.

Sergeant, E.P., 1964, Cocksfoot mottle virus, *Plant Pathology* **13**:23.

Shaw, J. G., Plaskitt, K. A., and Wilson, T. M. A., 1986, Evidence that tobacco mosaic virus particles disassemble cotranslationally *in vivo*, *Virology* **148**:326.

Shepherd, R. J., and Fulton, R. W., 1962, Identity of a seed-borne virus of cowpea, *Phytopathology* **52**:489.

Silva, A. M., and Rossmann, M. G., 1985, The refinement of southern bean mosaic virus in reciprocal space, *Acta Cryst.* **B41**:147.

Singh, R., and Singh, R., 1975, Studies on a mosaic disease of Urd bean (*Phaseolus mungo* L.), *Phytopath. Medit.* **14**:55.

Slack, S. A., and Scott, H. A., 1971, Haemolymph as a reservoir for the cowpea strain of southern bean mosaic virus in the bean leaf beetle, *Phytopathology* **61**:538.

Suck, D., Rayment, I., Johnson, J. E., and Rossmann, M. G., 1978, The structure of southern bean mosaic virus at 5Å resolution, *Virology* **85**:187.

Taliansky, M. E., Malyshenko, S. I., Pshannikova, E. S., and Atabekov, J. G., 1982, Plant virus-specific transport function. II. A factor controlling host range, *Virology* **122**:327.

Thomas, J. E., 1986, Purification and properties of ginger chlorotic fleck virus, *Ann. Appl. Biol.* **108**:43.

Toriyama, S., 1982, Cocksfoot mottle virus in Japan, *Ann. Phytopath. Soc. Jpn.* **48**:514.

Torrance, L., and Harrison, B. D., 1981, Properties of Scottish isolates of cocksfoot mild mosaic virus and their comparison with others, *Ann. Appl. Biol.* **97**:285.

Tremaine, J. G., 1966, The amino acid and nucleotide composition of the bean and cowpea strains of southern bean mosaic virus, *Virology* **30**:348.

Tremaine, J. H., and Hamilton, R. I., 1983, Southern bean mosaic virus, *CMI/AAB Descriptions of Plant Viruses* No. 274.

Tremaine, J. H., and Ronald, W. P., 1977, Assembly of particles from southern bean mosaic virus and sowbane mosaic virus components, *Can. J. Bot.* **55**:2274.

Tremaine, J. H., and Ronald, W. P., 1978, Limited proteolysis of southern bean mosaic virus by trypsin, *Virology* **91**:164.

Tremaine, J. H., Ronald, W. P., and Kelly, E. M., 1982, Intermediates of the *in vitro* assembly and disassembly of southern bean mosaic virus, *Virology* **118**:35.

Tremaine, J. H., Ronald, W. P., and McGauley, E. M., 1983, Effect of sodium dextran sulfate on some isometric plant viruses, *Phytopathology* **73**:1241.

Tremaine, J. H., MacKenzie, D. J., and Ronald, W. P., 1985a, Monoclonal antibodies as structural probes of southern bean mosaic virus, *Virology* **144**:80.

Tremaine, J. H., Ronald, W. P., and MacKenzie, D. J., 1985b, Southern bean mosaic virus monoclonal antibodies: reactivity with virus strains and with the virus antigen in different conformations, *Phytopathology* **75**:1208.

Tremaine, J. H., MacKenzie, D. J., and Ronald, W. P., 1966, Determination of affinity constants and the number of binding sites of monoclonal antibodies specific for southern bean mosaic virus, *Virology* **155**:452.

Uyemoto, J. K., 1980, Detection of maize chlorotic mottle virus serotypes by enzyme-linked immunosorbent assay, *Phytopathology* **70**:290.

Uyemoto, J. K., and Grogan, R. G., 1977, Southern bean mosaic virus: Evidence for seed transmission in bean embryos, *Phytopathology* **67**:1190.

Valverde, R. A., and Fulton, J. P., 1982, Characterization and variability of strains of southern bean mosaic virus, *Phytopathology* **72**:1265.

Veerisetty, V., and Sehgal, O. P., 1980, Proteinase K-sensitive factor essential for the infectivity of southern bean mosaic virus ribonucleic acid, *Phytopathology* **70**:282.

Walters, H. J., 1969, Beetle transmission of plant viruses, *Adv. Virus Res.* **15**:339.

Weintraub, M., and Regetli, H. W. J., 1970a, Electron microscopy of the bean and cowpea strains of southern bean mosaic virus within leaf cells, *J. Ultrastr. Res.* **32**:167.

Weintraub, M., and Ragetli, H. W. J., 1970b, Identification of the constituents of southern bean mosaic virus in crystals of infected cells, and their distribution within the virion, *Virology* **41**:729.

Wells, J. M., and Sisler, H. D., 1969, The effect of EDTA and Mg²⁺ on the infectivity and structure of southern bean mosaic virus, *Virology* **37**:227.

White, J. L., Wu, F. S., and Murakishi, H. H., 1977, The effect of low-temperature pre-incubation treatment of tobacco and soybean callus cultures on rates of tobacco and southern bean mosaic virus synthesis, *Phytopathology* **67**:60.

Wilson, T. M. A., 1985, Nucleocapsid disassembly and early gene expression by positive-strand RNA viruses, *J. Gen. Virol.* **66**:1201.

Wu, F. S., and Murakishi, H. H., 1978, Infection and synthesis rate of southern bean mosaic virus in soybean callus cells under selected cultural conditions, *Phytopathology* **68**:1389.

Wu, F. S., and Murakishi, H. H., 1979, Synthesis of virus and virus-induced RNA in southern bean mosaic virus-infected soybean cell cultures, *J. Gen. Virol.* **45**:149.

Wu, F. S., Hooper, G. R., and Murakishi, H. H., 1982, Cytopathology of soybean tissue culture cells infected with southern bean mosaic virus, *In Vitro* **18**:525.

Yerkes, W. D., and Patino, G., 1960, The severe bean mosaic virus, a new bean virus from Mexico, *Phytopathology* **50**:334.

Zaitlin, M., and Hull, R., 1987, Plant virus-host interactions, *Annu. Rev. Plant Physiol.* **38**:291.

Zaumeyer, W. J., and Harter, L. L., 1942, A new virus disease of bean, *Phytopathology* **32**:438.

Zaumeyer, W. J., and Harter, L. L., 1943, Two new virus diseases of beans, *J. Agric. Res.* **67**:305.

Zaumeyer, W. J., and Harter, L. L., 1944, A severe necrosis caused by bean mosaic virus 4 on beans, *Phytopathology* **34**:510.

CHAPTER 5

Tobacco Necrosis, Satellite Tobacco Necrosis, and Related Viruses

H. Fraenkel-Conrat

I. INTRODUCTION

Necrovirus is a recently coined name for a very small group, presently encompassing only tobacco necrosis virus (TNV) and possibly cucumber necrosis and melon necrotic spot virus. Although TNV was identified very early in the history of plant virology (Smith and Bald, 1935; Pirie *et al.*, 1938; Bawden and Pirie, 1945) and although it has a number of unusual features, this virus has been studied less frequently and in fewer laboratories than most of the long-known viruses. One of the particular aspects of TNV is that it is frequently associated with a satellite, tobacco necrosis satellite virus (TNSV), a very small particle unable to replicate in the absence of TNV (Kassanis and Nixon, 1960, 1961). Another cause for special interest is the mode of its specific transmission by a fungus, *Olpidium brassicae* (Teakle, 1962; Teakle and Hiruki, 1964). Although TNV can become replicated in many plant species it is not a widespread pathogen and is of little economic importance (Uyemoto, 1981). Its function as a helper for the replication of TNSV, however, makes it of great interest to the scientist concerned with the mode of viral replication.

Our understanding of the TNV–TNSV system was retarded almost 30 years by several factors. Among these were (1) the existence of several strains of TNV (113S) of limited serological interrelationship, (2) the pres-

H. FRAENKEL-CONRAT ● Department of Molecular Biology and Virus Laboratory, University of California, Berkeley, California 94720.

ence of a particle too small to be a competent virus, TNSV (50 S), and (3) the great stability of the latter and its tendency to aggregate to much larger particles, the predominant one being a dodecamer (240 S). Crystallizability was also a criterion that in the early days tended to confuse the issue. The actual relationship of the various components of TNV isolates was clarified by the work of Kassanis in the early 1960s, whose results are summarized in two reviews (Kassanis, 1978, 1981).

We now know that TNV is an isometric particle of 28-nm diameter and 7.2×10^6 daltons, consisting of one molecule of RNA and 180 molecules of protein, of 1.4×10^6 and 32×10^3 daltons, respectively. TNSV is also isometric and is 17 nm in diameter, with 1.7×10^6-dalton particle weight and an RNA of about 0.4×10^6 daltons (14 S), located within a 6-nm sphere with little contact with the shell of 60 coat protein molecules (which in different strains range from $22-24 \times 10^3$ daltons). Both the protein and the RNA of TNSV have been sequenced (see Figure 1).

II. GEOGRAPHICAL DISTRIBUTION, HOST RANGE, DISEASES, AND EPIDEMIOLOGY

TNV can be detected in the roots of a great variety of plants, mostly angiosperms, although usually not in the rest of the plant. It has been isolated all over the world from tobacco, bean, apple, pear, tulip, carrot, and so forth (Uyemoto, 1981) and has also been found in river water (Tomlinson et al., 1983).

When leaves are mechanically infected with Necroviruses, necrotic spots of varying size and color appear on the inoculated area. Such lesions may at times spread along leaf veins and lead to the death of a leaf (Babos and Kassanis, 1963c). However, generally no systemic disease results. Rare exceptions are the Augusta disease of tulip, stipple streak of bean, and cucumber necrosis. However, there are indications that systemic infection may spread in a symptomless manner throughout woody crops and other plants (Gama et al., 1982). Research on TNV has probably been handicapped by the failure to find a productive systemic host. The cryptic nature of TNV, occurring naturally almost only in the roots of plants, has also hindered epidemiological studies. All that can be said is that the virus may be ubiquitous, but it takes effort to demonstrate its presence; and since it does not affect plant growth no effort has been made to eradicate it (Uyemoto, 1981).

III. TRANSMISSION

The singular mode of natural transmission of TNV–TNSV by the chytrid fungus *Olpidium brassicae* has been the subject of considerable scientific interest in several laboratories (Teakle, 1962; Teakle and Hi-

ruki, 1964; Kassanis and Macfarlane, 1964a,b, 1965, 1968; and others).
Uniflagellate zoospores of the fungus can be shown to bind the viruses
(TNV and TNSV) *in vitro*. It appears that attachment to the axonemal
sheath rather than to the plasmalemma is the determining factor in vector
specificity. The axoneme (or flagellum) is then retracted into the cytoplasm. The cell is walled off and secretes an adhesive substance, which
attaches it to the epidermis of the plant root. The fungal protoplast can
then penetrate the host cell, with most of the membraneous matrial and
the zoospore's plasmalemma remaining outside. About 36 hr after penetration a thallus wall forms, but transmission of the virus probably takes
place during the intimate contact of the fungal ectoplast with the host
cytoplasm.

At various stages of these processes, the virus may become released
into the soil, and drainage water from symptomless plants can thus be
shown to contain infectious virus (Yarwood, 1960). It is believed that
zoospores released from the roots do not carry virus, but bind it and
become carriers in the soil water. The affinity of the virus for the zoospore
surface is remarkably high, and its binding is not reversible by washing.
It is for this and possibly similar reasons, that sterilized soil is an absolute
requirement for the culture of plants for the purpose of isolating and
studying any non-soil-transmitted viruses.

It should be particularly noted that tobacco callus tissue, not readily
infectible by most viruses, is easily infected by TNV in presence of the
fungus (Kassanis and MacFarlane, 1964b).

IV. PURIFICATION AND SEPARATION OF TNV AND TNSV

Most often TNV and TNSV are present together in the infected plant
material (e.g., Bawden and Pirie, 1950; Kassanis and Nixon, 1960); occasionally TNV but never TNSV is found alone. That TNSV does not
occur free from TNV is not surprising, since it can only be replicated in
the presence of TNV. The separation of the two types of particles of 50
S and 113 S should present no problem were it not for the great tendency
of TNSV particles to aggregate to hexamers, 12 mers, and even 24 mers
(Kassanis and Woods, 1968), thus being present at varying levels throughout the gradient during the customary rate zonal density gradient centrifugation. Isopycnic density gradient centrifugation is useless since both
viruses have about the same RNA : protein ratio and thus density.

To obtain the virus, the infected plant (e.g., tobacco or cowpea) leaves
are generally deep frozen. They are then ground to a powder and extracted
with a pH 7.5–8.0 buffer at 0.01 or lower molarity. We have found that
after several differential centrifugations of such extracts ammonium sulfate fractionation was a most useful tool. TNV is precipitated at 0°C at
0.2 saturation, and TNSV is precipitated at 0.33 saturation. This procedure
seems to dissociate TNSV aggregates so that upon sucrose gradient cen-

trifugation (5–30%, 3–5 hr at 4°C) the two components are obtained in seemingly pure form, peaking at tubes 8 and 14 of 24, respectively. This was judged by SDS polyacrylamide gel electrophoresis of their RNAs and proteins (about 1.4×10^6 versus 0.4×10^6, and 32,000 versus 22,000 daltons, respectively) (Fraenkel-Conrat, 1976; Salvato and Fraenkel-Conrat, 1977). Phenol extraction of the RNA required a pretreatment for 15 min with 2% sodium dodecyl sulfate (SDS) at 37°C to obtain good yields of RNA. The perchlorate method of Wilcoxon and Hull (1974) was also used successfully. Variants of this isolation procedure and other methods for the isolation, separation, and purification of the viruses have also been advocated (Uyemoto, 1981).

V. PROPERTIES OF THE VIRIONS, THEIR RNAs, AND THEIR COAT PROTEINS

Both TNV and TNSV are generally very stable viruses, not inactivated by temperatures of 80–90°C (Babos and Kassanis, 1963b), although structural changes occur between pH 5 and pH 7 (McCarthy *et al.*, 1980). Very unstable variants of TNV have also been observed, and it was shown that these represented uncoated or defectively coated viral nucleic acids (Babos and Kassanis, 1962; Kassanis and Welkie, 1963). TNSV remains active in terms of interacting with TNV even 10 days after applying it to the surface of leaves before TNV inoculation (Mossop and Francki, 1979). The amount of TNV in and around single necrotic lesions has been determined by the ELISA method (Roggero and Pennazio, 1984). Studies of the effects of antimetabolites on TNV replication have also been reported (Faccioli and Rubies-Antonelli, 1976). Many current studies, exemplified by a paper by Faccioli and Capponi (1983), deal with antiviral factors present in plants that are probably responsible for their ability to limit the spread of infection to necrotic spots.

The sedimentation rate of TNV is about 113 S; that of TNSV is 50 S, but aggregates of 6, 12, and 24 particles are often detected with S values of about 170, 230, and 330 (Kassanis and Woods, 1968). The diameters of the icosahedral particles are 28 and 17 nm, with weights of 7.2 and 1.7×10^6 daltons, respectively. The three-dimensional fine structure up to 3-Å resolution has been elucidated by a Swedish group (Fridborg *et al.*, 1965; Sjøberg, 1977; Liljas *et al.*, 1982) and by a neutron small-angle scattering study (Chauvin *et al.*, 1977). The low molecular weight for the RNA given by this method (0.28×10^6 daltons) is obviously erroneous, since sequencing has shown the presence of 1239 nucleotides in TNSV RNA.

TNV consists of an RNA molecule of about 1.4×10^6 daltons and 180 protein molecules of about 32×10^3 daltons (Lesnaw and Reichmann, 1969; Salvato and Fraenkel-Conrat, 1977)[*]; TNSV consists of an RNA

[*] The value of 22.6×10^3 frequently quoted is erroneous.

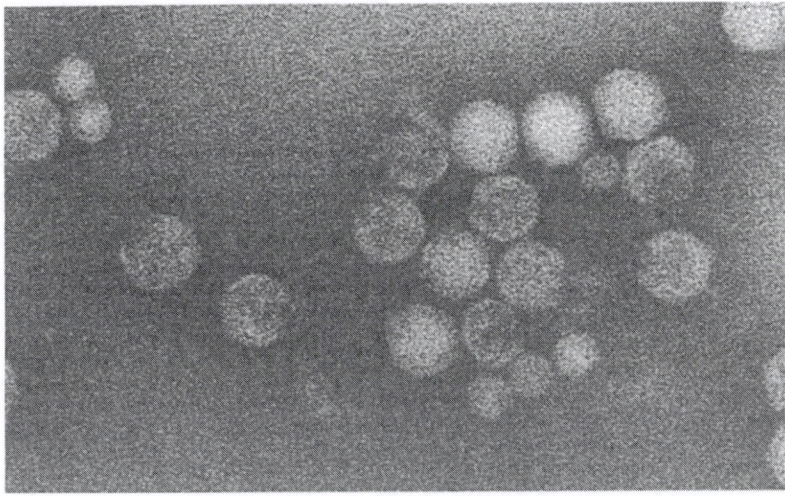

FIGURE 1. Electron micrograph of a mixture of TNV and TNSV particles. Isometric particles of 28 and 17 nm (0.9 and 0.5 cm on the image) are visible.

molecule of about 0.38×10^6 daltons, and 60 protein molecules of about 22×10^3 daltons. The latter value was less than earlier cysteine and end group analyses had indicated (Reichmann, 1964; Roy *et al.*, 1969; Roy and Fraenkel-Conrat, 1970). Both TNV and TNSV occur as several strains, not all of which are identical as far as these values or amino acid compositions are concerned (Rees *et al.*, 1970). Neither of the RNAs of these viruses show any of the terminal features of most other RNA plus strand plant and animal viruses. They lack 5'-terminal caps or covalently linked proteins, as well as 3'-terminal poly(A) or amino acid binding capability. In these respects both TNV and TNSV resemble the RNA phages. It may be regarded as significant that the RNAs of both viruses have the 5'-terminal sequence ppApGpUp − −, with TNSV RNA being partially triphosphorylated (Lesnaw and Reichmann, 1970; Horst *et al.*, 1971; Merregaert *et al.*, 1979; Wimmer and Reichmann, 1968; Wimmer *et al.*, 1968). It has been suggested that a 5'-terminal hairpin with seven base pairs plays a role in the recognition of the initiation codon at position 30-32 by the ribosomes (Leung *et al.*, 1979). The binding of initiation factor 2 to that site has also been studied (Kaempfer *et al.*, 1981). The 3' end of TNSV RNA, GACUACCCC3', is partly phosphorylated (Horst *et al.*, 1971). The complete sequence of that RNA is now known (Fig. 2) (van Emmelo *et al.*, 1980; Ysebaert *et al.*, 1980).

The coat proteins are the only known gene products of these viruses. The amino acid sequence of TNSV protein is known through direct analysis as well as from the nucleotide sequence (Ysebaert *et al.*, 1980). The 195 amino acids of this protein show a distribution typical also for other isometric plant virus coats (Chapters 2, 3, and 4 of this volume) with

```
AGUAAAGACAGGAAACUUUACUGACUAAC AUG GCAAAACAACAGACAACAACAGGCGAAAAUCCGCAACAAUGCGUGCAGUGAACGCCAUGAUAAAAUACACA      100

CUUGGAGCAUAAAAGGUUUGCACUGAUCAACACUCAGGGAACAACUCGUGGUACAGUACAAAAUCUGUCCAACGGUAUAAUCCAAGGAGAUGAU              200

AUCAACCAGAGAAGGUGGUGAUCAAGUGCGUAUAGUUUCACAUAAACUUCACAUUAAACCGUCGAGGCACUGCCAUCACCGUCAGCCUUUAGCAUUUAUCUGGU    300

UUCGUGAUAACAUGACCGUGGGACCCACUCUCCCAGGUUCUUGAGGUGUUGAACACUGCGAAUUUCAUGUCGCAGUAUAAACCCAAUCACGUGCAGCAAAA      400

GAGAUUUACAAUACUCAAGGAUGAUGAACUCUCAAUGGUCCCGGCCAAUAUUAUGUUGCAGAUUGGGCCAGAUGGGGAGCAUUAAAGAUCGGAUAAUUAACCUUCCAGGACAACUGGUGAACUAU  500

AAUGGAGCGACGGCUGUAGCAGCCUCCAAUGGUCCCGGCCAAUAUUUAUGUUGCAGAUUGGGCCAGAUGGGGAGCAUUAAAGAUCGGAUAAUUAACCUUCCUUGGUUGGGACUCCCUUUAUGAGG  600

CUGUGUACACAGAUGCA UAA UCCCAGAGGUUCACAAUGUUAGUGAUGGGCGCUGAAAGAUGCGUGAGCUCGUAACACCGAGGUCAUUGGGAAAGC             700

AGAAUCCAAGGGUACGGUGGUACGGUUGGCCUCGACGUGUCGAGGGGCCUAAGGAUUGGAACCCGCUUACACCGAGCUUAGGCUAAAGGUACUACCUUGCUCAU   800

UUGUAGUCUAAAUGACGUUGGCCUCGACGUGUCGAGGGGCCUAAGGAUUGGAACCCUCGUGUUUCCGUGUUUCCGAGUUAAAGCCGGU                  900

CUCUUGGUCAUAAUGCCAUUAGUAGGUCUAGCACUCAACGUAACUUCUUGCAACAAGAAUAUCCUCCGGUGUCAACAAGGUCAACAAGAUUAAAGCCGGU      1000

AUAUUAAGUGCGCCGGCUUAUCGUUGUUUGGACGUUUGGACCACGCCCCACGCCGGUUGGUACCCCUUCAGGCGGUCACAGGGCUUUAGGAGAUGAGAUAAG   1100

GUAUAGUUAUUAGACAAAUGCGGACAAACCUGAAAAGCUCGCUAGUGGUGGGGGCUGGCCAAGCGAAGAACCUCAUCCAGGUAUAGUUCUACAUGGGAAAUU  1200

UGGUACCAUCCAAACUUCUAUGAAGUCCUGACUACCCC
```

FIGURE 2. The nucleotide sequence of TNSV RNA. The initiation and termination codons for the TNSV protein are boxed in. The terminal phosphates are not shown.

many basic residues in the N-terminal part, presumably internal in the virion and in contact with the RNA, followed by many hydrophobic residues, and the acidic ones toward the C-terminus and the virion surface (Fig. 3). The gene for this protein is situated at the 5' end of the RNA, starting at the 33rd residue (30-32 = AUG). The 3' half of the RNA is apparently not translated (van Emmelo et al., 1980, 1984).

For the TNV coat protein only terminal sequences are now known (ALA N-terminal, and possibly Ser-Val-Val-Met C-terminal). Concerning the nature of the N-terminus, there remains some doubt. N-terminal alanine was reported for both viruses from Reichmann's laboratory (Lesnaw and Reichmann, 1969), but later studies by Uyemoto and Grogan (1969) failed to find a free amino end group. Unpublished results from that laboratory using gas chromatography indicated the presence of different N-terminal acyl groups in different strains of both viruses. Thus TNSV-C and TNV AC39 were reported to be terminally acetylated, while TNSV-B and TNV AC36 were butyrylated. Uyemoto suggests that the discrepancy between Reichmann's and their results might have been due to the former having studied different strains of both viruses (TNSV-A = TNSV-1) and the corresponding Urbana isolate of TNV.

It may also be considered whether the N-terminal alanine was an artifact due to the use of 1N HCl (24 hr, room temperature) as the isolation procedure of the proteins. The yield of terminal alanine was low in terms of the now-known molecular weights of the two proteins. The necessary assumption that an acid-sensitive bond precedes alanine in these two unrelated proteins throws doubt, however, on this hypothesis. There is now no question from sequence analyses that alanine is N-terminal in TNSV. However, in these recent studies the acid procedure was also used to isolate the protein. The possibility of an acid-labile acyl group at the N-terminus cannot be ruled out. The in vitro synthesized protein shows the expected methionine preceding the alanine (Rice and Fraenkel-Conrat, 1973; Klein et al., 1972, 1973; Leung et al., 1976, 1979). Obviously further studies of this question are needed.

Ala – Lys – Gln – Gln – Asn-Asn – Arg – Arg-Lys – Ser – Ala – Thr-Met – Arg – Ala-Val – Lys-Arg – Met-Ile – 20

Asn – Thr – His – Leu – Glu – His – Lys – Arg-Phe – Ala-Leu – Ile – Asn – Ser – Gly – Asn – Thr-Asn – Ala-Thr – 40

Ala – Gly – Thr – Val – Gln – Asn-Leu – Ser – Asn-Gly – Ile – Ile – Gln – Gly – Asp-Asp – Ile – Asn-Gln – Arg – 60

Ser – Gly – Asp – Gln – Val-Arg – Ile – Val – Ser – His – Lys – Leu-His – Val – Arg-Gly – Thr – Ala – Ile – Thr – 80

Val – Ser – Gln – Thr – Phe-Arg – Phe – Ile – Trp – Phe – Arg-Asp-Asn – Met-Asn-Arg – Gly – Thr-Thr-Pro – 100

Thr – Val-Leu – Glu – Val-Leu – Asn – Thr – Ala-Asn – Phe-Met-Ser – Gln – Tyr-Asn-Pro – Ile – Thr-Leu – 120

Gln – Gln-Lys – Arg-Phe – Thr – Ile – Leu-Lys – Asp – Val-Thr-Leu – Asn – Cys-Ser – Leu – Thr-Gly-Glu – 140

Ser – Ile – Lys – Asp-Arg-Ile – Ile – Asn-Leu-Pro – Gly – Gln-Leu – Val-Asn-Tyr – Asn – Gly – Ala-Thr – 160

Ala – Val – Ala-Ala-Ser-Asn-Gly – Pro-Gly – Ala – Ile – Phe-Met – Leu – Gln – Ile – Gly – Asp – Ser-Leu – 180

Val – Gly – Leu – Trp – Asp-Ser – Ser – Tyr – Glu-Ala-Val-Tyr-Thr-Asp – Ala   195

FIGURE 3. The amino acid sequence of the TNSV coat protein, the only known translation product of TNSV RNA (see Fig. 2).

It should be pointed out that nucleotide-sequence-derived amino acid sequence data are unable to solve the question of whether a protein is terminally acylated. Protein-chemical studies continue to be of essential value.

VI. SEROLOGY OF TNV AND TNSV

Uyemoto (1981) differentiated by gel diffusion plates and spur formation seven TNV strains among 11 isolates tested (CH, RO, RE, KD, AC36, AC39, AC43; see Table I). Of TNSV at least three serotypes were differentiated, two more closely related (SVA, SVB or SV1, SV2) and one clearly different (SVC). More recently a fourth TNSV strain was detected (Ameloot *et al.*, 1983). The former group is able to become activated by seven of the 11 TNV isolates, while TNSV-C is activated by the other four TNV strains (Grogan and Uyemoto, 1967). These differences are correlated with TNV host specificities, in that the TNSV-C-activated TNV strains, in contrast to the others, do not readily infect French bean and tobacco. However, the exact relationship of the Rothamstead strain and those studied by Kassanis (Babos and Kassanis, 1963a; Rees *et al.*, 1970), and the strains used by Uyemoto is not clear (SV1 to SV4 versus SVA to SVD).

VII. TRANSLATION AND REPLICATION OF THE GENOMES OF TNV AND TNSV

The most important aspect in the replication of the TNV–TNSV system is the absolute dependence of TNSV replication on the presence of TNV. This dependence is also quite specific in that certain strains of TNSV are only replicated in the presence of certain TNV strains (Uyemoto *et al.*, 1968; Kassanis and Phillips, 1970; Uyemoto, 1981). This dependence led to the assumption that TNSV, lacking information for a viral replicase, depended on the replicase of the helper virus. The questions concerning the replicase of TNV (and other plant viruses) have led to 12 years of studies of RNA-dependent RNA polymerases in virus-infected and uninfected plants in my laboratory (Fraenkel-Conrat, 1976; Stussi-Garaud *et al.*, 1977; Ikegami and Fraenkel-Conrat, 1978). I will briefly summarize the state of our present knowledge concerning these enzymes: (1) All plants have RNA-dependent RNA polymerases. (2) The amounts of these enzymes are increased by RNA virus (as well as viroid) infection, this effect varying from two- to 100-fold for different plants and viruses. (3) Most researchers in this field believe that viral RNA replication requires one or two viral gene products with or without a role being attributed to the presence of increased amounts of the host RNA-dependent RNA polymerase. The presumed viral RNA replicase components have,

TABLE I. Antisera

	RO[a]		AC 36		AC 43		KD	
TNV and spur order		Ppt. titer	TNV and spur order	Ppt. titer	TNV and spur order	Ppt. titer	TNV and spur order	Ppt. titer
CH	1[b]	1024[c]	AC39 5	256	CH 6	128	CH 4	256
KA	1	1024	Ch 4	512	KA 6	128	KA 4	128
RO	1	1024	KA 4	512	RO 6	128	RO 4	128
RE	2	512	RO 4	512	RE 5	128	RE 3	128
AC43	3	256	RE 3	512	AC39 4	256	AC39 2	128
KD	4	128	AC43 2	1024	NZ 3	2048	NZ 2	512
AC42	5	64	KD 2	2048	AC38 3	2048	AC38 2	512
AC39	6	256	AC42 2	2048	AC36 3	2048	AC36 2	256
NZ	7	128	NZ 1	2048	AC42 2	2048	AC43 2	512
AC38	7	128	AC38 1	2048	KD 2	1024	AC42 2	512
AC36	7	128	AC36 1	2048	AC43 1	2048	KD 1	512

[a] Antisera and virus isolate designations as follows: RO, Rothamsted culture; KA and KD, Kassanis strains A and D; RE, Reichmann culture; AC 36, 38, 39, 42, 43, American Type Culture numbers; NZ, New Zealand culture; CH, chrysanthemum culture.

[b] Numbers indicate order of spur formation with respect to the homologous reactions which are labeled 1. In each column antigens with smaller numbers formed spurs beyond the precipitin lines of any others with larger numbers. Antigens with the same numbers in each column produced reactions of identity with that antiserum.

[c] Precipitin titers expressed as reciprocals of the highest two-fold dilution of antiserum that produced a perceptible precipitin line in gel diffusion. (Modified from Uyemoto, 1981.)

singly or together, molecular weights ranging from 110,000 to 350,000, while the host enzymes are all about 130,000 (Fraenkel-Conrat, 1983).

We now return to TNV. Its largest gene product, both *in vivo* and *in vitro*, is reported to be about 65,000 daltons (Jones and Reichmann, 1973; Salvato and Fraenkel-Conrat, 1977). Since TNV lacks a large gene product, this author believes that RNA replication of both TNV and TNSV must involve the host enzyme, even though a viral gene product may well play an important role in this process. The presence in infected plant tissue of three double-stranded viral RNAs of 2.6, 1.05, and 0.95 daltons, nested and from the 3' end, has been reported (Condit and Fraenkel-Conrat, 1979). However, nothing definitive appears to be known concerning the mode of replication of these RNAs.

More is known about their translation, particularly that of TNSV. This RNA is a very effective messenger in both prokaryotic and eukaryotic *in vitro* systems (Clark *et al.*, 1965; Klein *et al.*, 1972; Lundquist *et al.*, 1972; Rice and Fraenkel-Conrat, 1973; Klein and Clark, 1973; Salvato and Fraenkel-Conrat, 1977). Even though the RNA lacks a cap as well as a 5'-terminal protein, either of which is generally a prerequisite for facile translation, ribosomes attach themselves from about the 10th to the 40th nucleotides and start translating TNSV coat protein at the AUG, located at position 30-32. At a termination codon 588 nucleotides downstream, the 195 amino acid-long TNSV protein is released. The next 622 nucleotides remain untranslated, and their function is unknown. The lack of a role of a cap is supported by studies using 7-methyl guanosine-5'-monophosphate, which is an inhibitor in the translation of capped RNAs (Hickey *et al.*, 1976; Roman *et al.*, 1976; Sonnenberg *et al.*, 1980). The absence of a cap is also borne out by the ability of the benzo[a]pyrene epoxide treated TNSV RNA translation to initiate normally, in contrast to capped mRNAs (Haas *et al.*, 1983).

The translations of TNV RNA *in vitro* and *in vivo* each appear to have been studied in only one laboratory and not at all recently. In addition, the lack of a particular 5'-terminal feature renders this viral RNA a priori unusual. What is quite surprising about TNV RNA translation is that, in contrast to all other RNA viruses, particularly with single genomes, no large gene product is detected and the coat protein is the predominant product, besides three to five minor proteins of 63, 43, 26 $\times$ 10^3 daltons, all adding up to approximately the total genome information content. This was observed by *in vitro* translation using the wheat germ system (Salvato and Fraenkel-Conrat, 1977) as well as *in vivo* (Jones and Reichmann, 1973). It is proposed that translation starts, stops, and restarts at initiation and termination codons in the same manner, as it does in RNA phages. There are a few other instances in the literature suggesting that eukaryotic ribosomes can find internal initiation sites, but at least one such instance has recently been shown to be in doubt and others are also dubious. The fact that the absence of a 5'-terminal feature does not prevent the effective translation of TNSV RNA makes the same behavior of TNV RNA, though carrying at least three initiation codons, easier to accept. In competition between the two RNAs *in vitro*, both were translated (Klein and Reichmann, 1970): TNSV RNA translation predominated over that of TNV RNA, but both occurred (Salvato and Fraenkel-Conrat, 1977). We have suggested that TNV may actually have evolved from an RNA phage, but further studies are needed to confirm the observations that have led to this suggestion.

Doubt is thrown on the internal initiation of TNV RNA translation by our later finding of intracellular virus-specific double-stranded RNAs of the three sizes, nesting and showing the same 3'-terminal RNA sequence and its complement (Condit and Fraenkel-Conrat, 1979). That again suggests processing of the minus strand copy of TNV RNA, with possible use of the 3'-terminal plus strand segments as templates for the proteins.

VIII. CYTOPATHOLOGY

Detailed studies on the cytopathological effects of TNV and TNSV are lacking (Francki *et al.*, 1985).

IX. OTHER NECROVIRUSES

The cucumber necrosis virus has been studied more than other potential members of the group. It appears that this virus is transmitted specifically by another fungus, *Olpidium radicale*. It is serologically distinctly different from TNV. A 50 S particle frequently found associated with it is probably not a satellite virus, since its protein is serologically related to that of the cucumber necrosis virus, whereas TNSV protein is unrelated to TNV protein (Dias, 1970; Tremaine, 1972; Temmink *et al.*, 1970). However, recent studies have shown that several unrelated viruses carry satellite viruses similar though not related to TNSV (Valverde and Dodds, 1987).

X. CONCLUSIONS

This chapter surely differs from all others in *The Plant Viruses* in that most of the references to TNV are prior to 1971. The current literature on this virus deals largely with its detection in many hosts and many lands, and the effect of infection on plant enzymes and biology. But no progress appears to have been made in our knowledge of TNV RNA replication or translation. The satellite, TNSV, has not been so neglected, but the nature of its dependence on TNV has also not been elucidated. It is hoped that drawing attention to this gap in our knowledge of plant virology may stimulate action. TNV is surely a singular virus in several regards, and it is worthy of more study.

REFERENCES

Ameloot, P., van Emmelo, J., and Fiers, W., 1983, SV-4, a new satellite tobacco necrosis virus variant, *Med. Fac. Landbouww, Rijksuniv. Gant* **48**(3):787

Babos, P., and Kassanis, B., 1962, Unstable variants of tobacco necrosis virus, *Virology* **18**:206.

Babos, P., and Kassanis, B., 1963a, Serological relationships and some properties of tobacco necrosis virus strains, *J. Gen. Microbiol.* **32**:135.

Babos, P., and Kassanis, B., 1963b, Thermal inactivation of tobacco necrosis virus, *Virology* **20**:490.

Babos, P., and Kassanis, B., 1963c, The behaviour of some tobacco necrosis virus strains in plants, *Virology* **20**:498.

Bawden, F. C., and Pirie, N. W., 1945, Further studies on the purification and properties of a virus causing tobacco necrosis, *Br. J. Exp. Pathol.* **26**:277.

Bawden, F. C., and Pirie, N. W., 1950, Some factors affecting the activation of virus preparation made from tobacco leaves infected with TNV, *J. Gen. Microbiol.* **4**:464.

Chauvin, C., Jacrot, B., and Witz, J., 1977, The structure and molecular weight of satellite tobacco necrosis virus: A neutron small-angle scattering study, *Virology* **83**:479.

Clark, J. M., Jr., Chang, A. Y., Spiegelman, S., and Reichmann, M. E., 1965, The *in vitro* translation of a monocistronic message, *Proc. Natl. Acad. Sci. USA* **54**:1193.

Condit, C., and Fraenkel-Conrat, H., 1979, Isolation of replicative forms of 3' terminal subgenomic RNAs of tobacco necrosis virus, *Virology* **97**:122.

Dias, H. F., 1970, The relationship between Cucumber Necrosis Virus and its vector, *Olpidium cucurbitacearum, Virology* **42**:204.

Faccioli, G., and Capponi, R., 1983, An antiviral factor present in plants of *Chenopodium amaranticolor* locally infected by TNV. 1. Extraction, partial purification, biological and chemical properties, *Phytopathol. Z.* **106**:289.

Faccioli, G., and Rubies-Antonelli, C., 1976, Action of antimetabolites on the biosynthesis of TNV in locally-infected cells of *Chenopodium amaranticolor, Phytopathol. Z.* **87**:48.

Fraenkel-Conrat, H., 1976, RNA polymerase from TNV infected and uninfected tobacco. Purification of the membrane-associated enzyme, *Virology* **72**:23.

Fraenkel-Conrat, H., 1983, RNA-dependent RNA polymerases of plants, *Proc. Natl. Acad. Sci. USA* **80**:422.

Francki, R. I. B., Milne, R. G., and Hatta, T. (eds.), 1985, *Atlas of Plant Viruses*, Volume I, CRC Press, Boca Raton.

Fridborg, K., Hjertén, S., Höglund, S., Liljas, A., Lundberg, B. K. S., Oxelfelt, P., Philipson, L., and Strandberg, B., 1965, Purification, electron microscopy and x-ray diffraction studies of the satellite tobacco necrosis virus, *Proc. Natl. Acad. Sci. USA* **54**:513.

Gama, M. I. C. S., Kitajima, E. W., and Lin, M. T., 1982, Properties of a TNV isolate from *Pogostemum patchuli* in Brazil, *Phytopathology* **72**:529.

Grogan, R. G., and Uyemoto, J. K., 1967, A D serotype of satellite virus specifically associated with a D serotype of tobacco necrosis virus, *Nature* **213**:705.

Haas, R., Pulkrabek, P., Takanami, Y., and Grunberger, D., 1983, Translation of satellite tobacco necrosis virus RNA modified by (±)-r-7,t-8-dihydroxy-t-9,10-epoxy-7,8,9,10-tetrahydrobenzo[a]pyrene is inhibited in a wheat germ cell-free system, *Carcinogenesis* **4**:221.

Hickey, E. D., Weber, L. A., and Baglioni, 1976, Inhibition of protein synthesis by 7-methyl guanosine-5'monophosphate, *Proc. Natl. Acad. Sci. USA* **73**:19.

Horst, J., Fraenkel-Conrat, H., and Mandeles, S., 1971, Terminal heterogeneity at both ends of the satellite tobacco necrosis virus ribonucleic acid, *Biochemistry* **10**:4748.

Ikegami, M., and Fraenkel-Conrat, H., 1978, The RNA-dependent RNA polymerase of cowpea, *FEBS Lett.* **96**:197.

Jones, J. M., and Reichmann, M. E., 1973, The proteins synthesized in tobacco leaves infected with tobacco necrosis virus and satellite tobacco necrosis virus, *Virology* **52**:49.

Kaempfer, R., van Emmelo, J., and Fiers, W., 1981, Specific binding of eukaryotic initiation factor 2 to satellite tobacco necrosis virus RNA at a 5'-terminal sequence comprising the ribosome binding site, *Proc. Natl. Acad. Sci. USA* **78**:1542.

Kassanis, B., 1978, *TNV Group Atlas of Insect and Plant Viruses* (Maramorosh, K., ed.), p. 281, Academic Press, New York.

Kassanis, B., 1981, Portraits of viruses: Tobacco necrosis virus and its satellite virus, *Intervirology* **15**:57.

Kassanis, B., and Macfarlane, I., 1964a, Transmission of tobacco necrosis virus by zoospores of *Olpidium brassicae, J. Gen. Microbiol.* **36**:79.

Kassanis, B., and Macfarlane, I., 1964b, Transmission of tobacco necrosis virus to tobacco callus tissues by zoospores of *Olpidium brassicae, Nature* **201**:218.

Kassanis, B., and Macfarlane, I., 1965, Interaction of virus, strain, fungus isolates, and host species in the transmission of tobacco necrosis virus, *Virology* **26**:603.

Kassanis, B., and Macfarlane, I., 1968, The transmission of satellite viruses of tobacco necrosis virus by *Olpidium brassicae, J. Gen. Virol.* **3**:227.

Kassanis, B., and Nixon, H. L., 1960, Activation of one plant virus by another, *Nature* **187**:713.

Kassanis, B., and Nixon, H. L., 1961, Activation of one tobacco necrosis virus by another, *J. Gen. Microbiol.* **25**:459.

Kassanis, B., and Phillips, M. P., 1970, Serological relationships of strains of tobacco necrosis virus and their ability to activate strains of satellite virus, *J. Gen. Virol.* **9**:119.

Kassanis, B., and Welkie, G. W., 1963, The nature and behavior of unstable variants of tobacco necrosis virus, *Virology* **21**:540.

Kassanis, B., and Woods, R. D., 1968, Aggregated forms of the satellite tobacco necrosis virus, *J. Gen. Virol.* **2**:395.

Klein, A., and Reichmann, M. E., 1970, Isolation and characterisation of two species of doublestranded RNA from tobacco leaves doubly infected with TNV and TNSV, *Virology* **42**:269.

Klein, W. H., and Clark, J. M., Jr., 1973, N-terminal sequence of the eucaryotic *in vitro* product made upon translation of satellite tobacco necrosis virus RNA, *Biochemistry* **12**:1528.

Klein, W. H., Nolan, C., Lazar, J. M., and Clark, J. M., Jr., 1972, Translation of satellite tobacco necrosis virus ribonucleic acid. I. Characterisation of *in vitro* procaryotic and eucaryotic translation products, *Biochem.* **11**:2009.

Lesnaw, J. A., and Reichmann, M. E., 1969, The structure of tobacco necrosis virus, 1. The protein subunit and the nature of the nucleic acid; 2. Terminal amino acid residues of the protein subunit, *Virology* **39**:729, 738.

Lesnaw, J. A., and Reichmann, M. E., 1970, Identity of the 5'-terminal RNA nucleotide sequence of the satellite tobacco necrosis virus and its helper virus: Possible role of the 5' terminus in the recognition by virus-specific RNA replicase, *Proc. Natl. Acad. Sci. USA* **66**:140.

Leung, D. W., Browning, K. S., Heckman, J. E., RajBhandary, U. L., and Clark, J. M., Jr., 1979, Nucleotide sequence of the 5' terminus of satellite tobacco necrosis virus ribonucleic acid, *Biochemistry* **8**:1361.

Leung, D. W., Gilbert, C. W., Smith, R. E., Sasavage, N. L., and Clark, J. M., Jr., 1976, Translation of satellite tobacco necrosis virus RNA by an *in vitro* system from wheat germ, *Biochemistry* **15**:4943.

Liljas, L., Unge, T., Jones, T. A., Fridborg, K., Lörgren, F., Skoglund, U., and Strandberg, B., 1982, Structure of satellite tobacco necrosis virus at 3.0 Å resolution, *J. Mol. Biol.* **159**:93.

Lundquist, R. E., Lazar, J. M., Klein, W. H., and Clark, J. M., Jr., 1972, Translation of satellite tobacco necrosis virus ribonucleic acid. II. Initiation of *in vitro* translation in procaryotic and eucaryotic systems, *Biochemistry* **11**:2014.

McCarthy, D., Bleichmann, J., and Thorne, J., 1980, Some effects of pH, salt, urea, ethanediol and SDS on TNV, *J. Gen. Virol.* **46**:391.

Merregaert, J., van Emmelo, J., and Fiers, W., 1979, 3'-terminal nucleotide sequence of satellite tobacco necrosis virus RNA, *Virology* **98:**182.

Mossop, D. W., and Francki, R. I. B., 1979, The stability of satellite viral RNAs *in vivo* and *in vitro*, *Virology* **94:**243.

Pirie, N. W., Smith, K. M., Spooner, E. T., and McClement, W. D., 1938, Purified preparations of tobacco necrosis virus (Nicotiana virus 11), *Parasitology* **30:**543.

Rees, M. W., and Short, M. N., and Kassanis, B., 1970, The amino acid composition, antigenicity and other characteristics of satellite viruses of TNV, *Virology* **40:**448.

Reichmann, M. E., 1964, The satellite tobacco necrosis virus: A single protein and its genetic code, *Proc. Natl. Acad. Sci. USA* **52:**1009.

Rice, R., and Fraenkel-Conrat, H., 1973, Fidelity of translation of satellite tobacco necrosis virus ribonucleic acid in a cell-free *Escherichia coli* system, *Biochemistry* **12:**181.

Roggero, P., and Pennazio, S., 1984, Quantitative determination by ELISHA of TNV from necrotic local lesions in tobacco, *J. Virol. Methods* **8:**282.

Roman, R., Brooker, J. D., Seal, S. N., and Marcus, A., 1976, Inhibition of the translation of a 40 S ribosome-met-tRNA,^{met} complex to an 80 S ribosome-met-tRNA,^{met} complex by 7-methylguanosine-5'-phosphate, *Nature* **260:**359.

Roy, D. D., and Fraenkel-Conrat, H., 1970, Reduction and alkylation of the coat protein of the tobacco necrosis satellite virus, *Virology* **40:**767.

Roy, D., Fraenkel-Conrat, H., Lesnaw, J., and Reichmann, M. E., 1969, The protein subunit of the satellite tobacco necrosis virus, *Virology* **38:**368.

Salvato, M. S., and Fraenkel-Conrat, H., 1977, Translation of tobacco necrosis virus and its satellite in a cell-free wheat germ system, *Proc. Natl. Acad. Sci. USA* **74:**2288.

Sjøberg, B., 1977, A small angle x-ray investigation of the satellite tobacco necrosis virus, *Europ. J. Biochem.* **81:**231.

Smith, K. M., and Bald, J. G., 1935, A description of a necrotic virus disease affecting tobacco and other plants, *Parasitology* **27:**231.

Sonnenberg, N., Trachsel, H., Hecht, S., and Shatkin, A. J., 1980, Differential stimulation of capped mRNA translation in vitro by cap binding protein, *Nature* **285:**331.

Stussi-Garaud, C., Leminus, J., and Fraenkel-Conrat, H., 1977, RNA polymerase from tobacco necrosis virus-infected and uninfected tobacco, *Virology* **81:**224.

Teakle, D. S., 1962, Transmission of tobacco necrosis virus by a fungus, *Olpidium brassicae*, *Virology* **18:**224.

Teakle, D. S., and Hiruki, C., 1964, Vector specificity in *Olpidium*, *Virology* **24:**539.

Temmink, J. H. M., Campbell, R. N., and Smith, P. R., 1970, Specificity and site of *in vitro* acquisition of tobacco necrosis virus by zoospores of *Olpidium brassicae*, *J. Gen. Virol.* **9:**201.

Tomlinson, J. A., Faithfull, E. M., Webb, M. J. E., and Fraser, R. S. S., 1983, Chenopodium necrosis: A distinctive strain of tobacco necrosis virus isolated from river water, *Ann. Appl. Biol.* **102:**135.

Tremaine, J. H., 1972, Purification and properties of cucumber necrosis virus and a smaller top component, *Virology* **48:**582.

Uyemoto, J. K., 1981, Tobacco necrosis and satellite viruses, in: *Handbook of Plant Virus Infections and Comparative Diagnosis* (E. Kurstak, ed.), p. 123, Elsevier/North Holland Biomedical Press, Amsterdam.

Uyemoto, J. K., and Grogan, R. G., 1969, Chemical characterisation of tobacco necrosis and satellite viruses, *Virology* **39:**79.

Uyemoto, J. K., Grogan, R. G., and Wakeman, J. R., 1968, Selective activation of satellite virus strains of tobacco necrosis virus, *Virology* **34:**410.

Valverde, R. A., and Dodds, J. A., 1987, Some properties of isometric virus particles which contain the satellite RNA of tobacco mosaic virus, *J. Gen. Virol.* **68:**965.

van Emmelo, J., Ameloot, P., Plaetinck, G., and Fiers, W., 1984, Controlled synthesis of the coat protein of satellite tobacco necrosis virus in *E. coli*, *Virology* **136:**32.

van Emmelo, J., Devos, R., Ysebaert, M., and Fiers, W., 1980, Construction and characterisation of a plasmid containing a nearly full-sized DNA copy of satellite tobacco necrosis virus RNA, *J. Mol. Biol.* **143:**259.

Wilcoxon, J., and Hull, R., 1974, The rapid isolation of plant virus RNAs using sodium perchlorate, *J. Gen. Virol.* **33**:107.

Wimmer, E., Chang, A. Y., Clark, J. M., Jr., and Reichmann, M. E., 1968, Sequence studies of satellite tobacco necrosis virus RNA, isolation and characterisation of a 5' terminal trinucleotide, *J. Mol. Biol.* **38**:59.

Wimmer, E., and Reichmann, M. E., 1968, Pyrophosphate in the 5' terminal position of a viral ribonucleic acid, *Science* **160**:1452.

Yarwood, C. E., 1960, Release and preservation of virus by roots, *Phytopathol.* **50**:111.

Ysebaert, M., van Emmelo, J., and Fiers, W., 1980, Total nucleotide sequence of a nearly full-size DNA copy of satellite tobacco necrosis virus RNA, *J. Mol. Biol.* **143**:273.

CHAPTER 6

Tymoviruses

L. Hirth and L. Givord

I. INTRODUCTION

The Tymoviruses are a major group of viruses, whose type member, turnip yellow mosaic virus (TYMV), was discovered by Markham and Smith in 1949. Since then, many other viruses have been discovered and classified in this group. For a long time Tymoviruses in general and TYMV in particular were known only in Europe, but later (Koenig and Lesemann, 1979), viruses of the Tymovirus group were discovered more or less everywhere in the world.

TYMV first became prominent because of its structure: It was the first plant virus for which the presence of noninfectious capsids was demonstrated in virus preparations. The existence of these capsids enabled Markham and Smith (1949) to suggest that the nucleic acid of TYMV was an RNA inside the virus, whose presence was related to the production of infection. However, they did not demonstrate the pathogenicity of isolated RNA, which was not achieved until seven years later by Gierer and Schramm (1956) with tobacco mosaic virus (TMV).

During the following 25 years TYMV was intensively studied, especially in molecular biology, because of several of its peculiarities. These included the formation of artificial capsids, the way in which the various proteins encoded by the RNA are expressed, the capacity of its RNA to bind valine at its 3'-end, and so forth; these peculiarities will be described below. In addition, in the last ten years research has been directed toward investigating viruses related to TYMV, which have been discovered mainly in tropical regions.

L. HIRTH and L. GIVORD • Institute of Cellular and Molecular Biology of CNRS, 67084 Strasbourg Cédex, France.

This chapter summarizes the major discoveries made in the field of the Tymoviruses and particularly on TYMV.

II. GEOGRAPHICAL DISTRIBUTION, HOST RANGES, DISEASES, AND SYMPTOMATOLOGY

Tymoviruses have been isolated from different countries in all continents (Table I). Each particular virus seems to be restricted to a certain geographic area, either a country or a continent. Three viruses are exceptions to this rule: BelMV, isolated in Europe and in North America (Paul, 1971; Moline and Fries, 1974; Lee *et al.*, 1979), PoiMV, isolated in North America (United States and Canada) and in Europe (Fulton and Fulton, 1980; Koenig and Lesemann, 1980; Brunt *et al.*, 1981; Chiko, 1983), and TYMV, isolated in Europe and Australia (Matthews, 1970; Guy and Gibbs, 1981). The extent of geographical spread of PoiMV is easily explained by the fact that its host, Poinsettia, an ornamental plant that reproduces vegetatively, is commercialized everywhere.

A strain of TYMV (TYMV-Cd) that is serologically identical to the classic strain has been isolated from a mountainous region of Australia (i.e., at a great distance from Europe) (Guy and Gibbs, 1981), which suggests the emergence of the same type of virus at two very distant points of the globe. In a further article (Guy and Gibbs, 1985), the authors assemble evidence that indicates that TYMV-Cd, as well as KYMV, has probably been in Australia for a considerable time rather than only having been introduced since European settlement.

The extent of geographical spread of different Tymoviruses can be extremely variable. For instance, a virus such as DuMV has only been isolated in Streastley, Bedfordshire, where it infected a quarter of all *"Solanum dulcamara"* plants; a search for this virus carried out in 31 other districts in the same area of the south of England (230 samples) was unsuccessful (Gibbs *et al.*, 1966). More recently, however, a strain of DuMV has been isolated in Hungary (Beczner *et al.*, 1976). A virus such as EryLV, although limited to only one country, was found to be distributed throughout East Germany in areas far distant from each other (Shukla *et al.*, 1975). On the other hand, APLV was found to be widespread throughout four different countries (Bolivia, Chile, Colombia, and Peru), which is easily explained since it is propagated by potato tubers. Although it is surprising that this virus remained restricted to South America, this is probably due to the lack of the vector in other countries and to quarantine measures.

Several explanations for the variable geographical distribution of different Tymoviruses have been given. Certain viruses that have been found only in a limited area are transmitted mainly by beetles (Coleoptera), i.e., by insects that never travel very far (Fulton *et al.*, 1980). In addition, the fact that certain viruses have been isolated in only one region does not

TABLE I. Some Properties of Definitive and Possible Members of the Tymovirus Group

Virus and abbreviations[a]	C.M.I./A.A.B. description number	Geographical distribution	Host plant families	Main diseases
Eggplant mosaic virus (EMV)	124	Trinidad, West Indies, Venezuela South America	Chenopodiaceae, Solanaceae	Mottling (Solanum melongena)
Tomato white necrosis virus (TWNV)		South America	Amaranthaceae, Chenopodiaceae, Compositae, Leguminosae, Malvaceae, Polemoniaceae, Solanaceae	White necrosis (Lycopersicon esculentum)
Andean potato latent virus (APLV)	124	South America	Chenopodiaceae, Solanaceae	Symptomless (Solanum tuberosum)
Physalis mosaic virus (PhyMV)		United States	Solanaceae	Interveinal chlorosis and mosaic (Physalis subglabrata)
Belladonna mottle virus (BelMV)	52	Europe, United States	Malvaceae, Scrophulariaceae, Solanaceae	Mild mottling, deformation (Atropa belladonna)
Dulcamara mottle virus (DuMV)		Europe	Solanaceae	Mild mottling (Solanum dulcamara)
Ononis yellow mosaic virus (OYMV)		Europe	Papilionaceae, Solanaceae	Chlorotic mosaic (Ononis repens)
Scrophularia mottle virus (ScMV)	113	Europe	Caryophyllaceae, Labiatae, Scrophulariaceae, Solanaceae, Umbelliferae, Valerianaceae	Mottling (Scrophularia nodosa)
Plantago mottle virus (PlMV)		United States	Aizoaceae, Leguminosae, Plantaginaceae, Scrophulariaceae, Solanaceae	Mild mottle (Plantago major), veinal chlorosis and mottle (Pisum sativum)
Wild cucumber mosaic virus (WCMV)	105	United States	Apocynaceae, Cucurbitaceae, Solanaceae	Mosaic (Marah macrocarpus and M. oreganus)
Turnip yellow mosaic virus (TYMV)	2	Europe	Cruciferae, Resedaceae	Mosaic (Brassica spp.)
Kennedya yellow mosaic virus (KYMV)	193	Australia	Leguminosae, Solanaceae	Blotchy chlorotic mosaic (Kennedya rubicunda)
Voandzeia necrotic mosaic virus (VNMV)	279	Africa	Leguminosae, Chenopodiaceae	Stunting, mosaic, necrosis, and distortion (Voandzeia subterranea)

(continued)

TABLE I. (continued)

Virus and abbreviations[a]	C.M.I./A.A.B. description number	Geographical distribution	Host plant families	Main diseases
Clitoria yellow vein virus (CYVV)	171	Africa	Leguminosae, Malvaceae, Solanaceae	Vein yellowing (*Clitoria ternatea*) and dark green mottle (*Abrus precatorius*)
Cocoa yellow mosaic virus (CYMV)	11	Africa	Apocynaceae, Bombacaceae, Chenopodiaceae, Solanaceae, Sterculiaceae	Yellow mosaic (*Theobroma cacao*)
Passiflora yellow mosaic virus (PasYMV)		South America	Passiflora genus	Yellow mosaic (*Passiflora* sp.)
Desmodium yellow mottle virus (DYMV)	168	United States	Papilionaceae	Yellow mottling, leaf deformity (*Desmodium* spp.)
Okra mosaic virus (OkMV)	128	Africa	23 families including all those listed for other viruses except Valerianaceae and Caesalpiniaceae	Mosaic (*Hibiscus esculentus*) and spotting (*H. rosa sinensis*)
Peanut yellow mottle virus (PeYMV)		Africa	Amaranthaceae, Apocynaceae, Bombacaceae, Chenopodiaceae, Cucurbitaceae, Euphorbiaceae, Leguminosae, Malvaceae, Solanaceae	Bright yellow mottle (*Arachis hypogea*)
Erysimum latent virus (EryLV)	222	Europe	Amaranthaceae, Caryophyllaceae, Cruciferae, Labiatae, Leguminosae, Resedaceae	Symptomless (*Erysimum* spp.)
Poinsettia mosaic virus (PoiMV)		North America and Europe	Euphorbiaceae, Solanaceae	Mosaic, leaves and bracts distortion (*Euphorbia pulcherrima*)

[a] The viruses are listed in order of their approximate serological relationships to each other (Koenig, 1976; Barradas, 1983; Crestani et al., 1984b; Lana, 1980; Desc. 279). EryLV is only distantly related serologically to the other viruses (Koenig and Lesemann, 1981). PoiMV is not serologically related to the other viruses and is considered as a possible member (Lesemann et al., 1983). Virus abbreviations are according to Van Regenmortel (1982). The data are based on C.M.I./A.A.B. Descriptions of Plant Viruses and on Barradas, 1983 (TWNV); Chiko, 1983 (PoiMV); Christie and Crawford, 1978 (WCMV); Crestani et al., 1984a (PasYMV); Debrot et al., 1977 (EMV); Fernandez-Northcote et al., 1982 (APLV), Fulton and Fulton, 1980 (PoiMV); Gibbs et al. 1966 (DuMV, OYMV); Gibbs and Harrison, 1969 (APLV, EMV); Granett, 1973 (PIMV); Hein, 1969 (ScMV); Horvath et al. 1976 (BelMV); Horvath, 1979 (BelMV); Koenig and Leseman, 1979 (Tymoviruses); Koenig and Lesemann, 1980 (PoiMV); Lana, 1980 (PeYMV); Lee et al., 1979 (BelMV); Lesemann et al. 1983 (PoiMV); Mamula, 1976 (BelMV); Moline and Fries, 1974 (BelMV); Peters and Derks, 1974 (PhyMV); Stefanac, 1974 (BelMV).

prove that they do not exist elsewhere. For a long time, TYMV was isolated only in Europe, although it was subsequently discovered to be present in Australia (Guy and Gibbs, 1981).

Furthermore, whether a virus is isolated in only one area in the world or in many different areas depends to a large extent on the economic importance of its natural plant host, on how long ago it was discovered, and on the number of virologists or phytopathologists working in the corresponding countries (out of 20 Tymoviruses, 12 were isolated in the United States or Europe, three in South America, two in Australia, and five in Africa).

The host range of Tymoviruses is usually very narrow, limited to either one family or to a few hosts in a very limited number of another family. In some cases a new host belonging to a different family from that of the original host has been discovered many years after the initial description of the virus: e.g., *Nicotiana benthamiana* for WCMV (Allen and Fernald, 1971; Christie and Crawford, 1978). Certain Tymoviruses, however, have a wide host range (Table I); OkMV, for instance, has been transmitted to plants belonging to 23 families (Givord, 1979).

Guy *et al.* (1984) published a taxonomic study of the host ranges of different Tymoviruses. They found that by separating Tymoviruses into two groups, one group preferring "crassi-nucellate species" and the other "tenui-nucellate species," they obtained approximately the two major divisions in the serological classification of Koenig (1976); i.e., one group of Tymoviruses clustering around TYMV and one around APLV. Koenig and Lesemann (1981) have attempted to distinguish between Tymoviruses on the basis of the reaction of 12 plant hosts commonly used in plant virology. This single criterion is, of course, insufficient for identifying with certainty a particular virus since host reactions depend on climatic conditions, on the cultivar used, and on the particular virus strains (Koenig and Lesemann, 1981) (Table II).

Macroscopic symptoms consist mainly of the appearance of a mosaic pattern on the leaves and sometimes of a stunting of the infected plants.

Stunting can reduce the size of TYMV-infected plants, *Brassica rapa* and *Brassica chinensis*, by a factor of 20 (Crosbie and Matthews, 1974a) depending on the virulence of the particular virus strain. In the case of *Brassica pekinensis* infection, the virus affects essentially the chloroplasts, in particular chlorophyll a, fraction I protein, and 68-S ribosomes (Crosbie and Matthews, 1974a). It seems that stunting of the plant is directly related to the reduction in synthesis of these three components.

Crosbie and Matthews (1974b) also showed that other factors linked to photosynthesis were affected by TYMV infection. This is the case for chlorophyll and the four major pigments that take part in photosynthesis (Neoxanthine, Violaxanthine, Lutein, and Carotenoids). The reduced levels of these substances are the result of a stop in synthesis rather than a degradation.

In addition to the permanently decreased levels of photosynthetic

TABLE II. Some Properties of Definitive and Possible Members of the Tymovirus Group (continuation)

Virus	Assay species	Particle diameter (nm)	Sedimentation coefficients [Top and bottom component ($S_{20,w}$)]	Juice extract	Thermal inactivation point	Storage at room temperature	Dilution end point
					Stability in sap		
EMV	Chenopodium amaranticolor	28–30	53;111	Similar results with 4 solanaceous species	78°C	Over 3 weeks	10^{-6}
TWNV	Unreported	24–25	49;109	Unreported		30 days	
APLV	None	28–30	53;111	Nicotiana glutinosa	90°	1 week	10^{-6}–10^{-7}
PhyMV	None	27	50;112	Nicotiana clevelandii			10^{-8}
BelMV	None	27	53;113	Nicotiana tabacum cvs. Samsun or Xanthi	80°	5 days to 3 weeks	10^{-6}–10^{-7}
DuMV	None	25–30	55;121	Nicotiana glutinosa	65°	1 week	10^{-3}–10^{-5}
OYMV	None	25–30	56;114	Pisum sativum cv. Onward	65°	2 weeks	10^{-3}–10^{-5}
ScMV	Vicia faba	26	54;116	Scrophularia nodosa, Datura stramonium	92–94°C	30 days	10^{-5}–10^{-6}
PIMV	None	26–27		Pisum sativum cv. Ranger	65°C	Unreported	10^{-7}–10^{-8}
WCMV	None	28	53;119	Cucurbita pepo	70°C	28 days	10^{-4}
TYMV	None	28		Brassica pekinensis $\dfrac{}{1+\theta^{-\alpha T-M}}$	70–75°C	A period of weeks	10^{-4}–10^{-6}
KYMV	None	28	54;110	Pisum sativum	65–70°C	10 days but not 100 days (20°C)	10^{-6}

VNMV	Chenopodium amaranticolor	28	Voandzeia subterranea	70°C	Less than 2 days (20°C)	10^{-2}	51,113
CYVV	Phaseolus vulgaris cvs. Long Tom or Premier	28	Nicotiana clevelandii	72°C	3 weeks (18–20°C)	10^{-7}–10^{-8}	50,109
CYMV	Chenopodium amaranticolor, C. quinoa	28	Theobroma cacao	60–65°C	16–32 days (25–30°C)	20^{-3}–20^{-4}	49,108
PasYMV	Unreported	30		55°C	8 days	10^{-5}	
DYMV	Desmodium tortuosum	30	Phaseolus vulgaris cv. Great Northern	70°C	38–44 days (20°C)	10^{-7}–10^{-8}	54,114
OkMV	None	28	Hibiscus esculentus	80°C	7–9 days (24°C)	10^{-6}	42,106
PeYMV	Chenopodium amaranticolor, C. quinoa	29	Unreported	70°C	23 days	10^{-7}	52,110
EryLV	Brassica napus cv. nappo brassica	27	Synapis alba	76–78°C	21 days (22°C)	10^{-5}–10^{-6}	59,113
PoiMV	None	26–29	Euphorbia cyathophora	60–65°C	8–10 days (24°C)	over 10^{-4}	55,117

[a] The viruses are listed in order of their approximate serological relationships to each other (Koenig, 1976, Barradas, 1983; Crestani et al., (1984b; Lana, 1980; Desc. 279). EryLV is only distantly related serologically to the other viruses (Koenig and Lesemann, 1981). PoiMV is not serologically related to the other viruses and is considered as a possible member (Lesemann et al., 1983). Virus abbreviations are according to Van Regenmortel (1982). The data are based on C.M.I./A.A.B. Descriptions of Plant Viruses and on Barradas, 1983 (TWNV); Dale, 1954 (EMV); Crestani et al., 1984a (PasYMV); Fulton and Fulton, 1980 (PoiMV); Gibbs et al., 1966 (DuMV, OYMV); Granett, 1973 (PiMV); Koenig and Leseman, 1979 (Tymoviruses); Koenig and Lesemann, 1980 (PoiMV); Lana, 1980 (PeYMV); Lesemann et al., 1983 (PoiMV); Peters and Derks, 1974 (PhyMV).

proteins, the activity of other enzymes is enhanced by TYMV infection in a more or less transitory manner. This is the case for instance with phosphoenolpyruvate carboxylase and aspartate aminotransferase. It would appear that the increased activity of these enzymes is specific for TYMV infection. It is likely that other enzymes are affected, but less so than in the case of viruses that induce the formation of local lesions (e.g., TMV). In plants infected with such viruses, the activity of the enzymes of phenylpropanoid metabolism, for example, is considerably increased (Legrand, 1983). On the whole, the metabolism of *Brassica pekinensis* plants infected with TYMV is reduced and this decrease can be measured by a drop in $^{14}CO_2$ incorporation (Bedbrook and Matthews, 1973).

The visible symptoms caused by TYMV on the leaves of *Brassica pekinensis* are a mosaic, i.e., a coalescence of dark green and light green patches. Usually, there are few dark green and many light green patches. During the course of infection, the symptoms change and the dark patches become more numerous as the leaves age (Matthews, 1981). These symptoms are sometimes accompanied by considerable deformations of the lamina. In the case of severe strains, local lesions are observed, the number of which increases with the concentration of the inoculum. The formation of these lesions does not prevent systemic infection of the plant. Ten to thirty virus particles are required to form one local lesion (Fraser and Matthews, 1979). It has not been possible to use the phenomenon of local-lesion formation by TYMV for developing a reliable assay to measure the virus concentration of an inoculum.

III. TRANSMISSION, EPIDEMIOLOGY, AND CONTROL

Tymoviruses are easily transmitted mechanically. Back-inoculation of purified virus to the original host plant has been achieved with all Tymoviruses except PoiMV (Fulton and Fulton, 1980; Lesemann *et al.*, 1983).

Three Tymoviruses have been studied for transmission by contact: APLV (Jones and Fribourg, 1977) and OkMV (Lana and Bozarth, 1975) could be transmitted in this manner whereas KYMV (Gibbs, 1978) could not. Clearly, part of the natural field transmission of Tymoviruses is assured by humans and other agents such as animals and wind that bring healthy leaves in contact with diseased plant material.

Vector transmission has been extensively studied using insects: Coleoptera, Homoptera (especially aphids), and Orthoptera, as well as cockroaches, bugs, collembola, cochineals, and even mites. Only Coleoptera were found to be efficient vectors of Tymoviruses. This argues against the idea that beetle transmission is a simple mechanical process akin to contamination not involving any biological specificity (Fulton *et al.*, 1980).

Among the Coleoptera, only Phytophagoïdea (Jeannel, 1949) were found to be vectors of Tymoviruses. In this supergroup, mostly the superfamily of Chrysomelids (in particular Halticids and Galerucids families) proved to be vectors of Tymoviruses with the exception of ScMV, which is transmitted only by the Curculionids superfamily (Weidemann, 1973). Guy and Gibbs (1985) reported that TYMV-Cd is transmitted by byrrhid beetles; this family belongs to the other supergroup of Cucujoïdea (Jeannel, 1949). Furthermore, this is the first time that byrrhids are reported as vectors of a plant virus (Guy and Gibbs, 1985).

In addition to Coleoptera, Orthoptera have also been found to be vectors. *Zonocerus variegatus* (Acrididae) was found to transmit OkMV, and *Leptophyes punctatissima*, *Locusta migratoria*, and *Stauroderus bicolor* (Acrididae) transmitted TYMV (Markham and Smith, 1949; Martini, 1957; Givord and Den Boer, 1980). *Forficula auricularia* (Dermaptera) has also been reported to transmit TYMV (Markham and Smith, 1949). In these instances, it is generally accepted that the transmission is of a purely mechanical nature.

Most transmission experiments have been carried out with adult insects, although larvae of the *Cionus* species (Curculionidae) have been found to transmit ScMV (Bercks, 1973), and larvae of Chrysomelidae, *Phaedron cochleriae*, and *Phyllotreta nemorum* have been found to transmit TYMV (Markham and Smith, 1949; Martini, 1957), as well as larvae of the butterfly *Pieris brassicae* (Martini, 1957).

Finally, only one case of transmission by Aleurode has been reported, i.e., the Nigerian strain of OkMV (N-OkMV) (Lana and Taylor, 1976). However, this needs to be confirmed as an OkMV strain from the Ivory Coast (IC-OkMV) is not transmitted by this insect (Givord *et al.*, 1972).

Seed transmission has been tested for many viruses and has been found to occur in only a few cases. Examples are the seed transmission of APLV in *Nicotiana clevelandii* and *Solanum melongena* (Gibbs *et al.*, 1966; Jones and Fribourg, 1977; Jones, 1982), of DuMV in *Solanum dulcamara* (Gibbs *et al.*, 1966), and recently of TYMV in *Camelina sativa* (Hein, 1984).

Dodder transmission was tested in the case of OkMV and found to be negative (Givord and Hirth, 1973).

Soil transmission was shown to occur in the case of N-OkMV (Lana *et al.*, 1978) and was due only to contact between infected plant debris and roots. It did not occur with CYMV (Brunt and Kenten, 1965).

As far as the spread of certain Tymoviruses from one location to another is concerned, it can be explained either by the efficacity of a particular vector (for instance in the case of BelMV) (Weidemann and Bode, 1973) or by the abundance and particular activity of a vector, e.g., EMV (Dale, 1954).

A thorough study of field transmission was carried out in the case of TYMV (Markham and Smith, 1949). A patch of turnip plants was artificially infected and the spread of the virus to adjoining fields was followed. The results showed that flea-beetles were the only factor responsible for the observed spread, which confirmed earlier experimental findings concerning vector transmission of TYMV by flea-beetles (Markham and Smith, 1949), as well as the fact that virus spread was linked to climatic conditions that favour the invasion of the beetles (Croxall *et al.*, 1953).

Field propagation has also been studied in detail in the case of EryLV (Proeseler and Schmelzer, 1977).

Apart from these two detailed studies, no proper epidemiological studies have been undertaken for Tymoviruses, although certain ecological studies have been reported. In the case of OkMV, which is found in the south of the Ivory Coast, the influence of cultivation practices and of the prevalence of certain reservoir plants on the spread of the disease has been studied (Givord, 1978).

In the case of other Tymoviruses, investigators have studied mainly the natural hosts that act as a reservoir for the virus. According to Provvidenti and Granett (1976), plantain is an important reservoir plant for PlMV. In California, Milne *et al.* (1969) observed that WCMV was only ever isolated from wild cucumber plants and never from commercial cucurbits growing nearby. Many Tymoviruses have been isolated from several natural hosts. The following four viruses, for instance, have been shown to infect naturally about a dozen different hosts: the EMV and APLV strains (Waterworth *et al.*, 1975; Jones and Fribourg, 1977; Debrot *et al.*, 1977), EryLV (Shukla *et al.*, 1975; Shukla and Schmelzer, 1975; Schmelzer, 1976), OkMV (Givord, 1978; Igwegbe, 1983; Atiri, 1984), and TYMV (Croxall *et al.*, 1953; Broadbent and Heathcote, 1958; Shukla and Schmelzer, 1974; Guy and Gibbs, 1981).

As far as the spread from one plant generation to the next is concerned, seed transmission and vegetative propagation of the primary host are particularly important. In addition, in most cases the virus survives from one year to the next because the host plant is a perennial plant (*Kennedya rubicunda* for KYMV, *Theobroma cacao* for CYMV, *Clitoria ternatea* for CYVV, *Desmodium* sp. for DyMV, *Passiflora* sp. for PasYMV, *Erysimum* sp. for EryLV, *Scrophularia nodosa* for ScMV, and *Atropa belladonna* for BelMV).

An interesting study by Gibbs (1980a) should be mentioned in this respect: it was found that herbivores tend to feed preferentially on healthy *Kennedya rubicunda* plants rather than on virus-infected ones, which increases the chances of survival of KYMV.

In the area of disease control, very few studies have been carried out. Two efficient methods for controlling PoiMV infections have been devised, one based on thermotherapy (Pfannenstiel *et al.*, 1982) and the other on cell culture (Preil *et al.*, 1982). Very few attempts have been made to obtain varieties resistant to infection by Tymoviruses. In the Ivory Coast, a variety of okra tolerant to OkMV has been found (Givord, 1980). In New York State, varieties of green peas resistant to PlMV are cultivated (Provvidenti and Granett, 1976). In this case, the genetics of the resistance were investigated and a single dominant gene in *Pisum sativum* was shown to be responsible for resistance to PlMV (Provvidenti, 1979).

IV. PURIFICATION

A. General Procedures

It is relatively easy to obtain large quantities of TYMV from infected turnips, but at present the host species most often used for growing TYMV is *Brassica chinensis*, variety Pe-Tsaï. For other members of the Tymovirus group, specific hosts displaying systemic infection are used and make it possible to obtain varying, but always quite large, amounts of virus (up to several hundred mg per kg of fresh leaves). Koenig and Lesemann (1981) (Table I) have listed the species that display a systemic infection with each of the Tymoviruses known at present. Whatever the type of Tymovirus examined, its method of purification is derived from one of the methods for TYMV, ultimately leading to a separation of virions and capsids.

The method described by Markham and Smith (1949) comprises, after an aqueous extract has been obtained from the infected plants, a stage of clarification with ethanol, then crystallization of the virus using ammonium sulfate. Several successive crystallizations yield a pure preparation. Examination of the crystals in the light microscope (under ultraviolet light) can distinguish two types of crystals: dark ones contain virions and light ones contain empty capsids. The two major components may be separated by centrifugation on a cesium chloride gradient.

Markham and Smith's (1949) technique is long and gives rise to virions and capsids associated with cellular contaminants. Matthews (1955) improved the method by taking the juice extracted from virus-infected plants to pH 4.8. The precipitate thus obtained is discarded and the supernatant is centrifuged at high speed; several cycles of centrifugation yield a particularly clean preparation of virus and capsids. The top and bottom components are separated by two cycles of sedimentation on a cesium chloride gradient.

A simpler method has been used by Leberman (1966). It consists of precipitating the virus from a crude extract of infected plants using appropriate concentrations of polyethylene glycol (PEG) and NaCl, which vary according to the virus being purified. The precipitate is resuspended and several cycles of alternate high- and low-speed centrifugation lead to the isolation of pure virus.

Virions can be separated from capsids, either on a sucrose gradient (Cosentino *et al.*, 1956) or, more effectively, on a preformed gradient of cesium chloride (Matthews, 1960). In order to separate virions from in-complete nucleoprotein particles, three or four cycles of density-gradient centrifugation are necessary.

B. Diversity of Purification Products

With TYMV, a more detailed analysis of cesium chloride gradients (Matthews, 1960, 1974; Johnson, 1964; DeRosier and Haselkorn, 1966) showed the existence, alongside the major bands representing top and bottom components, of minor bands relatively close to the bottom com-ponent, containing varying quantities of nucleic acid. These bands rep-resent four major groups of nucleoproteins containing various quantities of nucleic acid (from 20 to 36%). Each band may have a satellite, which probably has a slightly higher density. Nucleoproteins harvested from these bands are not infectious. However, RNA can be extracted from several of these bands, which can be translated in vitro and tested im-munologically to see if the coat protein has been produced (Pleij et al., 1977).

It seems that RNA extracted from the densest particles does not translate into coat protein, but RNA from intermediate density and top particles does. From this it may be deduced that the dense particles con-tain the complete viral RNA from which only the high-molecular-weight proteins (see Section VII.A) are translated. This observation is in agree-ment with the existence of a subgenomic RNA, coding for the coat protein alone and independently encapsidated (cf. Section VII.A). The situation is different in the case of EMV, where the genomic and the subgenomic RNAs are present inside the same capsid and also in the top (Szybiak et al., 1978). The mechanism of discrimination among the various RNAs is not known.

Various authors (Matthews, 1960; Johnson, 1964; Stols, 1964; De-Rosier and Haselkorn, 1966; Pleij et al., 1977) have shown that the major component (the densest one) of the TYMV may be converted into a sig-nificantly less dense one, and that this conversion is reversible. It seems that cations exchange (between the interior of a particle and the external medium) in various conditions of pH and temperature. The Mg^{2+} ions and polyamines associated with the RNA may exchange with Cs^+ ions, which increases the density of the particle. The intermediate components containing small quantities of RNA are not sensitive to this type of change (Noort et al., 1982).

Similar changes in density have also been observed for the particles of cowpea mosaic virus (Wood, 1971) and of poliovirus (Wiegers et al., 1978). On the other hand, many other viruses do not display this phe-nomenon. If one accepts that these changes in buoyant density are related to exchanges of ions between virions and the external medium, one may imagine that the viruses that display this phenomenon have a capsid that can allow ions to pass only under the described experimental conditions; those viruses whose apparent buoyant density remains constant have either freely permeable or completely impermeable capsids.

V. PROPERTIES AND CONSTITUENTS OF PARTICLES

Although only TYMV has been studied in depth, there are good reasons to think that probably all Tymoviruses are built on the same basic model (Table II).

In a review in 1959, Markham reported the results of his earlier work, showing that the nucleic acid contained in the capsid was RNA and that its base composition was very uneven. Since this initial observation, many investigations (cf. Matthews, 1955) have confirmed that the infectious entity is, for TYMV and the majority of viruses of the same group, an RNA of mean mol. wt. 1.9×10^6.

A. The RNAs

As already mentioned, in addition to the genomic RNA there is an RNA that encodes the coat protein. This represents a redundancy of 695 nucleotides next to the 3'-extremity. It has not been ruled out that RNA molecules derived from partial replication are enclosed in particles of lower density than the infectious particles, but no proof has been offered for this assertion.

Bouley et al. (1976) demonstrated the existence of transfer RNAs in the top component of eggplant mosaic virus. These tRNAs, of which the main one was lysine tRNA, are not contaminants but are inside the top component. Szybiak et al. (1978) demonstrated that in addition to these tRNAs there was a 200-Kd RNA that could be translated in vitro into a protein that reacted with an antiserum prepared against the coat. Similar subgenomic RNAs have also been found by these authors in the top components of OkMV and WCMV. TYMV, whose top component contains no RNA, appears to be a special case among Tymoviruses in this respect.

The base composition of the RNA of all the Tymoviruses is characterized by a high content of C and a low content of G. The mean percentage base compositions lie within the following ranges: G 15.2 to 17.5, A 17 to 23.8, C 31.9 to 42.1, and U 22.1 to 29.4. Cytidine residues seem to be distributed in groups of up to six approximately homogeneously along the whole RNA molecule; this is the case for the subgenomic RNA of TYMV, whose complete sequence has been established (Guilley and Briand, 1978).

The sequence of the 110 nucleotides closest to the 5' end of the genomic RNA is slightly poorer in G than the rest of the molecule. The

5' and 3' ends contain two complementary sequences of nine nucleotides, which explains previous results (Strazielle *et al.*, 1965) suggesting that in certain conditions of pH and temperature, the RNA molecule has a circular structure. Such a structure has also been found among Bromoviruses. Its significance is not known.

B. The Capsid Protein

Degradation of the capsid or the nucleocapsid in a denaturing medium gives rise to a unique protein of 20,133 d in the case of EMV and of 19,809 d in the case of TYMV (molecular weights are deduced from their protein subunit sequences: Peter *et al.*, 1972; Dupin *et al.*, 1984).

For viruses whose coat protein sequence has not been determined, the molecular weights are of the order of 20,000. The coat protein molecular weights seem to be very similar for all Tymoviruses. The same is not true for the amino acid composition and certainly not for the sequence, since the serological relationships among the various types of Tymovirus are often very remote or nonexistent (Koenig, 1976). Differences have been shown between TYMV and EMV, whose sequences are known: 32% of the amino acids of these two viruses lie within sequences common to both, of which the longest consists of four adjacent amino acids; an addition at the N-terminus and a deletion at the C-terminus are among the differences (Dupin *et al.*, 1985).

From the sequence of EMV protein and the methods of Chou and Fasman (1978) it is possible to propose a model for the organization of its nucleocapsid quite similar to that of TYMV (see below). Although little is known about the capsid proteins of other Tymoviruses, it is likely, considering their amino acid composition, that they have a similar distribution of hydrophobic amino acids to that in the two viruses just described and that their capsids have the same fundamental structure (Paul *et al.*, 1980).

C. Polyamines

Johnson and Markham (1962) first showed that polyamines were present in purified TYMV particles. This was confirmed by Beer and Kosuge (1970) and investigated in more detail by Cohen *et al.* (1981) who showed that there were less tetramines (spermine) than triamines (spermidine). In infected plants there is also putrescine besides these two polyamines. More recently, Cohen and Greenberg (1981) have shown that the presence of polyamines in TYMV is not a matter of contamination (that depends on the method of purification of the virus), but that most of the polyamines are contained inside the virions. Depending on the preparation, the polyamine content varies from 200 to 700 molecules of polyamine per virion, and 5 to 22% of these molecules are spermine, the rest being spermidine. These polyamines are associated with the RNA and cannot come out of the virions except in conditions such as relatively low pH.

In general, polyamines are tightly bound to the RNA and not exchangeable, even in media with a high concentration of salt (Cohen and Greenberg, 1981). It seems likely that polyamines serve to neutralize the negative charges of RNA, which lies inside a capsid with only a limited number of basic amino acid side-chains. It is probable that Tymoviruses related to TYMV must contain polyamines, but observations on BelMV by Virudachalam et al. (1983b) showed that this Tymovirus seems not to contain polyamines. It may be interesting in this context that at neutral pH TYMV migrates rapidly toward the anode, whereas BelMV migrates towards the cathode (Koenig, 1976).

Nickerson and Lane (1977) showed that viruses other than Tymoviruses may also contain polyamines: this is the case with cucumber mosaic virus (CMV), whose coat protein does, however, contain enough basic amino acids to neutralize the negative charges of RNA. On the other hand, brome mosaic virus (BMV) does not contain polyamines even though the amino acid composition of its coat protein is close to that of CMV. The reasons that lead two viruses with similar overall amino acid compositions to neutralize the negative charges of their nucleic acids by different means is still unknown.

D. Structure of the Capsids of TYMV

In 1957, Klug et al. proposed a structure of 60 identical subunits associated in the form of a dodecahedron. Later, using electron microscopy, Finch and Klug (1960, 1966) described a model based on 180 identical subunits organized into an icosadeltahedron of order $T = 3$, i.e., with 12 vertices having fivefold symmetry and 20 vertices with local sixfold symmetry.

The structure of TYMV, together with that of tomato bushy stunt virus, was what enabled Caspar and Klug (1962) to draw up rules for the organization of viruses with cubic symmetry (Caspar, 1956; Klug et al., 1957). These rules were derived from the initial observation by Crick and Watson (1956) on the relationships between the coding capacity of a viral RNA and the organization of the capsid. The RNAs of TMV and TYMV are not large enough to code for the whole of the capsid as a single protein, so the idea arose that the capsid is made up of an association of identical protein subunits organized symmetrically. Although some viruses display an organization that is a little different from that of TYMV, the rules of Caspar and Klug (1962) apply in general to all $T = 3$ cubic-symmetry viruses. Recently, Rayment et al. (1982) have shown that in the case of polyoma virus all the subunits are associated in pentamers. This means that the number of subunits calculated from the classical organization of order T = 7 is overestimated. This type of organization seems general for the T = 7 viruses.

E. Protein-RNA Interactions inside TYMV

An initial observation of Finch and Klug (1966) and Klug *et al.* (1966) based on analysis of X-ray diffraction diagrams of TYMV crystals strongly suggested that the viral RNA was deeply interdigitated with the protein subunits (Figs. 1A, B). The validity of this diagram was rapidly put in doubt by observations that led to the idea that the viral RNA could be released easily from the protein subunits without the capsid necessarily being destroyed.

F. Formation and Properties of Artificial Capsids

Natural capsids of TYMV were studied more than 35 years ago by Markham and Smith (1949) right after their discovery in virus preparations. Although the structure of the virus was unknown, these authors suggested that the capsids had a protein structure similar to that of the virions. This point of view was confirmed by Finch and Klug (1960) and by Longley and Leberman (1966), who showed that the dimensions of the capsids and the organization of their subunits were as in the virions, thus confirming the observations of Cosentino *et al.* (1956) and of Fraser and Cosentino (1957).

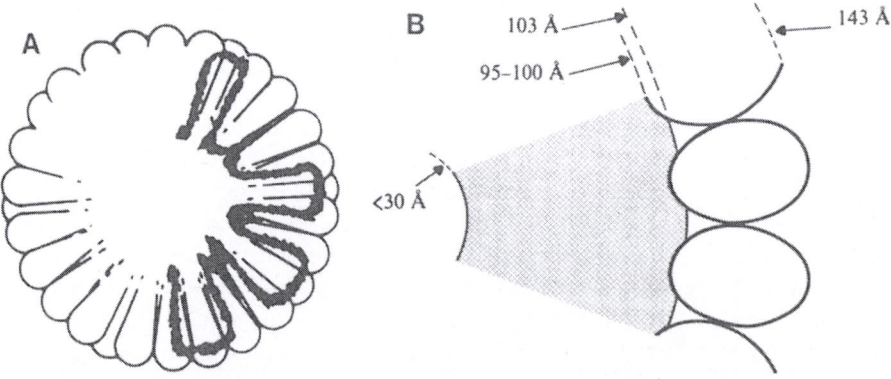

FIGURE 1. (A) Interaction RNA–protein in the nucleocapsid of TYMV determined by X-ray diffraction. The RNA seems deeply embedded in protein subunits (Klug *et al.*, 1966). This contrasts with Fig. 1B. (B) A model of organisation arising from neutron diffraction studies and showing that RNA interacts only with the inner part of the protein subunits (Jacrot *et al.*, 1977). The discrepancies between these models arise probably from the use of high salt concentration for X-ray diffraction.

The degradation of virions and of capsids of Tymoviruses by means as diverse as alkaline pH, urea, formamide, heat, and mercury compounds, has been very useful for understanding protein–protein and protein–RNA interactions. These investigations have shown the great importance in Tymoviruses of protein–protein interactions, especially hydrophobic interactions.

Kaper (1960a,b) showed that the virions of TYMV could be dissociated into RNA and protein in a pH 11.5 phosphate buffer and in media of high ionic strength. The RNA was then degraded. At pH 4.8 the capsids were degraded, provided the ionic strength was relatively low. Kaper (1960a,b) showed that capsids obtained from virions at high pH could be crystallized, and Longley and Leberman (1966) showed that natural and artificial capsids were identical. It should be noted that at pH 10.5 a particular form of the virion appears (Pleij, 1973) containing an RNA molecule sedimenting at 38 S, which is not infectious since this RNA is in fact a product of internal aggregation of RNA fragments of the order of 10 S.

Jonard et al. (1967) and Jonard (1972) showed that 8 M urea at pH 7 and low ionic strength completely degraded virions of TYMV into dissociated protein and degraded RNA. No intermediate product was found. The observed rates and completenesses of the reactions depended on the variations in concentration of urea and the pH. In fact, the protein isolated after the action of urea (which was then eliminated) had a sedimentation rate of 7 S and represented a protein aggregate that was stable because of disulfide bridges. Purified natural capsids are, in the conditions indicated, not sensitive to the action of urea; on the other hand, at pH 4.8 in the presence of urea, they degrade rapidly. This quite curious phenomenon may be due to forces of repulsion resulting from the protonation of three histidines on each subunit (Jonard, 1972).

In the presence of 8 M urea at pH 7 and high ionic strength, artificial capsids and a 7 S protein fraction are obtained. Purification and investigation of the capsids revealed that the capsids formed had practically all the properties of natural capsids and in particular those of capsids prepared by alkaline degradation. If the reaction is carried out at pH 4.8, the capsids formed remain relatively stable. The RNA released during the degradation with urea is infectious and therefore has all the properties of a viral RNA extracted from virions by the phenol method (Jonard et al., 1967).

Dissociation–reassociation experiments led to a definition of the nature of the protein–protein and RNA–protein interactions in the case of TYMV (Jonard, 1972; Briand et al., 1975; Bouley et al., 1975) and to a suggestion for the mechanism of dissociation, which has been significantly modified by recent experiments involving a system of freeze-thawing of virus preparations or by experiments on decapsidation in vivo. Nevertheless, protein–protein interactions are much decreased at pH 7 and in the presence of urea at moderate ionic strength, and the whole viral architecture collapses (Jonard, 1972). On the other hand, at low pH (4.7 to 4.2) hydrophobic interactions are reinforced and, if the ionic strength is increased (2 M NaCl), protein–RNA interactions are decreased, which allows the release of RNA by a mechanism that is ill-defined in these conditions, but which may be similar to that shown by freeze-thaw experiments (Katouzian-Safadi et al., 1980; Katouzian-Safadi and Berthet-Colominas, 1983).

The stability of capsids at low pH and high ionic strength demonstrated the importance of hydrophobic interactions in the stability of the capsid and also makes it possible to understand why, in the stability of the nucleocapsid, the forces of repulsion due to histidine may be neutralized by the phosphate groups of the RNA. Nevertheless, the degradation of capsids at low pH and low ionic strength was interpreted by Kaper (1969) and Jonard (1972) as indicating that the histidines of the capsid interact with the bases of the viral RNA, which may show that there are interactions other than salt bridges in the case of TYMV.

The reconstitution of a simple virus from its constituents (RNA and protein subunits) provides some basic understanding of the nature of the RNA–protein and protein–protein interactions. This is the case with TMV (Lebeurier et al., 1977; Butler et al., 1977) and with BMV (Bancroft, 1970), for example. All attempts in many laboratories to obtain TYMV coat protein in a nondenatured, reassociable form have failed, probably because of the strong hydrophobic interactions that stabilize the protein–protein interactions.

It has been possible to obtain 7 S aggregates, but specific conditions for producing these aggregates have not been discovered (Jonard, 1972; Kaper, 1975). While performing dissociation–reassociation experiments, Jonard et al. (1972) showed that it was possible to obtain a specific reassociation between TYMV RNA and "nascent" artificial capsids. This reassociation, performed with labeled RNA and protein, led to the formation, in the presence of Mg^{2+} and/or spermidine, of nucleoprotein complexes that were sensitive to RNase and were not infectious. Thus, this was not an authentic reconstitution, but the results obtained suggested that bases of the RNA chain were recognizing certain regions of the protein subunits.

Complementary experiments (Briand et al., 1975) in which synthetic polynucleotides were added to the dissociation medium (already containing Mg^{2+} and spermidine) showed that poly-C associated specifically with artificial capsids formed in the conditions described above. From this it was deduced that repetitive poly-C sequences follow one another in a more or less defined order along the viral RNA molecule. Indeed, it was shown that labeled RNA from either BelMV or EMV was capable of interacting with "nascent" capsids of TYMV obtained by the action of 8 M urea at low pH.

It should be noted that the proportion of cytosine in the RNAs of these two viruses is 32% and 38%, respectively; however, the proportion of cytosine is not the only factor involved, since TYMV RNA treated with bisulfite (which converts C to U) no longer binds to nascent capsids, even if its C content is still 32%. This demonstrates that there are sequences involved in preferential association, which when transformed into U no longer allow the association between the pyrimidine residues of cytosine and the viral protein. The complete sequence of the subgenomic RNA coding for the TYMV coat protein is, however, not in agreement with this structure. Only the complete sequence of TYMV RNA, which is not yet published, will make it possible to determine the significance and the arrangement of the 38% C in TYMV RNA.

Combination between TYMV RNA and heterologous protein has been investigated by mixing TYMV RNA and protein from TMV in various conditions of pH and ionic strength (Matthews, 1966; Fritsch et al., 1973). In conditions of pH near 6 and high protein concentration, a few relatively short particles are formed. In reality, in the light of what is known about the nature of the protein aggregate that recognizes the RNA initiation-site and the subsequent mechanism of the assembly of TMV (Durham and Klug, 1971; Lebeurier et al., 1977; Butler et al., 1977), it is likely that the rods observed (Matthews, 1960; Fritsch et al., 1973) are nonspecific products of the aggregation of short protein segments with unidentified sequences of the viral RNA. All this suggests that these experiments are rather artifactual. One should therefore consider that the mechanism of TYMV's morphogenesis is still unknown.

Attempts to relate observations on the morphogenesis of Picornaviruses with those performed on Tymoviruses remain at present inconclusive, despite the preliminary observations of Jonard et al. (1972). One question that has still not been definitely answered is whether the capsids are formed by the association of many protein subunits, or whether they arise by a loss of RNA from virions. The latter hypothesis is quite unlikely, considering that the accumulation of capsids is observed when the virus multiplies in the presence of thiouracil, which inhibits the replication of RNA (Francki and Matthews, 1962). It is possible that the synthesis of protein subunits is not linked in any quantitatively well-defined way to that of the molecules of viral RNA (cf. below).

G. Stability and Structure of Tymoviruses Other than TYMV

The majority of published observations have been on TYMV, but EMV and BelMV have been studied in some depth (Bouley et al., 1977; Virudachalam et al., 1983a,b). Despite the systematic distance between some Tymoviruses in the classification proposed by Koenig (1976) it is likely that the properties described for TYMV are on the whole valid for other viruses in this group.

EMV, which contains the two classic major components (top and bottom), has been tested with regard to its stability as a function of temperature, urea concentration, pH, and ionic strength of the medium. These studies show that its capsid and nucleocapsid are less stable than that of TYMV. This is true in particular for the hydrophobic interactions; likewise, the specific RNA–protein interactions described at low pH for TYMV have not been found in the case of EMV. The RNA destabilizes the capsid at all pH values, whereas it does not at neutral or high pH for TYMV.

Analysis of the putative tertiary structures of the protein subunits of TYMV and EMV performed by Dupin *et al.* (1985) according to the rules laid down by Chou and Fasman (1978) suggested reasons for these differences in the behavior of the two Tymovirus capsids, but things will become clear only when the tertiary structures of the capsid and nucleocapsid of these two viruses are known. The difficulties encountered by crystallographers in determining these structures derive from the size of the asymmetric unit in the crystals obtained. All the same, substantial progress in methods of interpreting X-ray diffraction diagrams (cf. review by Bricogne, 1984) suggests that this type of problem will be solved in the near future.

BelMV has been studied recently. Using proton nuclear magnetic resonance spectroscopy, Virudachalam *et al.* (1983b) showed that it is, like all the Tymoviruses, made up of capsids and nucleocapsids and that its RNA can be released at pH values much lower (above 7) than for TYMV (above 11.5). The method used showed that polyamines were present in TYMV and absent from BelMV. It is interesting to note that the addition of polyamines to a suspension of BelMV increases the stability of the virions, and that its RNA is not released until high pH values such as those observed in the case of TYMV.

The reason why some Tymoviruses (and also other plant viruses) contain polyamines is still unknown. Perhaps it is related to the nature of the host plant or the physiological conditions of growth. Nevertheless, BelMV is, like EMV, less stable than TYMV (Virudachalam *et al.*, 1983b). It is worth mentioning that the release of intact RNA has not been obtained by classic physicochemical techniques, but only by the use of urea. The significance of this observation and the mechanism(s) involved have been the subject of much study: some old (Jonard *et al.*, 1967), some new (Matthews and Witz, 1985).

H. Decapsidation of TYMV

In 1972, Kurtz-Fritsch and Hirth showed that the inoculation of purified nucleocapsids into sensitive plants induced the formation of a non-negligible quantity of capsids, and that this phenomenon was undoubtedly due to the release of infectious RNA, since the inoculated plants rapidly displayed symptoms. Similar results were obtained by incubating insensitive plants such as tobacco. These experiments suggested that the release of RNA did not involve destruction of the capsid, and the mechanism of the infection could be likened to the *in vitro* observations on the formation of artificial capsids.

In 1985, Matthews and Witz showed that the experiments of Fritsch *et al.* (1973) were reproducible and that the capsids formed in the course of the inoculation were incomplete; five to six subunits had been released, leaving a hole in the capsid. It is possible that the phenomenon takes place at cell structures, such as membranes, but the site of decapsidation remains unidentified. These *in vivo* observations are in agreement with observations on decapsidation *in vitro*. Keeling and Matthews (1982) showed that the empty shells formed by the effect of high pH differed from the natural capsids in having a lower sedimentation coefficient and by the presence in the suspension of a protein aggregate formed from five or six protein subunits. The release of RNA at high pH is therefore accompanied by the formation of a "hole" in the capsid. The subunits released remain associated with small quantities of viral RNA that can easily be eliminated with pancreatic ribonuclease. It should be noted that the loss of a hexamer or a pentamer is limited to the nucleocapsid: the same experiments performed on natural capsids had no effect (Keeling and Matthews, 1982). The loss of protein subunits is therefore related to the presence of RNA.

Using different methods, Katouzian-Safadi *et al.* (1980) showed that freeze-thawing of TYMV suspensions in defined conditions of pH and ionic strength led to the formation of artificial capsids and high-molecular-weight RNA. This experiment was inspired by earlier work of Kaper and Alting-Siberg (1969a) who showed that the virions were degraded into protein aggregates of low molecular weight but into high-molecular-weight RNA (1969b). However, this degradation could be avoided by the addition to the incubation medium of various chemicals (methanol, ethanol, etc.).

The results of Katouzian-Safadi *et al.* (1980) differed from those of Kaper (1975) by the formation of artificial capsids. The artificial capsids formed (Katouzian-Safadi and Berthet-Colominas, 1983) had a slightly lower sedimentation coefficient than natural capsids. In addition, their analysis by means of X-ray diffraction and their orientation in a magnetic field showed that they differed from natural capsids by the loss of five to nine subunits per particle, creating a hole through which high-molecular-weight RNA may escape.

Interpretation of these results leads to the hypothesis that there is some particular sequence in the viral RNA that destabilizes a particular region of the viral protein. It seems likely that the same process occurs *in vivo* in contact with specific cell structures (Fritsch *et al.*, 1973; Matthews and Witz, 1985). It would be very interesting to know which regions of the RNA are involved. It is worth mentioning that other systems for degrading virions have been used, as described by Kaper (1975), but have not revealed much beyond what has just been discussed.

VI. SEROLOGY AND IMMUNOCHEMISTRY

Tymoviruses are excellent immunogens. By means of a number of serological techniques their presence in infected plants can be detected and their concentration can be estimated.

Relationships between members of the group as determined by serological methods range from very close to very distant or not detectable. A classification scheme based on serological differentiation indices (Van Regenmortel and von Wechmar, 1970) has been worked out by Koenig and Givord (1974) and Koenig (1976). It shows that Tymoviruses that are distantly related to one another, are stepwise interconnected by other viruses that are more closely related. Figures 2 and 3 taken from data of Koenig (1976) illustrate these relationships and show a "loop structure" in the classification of Tymoviruses.

In Table I, Tymoviruses are listed in order of their approximate serological relationships to each other. This order is based on the classification of Koenig (1976) and earlier descriptions of viruses (see note to Table I). The case of the "strain" of BelMV described in *Physalis* in the United States (Iowa and Kansas) deserves special mention. This "strain" is so different from that of Europe that a more detailed study would be necessary in order to decide whether the United States "strain" is in fact a strain of BelMV or a distinct virus. First, the comparative serological

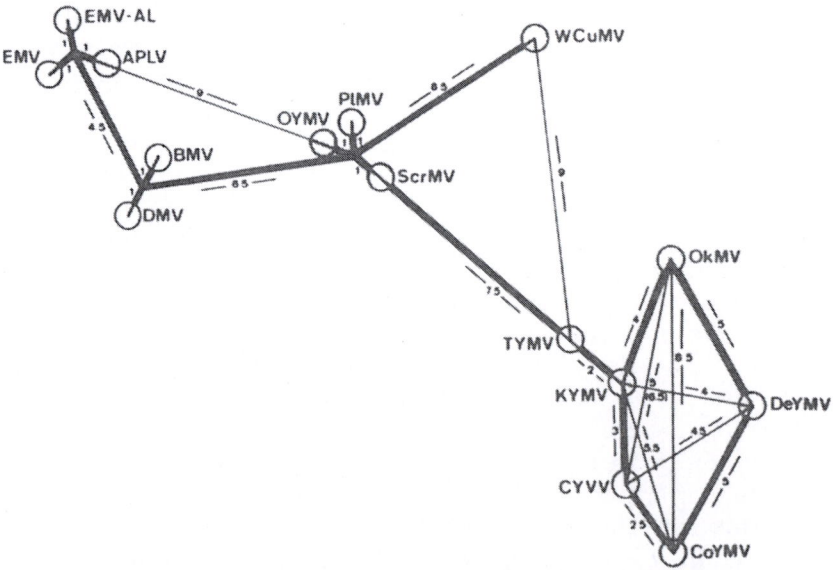

FIGURE 2. Serological classification of Tymoviruses on the basis of the average of serological differentiation indices (SDIs) of reciprocal tests (RT-SDIs), which are depicted as length units. Average RT-SDIs and measured distances are rounded off to nearest integers or to half integers, e.g., 4, 4.5, 5, 5.5. Figure and legend from Koenig (1976).

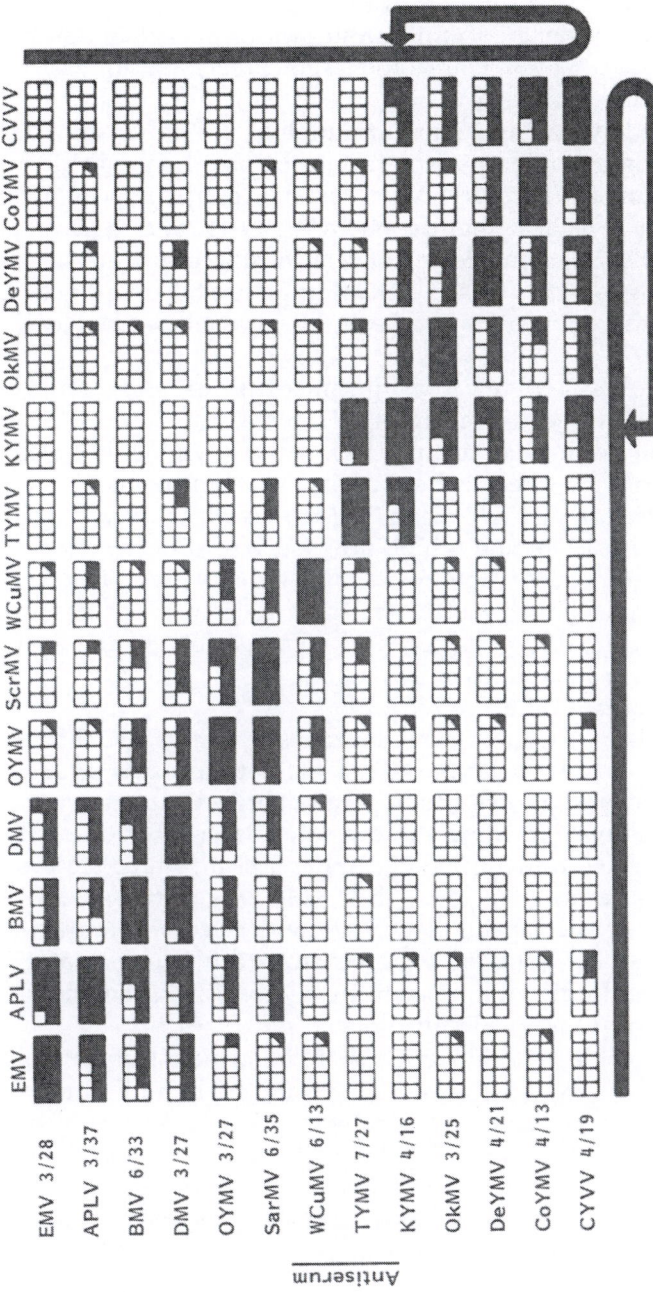

FIGURE 3. Serological classification of Tymoviruses on the basis of the average of serological differentiation indices which are represented by the number of white squares. The SDI can be deduced by counting of the remaining white squares. On the left of the figure the first number following each antiserum indicates the number of rabbits immunized and the second the number of the bleeding that was tested. From Koenig (1976).

studies that have been carried out are not convincing (Moline and Fries, 1974; Lee *et al.*, 1979) and, second, the amino acid composition of the two "strains" are quite different (Lee *et al.*, 1979) [the amino acid composition of the European strain (Jankulowa *et al.*, 1968) has been confirmed in Strasbourg by A. Dupin and R. Peter (personal communication)].

Figures 2 and 3 also show that viruses with similar physicochemical properties can have no apparent immunological relationship. This indicates that virus classification should not be based on a single criterion. Without entering a debate that is of interest primarily to people interested in systematics, it is clear that the choice of different criteria may lead to variations in the apparent degree of relationship between viruses. Some of the problems that may arise when attempts are made to detect different strains of a virus, e.g., APLV and EMV, have been described by Koenig *et al.* (1979).

It nevertheless remains that for practical purposes, serology is the most useful technique for identifying viruses. The classification of Koenig shows that the serological distance between two viruses can also parallel the degree of physicochemical similarity between them.

This is the case for EMV and TYMV, which are very distant serologically (Koenig, 1976) and which are also very different in their physicochemical properties (Bouley *et al.*, 1976, 1977). The serological classification also largely correlates with classifications based on host ranges (Guy *et al.*, 1984), amino acid compositions of the coat protein (Paul *et al.*, 1980), and base sequences in the tRNA-like structures (Van Belkum *et al.*, 1986).

The fact that immunological properties of viruses owe their existence to continuous or discontinuous antigenic determinants situated at the surface of the capsid allows one to predict partly the tertiary structure of the protein subunits forming the capsid, at least when the polypeptide chain sequence is known. In 1967, Anderer *et al.* suggested for the first time that the C-terminal region of the TYMV polypeptide chain is situated on the outside of the capsid, by analogy with what was known to be the case for TMV. This question was reexamined by the group of Van Regenmortel, who localized four antigenic regions in the polypeptide chain constituting the TYMV subunit (Pratt *et al.*, 1980).

The chain of 189 amino acids that constitutes this subunit can be cleaved into as many as 12 peptides by cyanogen bromide and by trypsin. The inhibition of the antigen–antibody reaction by synthetic peptide corresponding to defined regions of the polypeptide chain showed that the region 46-67 of the chain was situated on the outside.

These observations are derived from a comparison of the results obtained using antibodies prepared against virions or capsids, and against a 7 S aggregate in which the whole of the surface of subunits was exposed. These results do not agree with those of Re and Kaper (1975) showing that the central part of the polypeptide chain was folded toward the inside of the capsid.

Using synthetic peptides corresponding to the N-terminal region of the polypeptide chain, Quesniaux *et al.* (1983a) confirmed that the N-terminal peptide 1-12 is situated on the outside of the capsid. As far as the C-terminal region of the polypeptide chain is concerned, the techniques used by Quesniaux *et al.* (1983a,b) (inhibition of binding of antibody to synthetic peptides corresponding to different regions of the polypeptide chain) allowed the authors to suggest that the residues 183-189 and 187-189 (see Fig. 4) were situated on the outside of the capsid. The same pertained to residues 57-64, for which the position was established by complement fixation tests.

The use of immunological techniques has thus made it possible to localize three epitopes on the surface of the virus, and has further led to the suggestion that three other peptides can interact with RNA.

This proposed structure does not agree with that suggested by earlier cross-linking experiments between RNA and protein subunits of TYMV. Ehresmann *et al.* (1980), using UV irradiation of viral particles at pH 4.8 or bisulfite treatment of particles at pH 7.3, studied the binding sites of RNA in the protein. Figure 4 (Ehresmann *et al.*, 1980) shows which peptides were found to interact with the RNA. It is clear from this figure that the N-terminal peptide (12 amino acids) interacts with the RNA and thus that it should be considered an internal peptide; furthermore, the peptide immediately in front of the peptide 46-66 (defined as external by immunological detection) is also internal. It would of course be surprising if the adjacent peptide (residues 33-45) could be located externally.

It is difficult to take sides between the two proposed structures. It is clear that if one agrees with the point of view expressed by Quesniaux *et al.* (1983a), then the C- and N-termini of the polypeptide chain are situated on the outside of the nucleocapsid, a location similar to that described for TMV (Klug, 1979). In potato virus X (Koenig *et al.*, 1978) and Potyviruses (Dougherty *et al.*, 1985) the N-termini are also situated at the surface. It should be noted however that if TYMV follows the rules recently established for other nonenveloped isometric plant viruses, it should have the N-terminal region of the polypeptide chain inside the capsid and interacting with RNA, as was found for tomato bushy stunt virus (TBSV), southern bean mosaic virus (SBMV), and satellite tobacco necrostics virus (STNV) (for review, see Harrison, 1983).

Furthermore, recent results obtained with two animal viruses, poliovirus (Hogle *et al.*, 1985) and rhinovirus (Rossmann *et al.*, 1985), also show internal location of the N-terminal extremity of the polypeptide chain.

It would appear that isometric viruses are built according to the same general principles (for review, see Harrison, 1985) and it would be surprising if Tymoviruses did not conform to this pattern.

Whatever the case, only crystallography will be able to unambiguously resolve the apparent contradictions that remain. It should be noted, however, that the structure of rhinovirus established by crystallography (Rossmann *et al.*, 1985) demonstrated the existence in the tertiary struc-

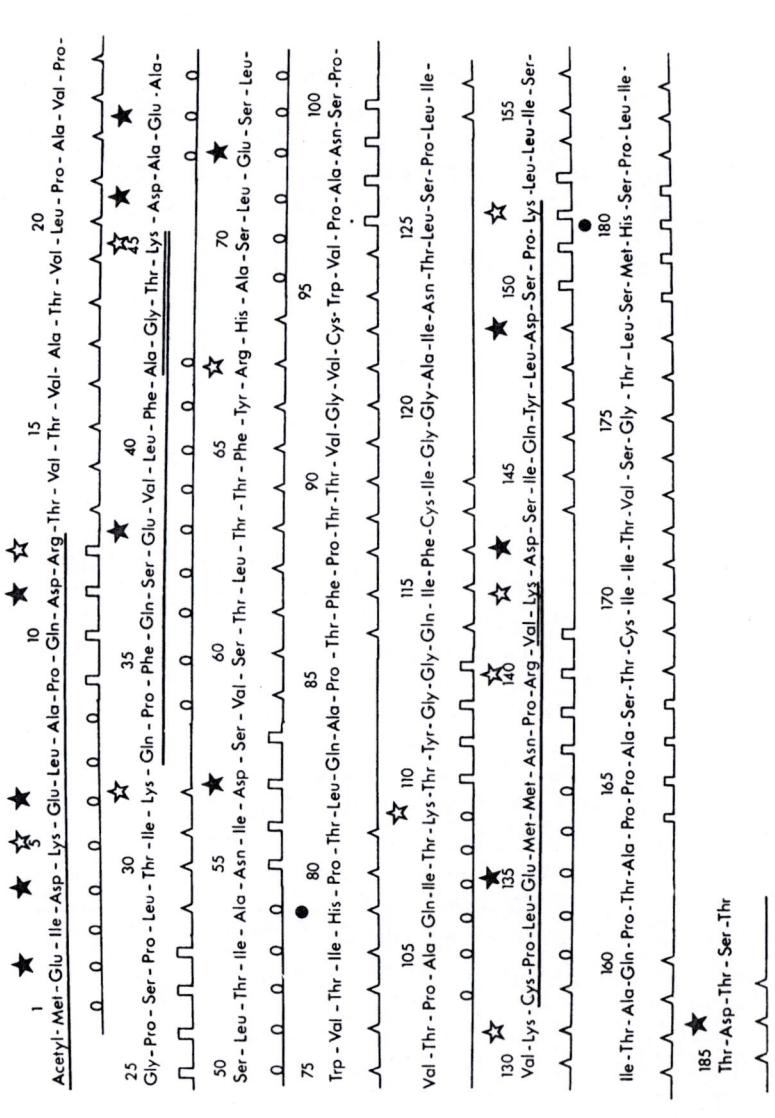

FIGURE 4. Primary structure of TYMV coat protein (Peter *et al.*, 1972). Dicarboxylic and basic amino acids are indicated by filled and empty stars, respectively. Histidine residues, which are positively charged at low pH, are indicated by filled circles. o-o-o, putative α-helix; ∿∿∿, β sheet; ⊓⊔⊓, β turn; ——, irregular. Peptides that were found to bind covalently to the RNA are underlined. Peptides underlined twice remained bound even after drastic hydrolysis. According to Ehresmann *et al.* (1980).

ture of the subunits, of a kind of groove that makes antibodies accessible to regions of the polypeptide chain that are, in fact, internally located.

In conclusion, it seems that only new crystallographic techniques will succeed in unraveling the structure of TYMV and of other viruses of the same group.

VII. GENOME PROPERTIES AND REPLICATION

A. *In vitro* Messenger Properties of TYMV RNAs

The messenger properties of plant viral RNAs were discovered very early on, and the RNA of TMV was described as being able to stimulate the activity of the *in vitro* protein synthesis developed by Nirenberg and Matthaei (1961). Later, improvement of the *in vitro* protein synthesis system (from wheat germ or rabbit reticulocytes) made it possible to disentangle the genetic organization of RNA viruses with a single positive strand (i.e., with direct messenger properties). In 1976, Hunter *et al.* and Seigel *et al.* showed that translation of the genomic RNA of TMV yielded only high-molecular-weight proteins but not the coat protein. The latter was produced only from a subgenomic RNA that was not encapsidated but was present in infected plants (Beachy and Zaitlin, 1975).

Initial results obtained with TYMV RNA by Bénicourt and Haenni (1976) tended to show the existence of three protein bands on the poly-acrylamide gels of translation products: the 195-Kd and 150-Kd bands derived from the expression of genomic RNA, less the coat protein represented by the 20-Kd band. Klein *et al.* (1976) and Pleij *et al.* (1976) showed that in reality the preparations of TYMV RNA contained, in addition to the infectious RNA of molecular weight 2000 Kd, one RNA of 280 Kd (Klein *et al.*, 1976) and expressed *in vitro* in the form of a polypeptide of molecular weight about 20 Kd. In contrast to the 195-Kd and 150 Kd proteins, this polypeptide reacted with an antiserum against coat protein.

Analysis of translation products by Higgins *et al.* (1978) suggested that the densest particles from a preparation of TYMV that was centri-fuged on a cesium chloride gradient, contained the genomic RNA with or without subgenomic RNA. In addition, some intermediate particles between the top and bottom components contained only the subgenomic RNA coding for the coat, or RNA of various sizes but significantly larger than the subgenomic RNA of 280 Kd.

A more detailed analysis by Mellema *et al.* (1979) of the translation products of the RNAs isolated from intermediate products of TYMV led to the conclusion that the majority of dense particles contain only the genomic RNA, which can give major proteins of 150 Kd and 195 Kd, but no protein corresponding to the coat protein. The least dense nucleoproteins contain several molecules of subgenomic RNA coding for the coat protein, and the intermediate nucleoproteins contain RNA molecules of various sizes with a common 5' extremity, whose *in vitro* translation products are related proteins of various lengths. The noninfectious character of the intermediate products (from 800 to 130–140 Kd) was described by DeRosier and Haselkorn (1966) and Faed *et al.* (1972).

The determination of the primary structure of the subgenomic RNA coding for the coat protein has given rise to discussion about the very effective translation of this RNA. One may imagine that the short 19-nucleotide leader sequence of the subgenomic RNA helps it to be read rapidly and repeatedly, especially as this leader sequence has not, of itself, any major secondary structure. In addition, Hirth *et al.* (1965) and Strazielle *et al.* (1965) reported that the RNA from TYMV may have a circular structure. It is now known that this may be due to pairing of nine nucleotides situated close to the 3' extremity, with nine complementary nucleotides close to the 5' end. Blocking of the 5' extremity of the genomic RNA by subgenomic RNA might play a role in regulating the expression of these two genes, but no experimental proof for this exists yet. Other hypotheses have been proposed, but none has been confirmed and it is worthy to note that the majority of plant and animal viruses display systems analogous to those just described for the expression of genes situated on a polycistronic RNA (for the mechanism of expression of polycistronic RNAs in eukaryotes, see Kozak, 1978).

The mechanism of formation of subgenomic RNAs is still not completely clear, and the relatively recent discovery that, in the case of TMV and other RNA viruses, truncated double-stranded forms exist in infected plants and might give rise to RNAs with identical 3' extremities that are translated into proteins with analogous C-terminal sequences, suggests a type of mechanism (at least for the subgenomic RNA of the coat protein) of a partial replication (Beachy and Zaitlin, 1975).

In the case of TYMV, besides the classical translation products derived from the genomic RNA (195 and 150 Kd), two other major proteins are found at 120 and 78 Kd. Although it is difficult to distinguish artifactual cleavages from specific cleavages, Morch *et al.* (1982) suggested that the formation of these proteins corresponds to specific cleavage products. The relative insensitivity of the system to the dilution of the 195-Kd protein suggests an intramolecular mechanism of maturation. The 150-Kd protein does not show any intramolecular proteolysis (Morch *et al.*, 1982). This observation is in good agreement with what is known about some Picornaviruses (Hanecak *et al.*, 1982) and Comoviruses (Peng and Shih, 1984). However, major differences exist between these viruses and the Tymoviruses, and the involvement of a protease belonging to the translation system, although unlikely, cannot be completely ruled out.

Hitherto, the role of products of cleavage of this 195-Kd protein has not been elucidated, although the C-terminus of the 195-Kd protein has been suspected of having proteolytic properties (Morch and Bénicourt, 1980). It is likely that the nonstructural proteins of TYMV play a role in the replication of the viral RNA. Recently, Mouches *et al.* (1984) suggested that the viral fraction of the RNA replicase has a molecular weight of 115 Kd and might originate from a 120-Kd protein described by Morch *et al.* (1982). All the same, no definite proof of this identity has been provided yet. The origin of the nonstructural proteins present in the translation products of TYMV RNAs remains an unresolved problem, which is also true for many other viruses.

It is worth mentioning that information about the existence of non-structural viral proteins is mainly based on *in vitro* translation. It would be very useful to know whether they exist *in vitro*, and if so where; they might be detected immunologically by antibodies obtained from synthetic peptides whose composition could be deduced from the sequence of the genomic nucleic acid, as in the case of cauliflower mosaic virus (Xiong *et al.*, 1984; Ziegler *et al.*, 1985) and of alfalfa mosaic virus (AMV) Berna *et al.*, 1984).

Structural investigation of the genomic and subgenomic RNAs has shown that the 5' leader sequence of the genomic RNA is 95 nucleotides long and relatively poor in G (Briand *et al.*, 1978), and that the 5' leader sequence of the subgenomic RNA is short (19 nucleotides) (Guilley and Briand, 1978). It is tempting to see in the shortness of the subgenomic RNA and its leader sequence a key factor in the effectiveness of its trans-lation, especially as the same situation holds for TMV. Nevertheless, some viral RNAs of the subgenomic type that are very effectively trans-lated have a relatively long leader sequence: this is the case with RNA 4 of AMV (Bol *et al.*, 1976; Pinck *et al.*, 1981). It is difficult at present to formulate general rules concerning the reasons for greater or lesser effec-tiveness of translation of a messenger RNA, especially in the case of plant viruses.

The major 195-Kd and 150-Kd proteins are expressed at different levels, even though they have the same leader sequence and the same reading phase (see above). Morch *et al.* (1982) attributed the synthesis of the 195-Kd protein to a readthrough of a weak stop codon read by a suppressor tRNA. This readthrough mechanism was suggested in 1976 by Pelham and Jackson in the case of the translation of two major proteins corresponding to the protein of TMV. The composition and the quantity of the 120-Kd and 78-Kd proteins may therefore depend on the formation of the 195-Kd protein resulting from a readthrough. Of course, as men-tioned above, *in vitro* identification, using synthetic peptides, of the var-ious products identified *in vitro* remains the fundamental question to be resolved.

Recently, Zagorski *et al.* (1983) showed that the messenger properties of TYMV RNA could be expressed in various protein synthesis systems (in addition to rabbit reticulocytes), such as extracts of wheat germ and the cell-free Ehrlich ascites system. These two latter systems are very effective for expression of the 195-Kd protein, but specific maturation processes are either absent or weakly expressed. This is particularly the case for the wheat germ system.

Szczesna and Filipowicz (1980) showed that various messengers from plant viruses, including TYMV, could be expressed in a cell-free system derived from *Saccharomyces cerevisiae.* Nevertheless, the effectiveness of this system was less than that of the systems just mentioned and it is little used.

Finally, in many cases, TYMV RNA has served as a messenger for testing the preferential effectiveness of translation of a system (Asselbergs *et al.*, 1980) without the specific understanding of the expression of TYMV RNA being much advanced in the process.

B. tRNA-like Structure at the 3'-OH Extremity of TYMV RNA

Although the RNAs of many plant viruses display the peculiarity of binding a specific amino acid at their 3' extremity, we shall be relatively brief on this subject, which has often been reviewed, especially in the case of TYMV (Hall, 1979; Haenni *et al.*, 1982). In 1970, Pinck *et al.* and Yot *et al.* showed that valine could bind to the 3' extremity of TYMV RNA, via the intermediate of a valyl tRNA-synthetase. This extremity could also recognize a large range of tRNA enzymes, such as CTP/ATP-nucleotidyl transferase, EF-Tu and EF-T elongation factors, a peptide hydrolase, and perhaps the RNase-P that can recognize specifically a site situated close to the 3' extremity of TYMV RNA. However, contrary to what has been suggested, the 3' extremity of TYMV RNA does not have the property of participating in the elongation of a polypeptide chain by addition of valine.

A fragment of 159 nucleotides containing the 3' extremity of TYMV RNA has been sequenced (Briand *et al.*, 1977; Silberklang *et al.*, 1977) and its folding proposed (Briand *et al.*, 1977; Silberklang *et al.*, 1977). The secondary structure proposed could have some analogy with the cloverleaf structure of tRNA in particular with the valine anticodon region, but presents also new structural features not found in canonical tRNA: a new stem and loop near the CCA-accepting end of the molecule.

Rietveld *et al.* (1982) and Florenta *et al.* (1982) proposed a secondary structure for the 86 and 159, respectively, 3'-terminal nucleotides of TYMV based on their digestion with ribonuclease T1, nuclease S1, and cobravenom ribonuclease (which digests double-stranded regions) and on their reactivity to chemicals. The secondary structure deduced from these works resembled, in some respects, the structure of a classic tRNA, but did not display the acceptor arm generally present in all the prokaryotic and eukaryotic tRNAs. The amino acid receptor arm proposed could be found for all tymoviral tRNA-like ends sequenced today (Van Belkum *et al.*, 1987).

The most recent data (R. Giegé, personal communication) show that steadily shorter and shorter fragments from the 3' extremity of TYMV RNA can be aminoacylated. Similar observations have been made with the 3' extremities of other Tymoviruses, and it seems in fact that all Tymoviruses can bind valine at their 3' extremity.

1. Role of the Valylation of TYMV RNA

The question that arises from this observation is this: what is the role of valylation? And can the observations made *in vitro* be extrapolated to the situation *in vivo*? In 1978, Joshi *et al.* showed that the introduction of TYMV RNA into oocytes of *Xenopus laevis* resulted in the appearance of 4–5 S fragments of aminoacylated and valylated TYMV RNA. It is likely that TYMV RNA is degraded by cellular ribonucleases into fragments of 100 to 120 nucleotides, of which some contain the 3' extremity of the viral RNA, which can be valylated *in vitro* (Joshi *et al.*, 1982).

The valylation of viral RNA fragments in *Xenopus* oocytes might be artifactual in character, but recently Joshi *et al.* (1982) reported the presence of valylated viral oligonucleotides in leaves of infected Chinese cabbage. The length of these valylated fragments was, however, relatively small and it is not clear whether the complete TYMV RNA or the subgenomic RNA were valylated. It should be noted in this context that encapsidated molecules of genomic and subgenomic RNA are never valylated. The reasons why the 3' extremities contained in the virions are not charged is unknown. In addition, Joshi *et al.* (1982) estimated that 21 to 38% of the viral RNA present in an infected plant was valylated. This observation is somewhat difficult to reconcile with the large number of nonvalylated molecules present in the virions contained in an infected leaf and the apparently low amount of free viral RNA molecules in the same tissue.

Many questions remain to be resolved, including the main one: what is the role, if any, of the charging of this 3' extremity of Tymoviruses? Or, at least, what is its significance (vestigial structure, see below)? Note that observations similar to those reported for Tymoviruses have been reported for Cucumoviruses and Bromoviruses (binding tyrosine) as well as Tobamoviruses (binding histidine).

Observations suggesting that the RNA of animal viruses bind amino acids [encephalomyocarditis, by Lindley and Stebbing (1977)] have not given rise to any further developments and remain ambiguous.

2. Tertiary Structure of the 3' Extremity of TYMV RNA

It must be stressed from the start that all structures proposed are hypothetical. In fact, until now, the 3' fragments of various lengths obtained from the products of digestion of the genomic or subgenomic RNA have not been crystallized and no tertiary structure has been established experimentally. However, major efforts have been made to suggest struc-

tures that may be in accord with the observations on the action of enzymes mentioned above and the accessibility of some nucleotides and
phosphates to specific chemical probes. Consequently, Rietveld *et al.*
(1982) proposed a tertiary structure for the 42 terminal nucleotides for
TYMV RNA, which differed significantly from the tertiary structure established for the Phe and Asp tRNAs of yeast. The proposed folding
includes a pseudoknot whose occurrence is strongly supported by graphic
modeling of the 86 last nucleotides as shown by Dumas *et al.* (1987).

Rietveld *et al.* (1983, 1984), on the basis of chemical and enzymatic
accessibility studies, proposed models for the tertiary structure that were
very similar to canonical tRNAs. This is particularly true for the 3′ extremity of TYMV RNA, as the structures of the 3′ extremities of other
plant viruses that bind amino acids (BMV and TMV) are significantly
more remote, even if in all the cases examined the L structure characteristic of all tRNAs is found.

Also, the results of Rietveld *et al.* (1984) and of Joshi *et al.* (1983)
show that specific regions are involved in the recognition of certain enzymes. In the case of TYMV, a sequence of 86 3′ nucleotides is enough
to ensure both the addition of an A to the 3′ extremity of the RNA, and
the valylation of the RNA. Shorter fragments (47 nucleotides) can bind
to the EF-Tu (elongation factor) complex with GTP and, so it also seems,
to tRNA nucleotidyl-transferase. The other enzymes mentioned above
do not recognize this type of fragment.

Not all the tRNA-like extremities of plant viruses have the same
properties. Thus the 3′-OH extremity of TMV RNA must be about 100
nucleotides long to be recognized by ATP-CTP nucleotidyl transferase,
(Rietveld *et al.*, 1984), and thus longer than that of TYMV. However, this
point of view has been disputed recently by Joshi *et al.* (1985), whose

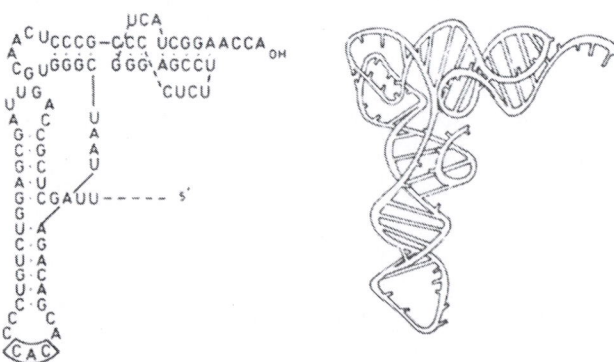

FIGURE 5. L arrangement of three-dimensional folding of the 3′ end of TYMV RNA. It is
clear that the structure is very similar to that proposed for yeast phenylalanine tRNA by
Kim *et al.* (1974).

experiments tend to show that the 55 nucleotides at the 3' extremity of TMV RNA can bind histidine and add an A to the RNA if it has had it experimentally removed. These differences show that crystallization of 3' segments of aminoacylatable viral RNA and determination of their structure of X-ray crystallography is needed to remove the ambiguities.

There is, however, a consensus to attribute to these tRNA-like viral RNAs and L structure quite close to that of classic tRNAs. In this context it should be remembered that no modified bases have been discovered in viral tRNAs, and their absence does not alter their capacity for amino-acylation. Figure 5 illustrates the preceding text.

3. Role of the tRNA-like Structures in Plant Viruses

The role of these structures has been the subject of many hypotheses, most of which are summarized in the reviews by Haenni and Chapeville (1980) and by Florentz et al. (1984). Among them, one is based on the possibility of an interaction between complementary sequences at the 3' and 5' ends of the viral RNA, giving rise to a circular structure. This structure might act on the effectiveness of translation of the genomic RNA, which might not be expressed in its circular form. However, there is no proof on this subject and anyway, aminoacylation of the 3' extremity of the circular form is theoretically not necessary for regulation to occur.

The hypothesis most favored at present is that the tRNA-like region serves as a recognition structure for the viral RNA polymerase. However, Mouchès et al. (1984) investigated the structure and the function of the viral RNA polymerase extracted from Chinese cabbage plants infected with TYMV, and could not link this structure with that of the tRNA-like structure of the RNA. The information came from the work of Hall (1979) on BMV, which has three RNA molecules necessary for infection with a 3' extremity that is aminoacylable by tyrosine.

Using site-directed mutagenesis on cloned cDNA of the 3' end of BMV RNA 3, followed by in vitro transcription, Dreher et al. (1984) showed that it was possible to dissociate the replicase activity from the aminoacylation activity. This point of view has been reinforced by the use of deleted versions of this cDNA, which lacked the final 3' nucleotides (Bujarski et al., 1985). Thus a deletion in arm C (Fig. 6) of the tRNA-like structure abolishes the replication of RNA, but not its capacity for ami-noacylation; deletion of arm B has the exact opposite effect. One may therefore assert that a sequence in the tRNA-like structure of BMV is recognized by the replicase.

New details were provided by Ahlquist et al. (1985), who showed that the 39 final nucleotides of the tRNA-like structure of RNA 3 of BMV are essential to recognition of the replicase, but that hybridization of sequences situated 200 nucleotides from the 3' extremity of the RNA with the corresponding cDNA did not inhibit the recognition by the

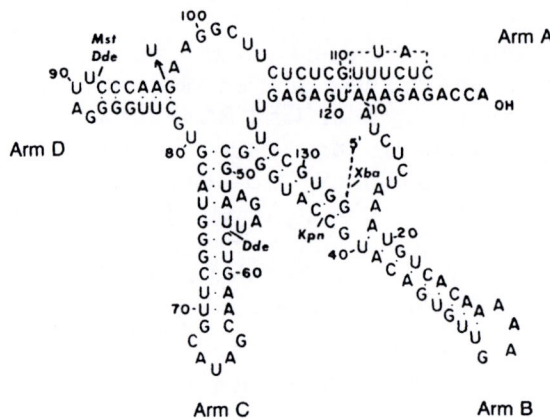

FIGURE 6. tRNA-like structure at the 3' terminus of BMV RNA-3. Nucleotides are numbered from the 3' end. This schema illustrates the text concerning the role of the various parts of the sequence.

replicase. Therefore there are two distinct regions in the tRNA-like structure concerning the aminoacylation and the recognition by the replicase.

Paradoxically, it seems that in BMV the anticodon region rather than the tyrosyl tRNA-synthetase-binding region acts as replicase-recognition region. It is still not clear if the 3' extremity needs to be charged in order to be recognized by the replicase, or not. It is possible that the remarks made about this virus may be extended to TYMV. Also, Meshi et al. (1981) showed that the cowpea strain of TMV, which binds valine at its 3' extremity, has a 3'-terminal sequence of 46 nucleotides analogous with that observed in TYMV, and its 120 C-terminal nucleotides may be folded in the same way as the 3' extremity of TYMV. The origin of this analogy is not clear, and the hypothesis of Gibbs (1980b) that some recombination between TMV and TYMV is responsible for this analogy is highly speculative. In any case, research on the Tymoviruses is currently directed mainly towards the functions of the tRNA-like structures, although structural studies on the virus, by means of X-ray diffraction, are beginning to be seriously undertaken, notably on BelMV (M. Rossmann, personal communication).

C. Replication

The replication of TYMV RNA was the starting point for in-depth research on the replication of other plant virus RNAs. The initial research in this field was by Astier-Manifacier and Cornuet (1965) and Bové et al. (1965), who showed that extracts from infected plants could synthesize a fragment of viral RNA using TYMV RNA as matrix. The base composition of this fragment suggested that it was complementary to the 3' part of the matrix. It therefore appeared that, as for bacteriophage Qβ (Weissmann, 1965; Weissmann and Feix, 1966; Mills et al., 1966) and animal viruses (Baltimore et al., 1963; Montagnier and Sanders, 1963), TYMV replicates from a positive strand using a specific replicase that is thought to be encoded by the viral genome. The work of Bové (1967), Laflèche and Bové (1971), and Laflèche et al. (1972) confirmed this point of view by isolating, from infected plants, double-stranded RNA molecules and localizing the replication complex to the chloroplast membranes.

From 1971 to 1983, the coding and in vitro function of the replication complex of several RNA viruses was a subject of much discussion, especially concerning the involvement of a (soluble) cytoplasmic RNA polymerase found in healthy cabbage leaves, but also in other plants sensitive to infection with other viruses (Astier-Manifacier and Cornuet, 1971; Duda et al., 1973; Romaine and Zaitlin, 1978). The same type of enzyme was also found on membrane structures (Bol et al., 1976; Fraenkel-Conrat, 1976; Stussi-Garaud et al., 1977). In 1983, Fraenkel-Conrat suggested that this type of enzyme (which is apparently universal) is responsible for the replication of RNA viruses.

The work of Mouchès et al. (1984) led to the isolation of purification of the TYMV replicase. It is made up of two major subunits of 115 and 45 Kd, respectively. The 115-Kd subunit is of viral origin and may arise from maturation of the 198-Kd protein being coded by the 5' extremity of the RNA. The 45-Kd protein may be of cellular origin, and appears to be essential to the functioning of the enzyme. Although its nature has not yet been determined, Mouchès (1984) suggested that it might be responsible for the host specificity of the virus; there is no evidence yet to justify this point of view. It is not absolutely certain, either, that the replicase (or rather the replication complex) consists of only these two major proteins (Mouchès et al., 1984). It is likely that the operation of the enzyme requires cofactors that have not yet been identified.

Substantial progress has been made in the case of BMV (Ahlquist et al., 1985) and cowpea mosaic virus (Dorssers et al., 1984), where the proteins of the replication complex have been identified. The polymerase activity for this latter virus seems to be associated with a protein (110 Kd) derived from the maturation of the polyprotein encoded by RNA 1. The role (if there is one) of the 68-Kd and 57-Kd proteins associated with the replication complex, which are probably of cellular origin, has not been defined, and here, too, these proteins have not been identified.

It is interesting to note that, for all the RNA viruses with a positive RNA strand a Glu-Asp-Asp peptide sequence is present in the fraction of the polyprotein coded by the RNA (poliovirus) or in the protein coded by RNA 1 in the case of multiple-component viruses (BMV, AMV, and beet necrotic yellow vein virus) (Kuszala *et al.*, 1986). It is very regrettable that the complete sequence of TYMV RNA has still not been determined, but it is likely that the codons for this acidic tripeptide will be found in it. The isolation and the purification of the replicase of plant viruses with a positive RNA strand remains a very open question.

VIII. RELATIONS WITH CELLS

A. Cellular Symptomatology

At the cellular level, the most important changes occur in the chloroplasts and in the nuclei, which accumulate large masses of "empty"-appearing capsids. Laflèche and Bové (1969), Ushiyama and Matthews (1970), Hatta and Matthews (1974), and Matthews (1973, 1981) showed the formation of vesicles arising from the double membrane of chloroplasts which become invaginated toward the inside of the organelle. These vesicles and other cytological changes, such as rounding and clumping of chloroplasts, vacuolization, characteristic advanced stages of chloroplast disorganization, formation of myelinlike structures, fragmentation of stroma parts, and accumulation of material with low-electron density (presumably masses of empty virus protein shells) in the nuclei, induced by various Tymoviruses have been studied in detail by Lesemann (1977) (for review see also Koenig and Lesemann, 1981). PoiMV, a possible Tymovirus, induces vesicles in chloroplasts that are bounded only by the inner of the two membranes (Lesemann *et al.*, 1983).

Electron microscopy has shown that the vesicles induced by TYMV are the seat of RNA replication (Laflèche and Bové, 1971). Matthews (1981) has proposed (Fig. 7) a model of virus assembly according to which pos-

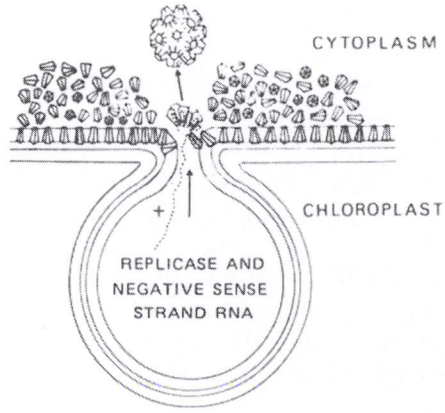

FIGURE 7. Putative model for the assembly of TYMV according to Matthews (1981). The protein subunits are synthesized in the cytoplasm and some of them are inserted in the outer chloroplast membrane. These subunits are oriented in such a manner that they can trap the + viral RNA molecules emerging from the vesicles in which these RNA molecules are synthesized. The first cluster of protein subunits interacting with the RNA could be a pentamer and assembly proceeds by the addition of other pentamers and hexamers.

itive-strand RNA excreted from the vesicle is incorporated into protein subunits present in the external membrane of the chloroplast. However, this model has not yet been experimentally confirmed.

B. Modifications in Protoplasts Infected by TYMV

Protoplasts of *Brassica* (*pekinensis* or *rapa*) plants have been shown to be an excellent material for studying cytological modifications in TYMV-infected cells, as well as for following virus multiplication. Under the influence of virus infection, chloroplasts have been found to fuse together to form polyplasts (Laflèche *et al.*, 1972; Garnier *et al.*, 1980), and this makes it easy to distinguish infected from noninfected cells using an optical microscope.

Renaudin *et al.* (1975a) established the conditions of pH, poly-L-ornithine, and virus concentration necessary for infecting the protoplasts of *Brassica sinensis*. These authors also studied changes occurring during virus multiplication (Renaudin *et al.*, 1975b). They showed that the formation of polyplasts coincides with the appearance of virus in the protoplasts (determined by immunofluorescence), and the visualization of polyplasts makes it possible to determine the number of infected protoplasts (up to 90%). Fraser and Matthews (1979) showed that the formation of polyplasts as well as the fragmentation and/or deformation of chloroplasts depends on the particular strain of TYMV used.

The number of infected protoplasts after inoculation can also be determined by autoradiography after incubation with [^{32}P]orthophosphate or [^{14}C]leucine (Cocker and Cassells, 1983). The latter method makes it possible to follow the first steps of virus multiplication. It has been well established that the multiplication of TYMV in protoplasts occurs in light conditions and that it therefore depends on photosynthesis. Nevertheless, in 1980, Fernandez-Gonzalez *et al.* showed that protoplasts obtained from hypocotyls of *Brassica sinensis* seedlings were able to sustain the multiplication of TYMV although they were totally devoid of chloroplasts. Virus multiplication could occur in darkness, indicating that the necessary energy was of respiratory origin. In fact, if respiration is blocked (multiplication in the presence of nitrogen) there is no virus production. In this case it is possible that virus multiplication takes place in undifferentiated plastids.

The protoplasts of *Brassica sinensis* represent an excellent material for studying the synthesis of polyamines (spermine and spermidine) during infection by TYMV (Cohen *et al.*, 1981) as well as for obtaining fragmented chloroplasts.

It must be emphasized that the behavior of chloroplasts is different in the cells of leaves and in the protoplasts obtained from the same infected leaves. In the case of leaves, the chloroplasts become fragmented in an approximately synchronized fashion. In the case of protoplasts, the intact form is present together with modified and fragmented forms (Matthews, 1981).

IX. CONCLUSIONS

TYMV is the type member of the group of Tymoviruses. The main characters of TYMV provided the criteria that subsequently made it possible to group together viruses which on an immunological basis would have been considered unrelated. In fact, the existence of two major components (top and bottom) and of intermediate, noninfectious components represents one of the crucial discoveries made in the early days of general virology. Nevertheless, it must be noted that the origin of the presence of empty particles (top) is not yet very clear except perhaps in case of *in vivo* as well as *in vitro* decapsidation (see Section V.H).

TYMV and its capsid are extremely stable, a characteristic shared by many other Tymoviruses although to different degrees, as shown for instance in the case of EMV and BelMV.

The presence of low molecular weight RNAs in top components makes it possible to differentiate between Tymoviruses. The most interesting properties of Tymoviruses relate in fact to their RNA composition—two or more characteristic properties of the viral RNA set Tymoviruses apart from many others. First, is the large dissymmetry in nucleotide composition (32–38% of C depending on the virus), the significance of which may become clear when the complete nucleotide sequence is established.

The second characteristic of all Tymovirus RNAs is their ability to fix valine at their 3' end. The structure of this extremity, which has been described above, presents an obvious evolutionary interest. In fact, whatever ideas one may have concerning the origin of RNA viruses (Matthews, 1981; Hunter *et al.*, 1983), it is evident that the Tymoviruses, as with all other RNA viruses that fix an amino acid, such as TMV, (Oberg and Philipson, 1972), BMV (Hall *et al.*, 1972) and CMV (Kohl and Hall, 1974), could not have appeared in their present form until relatively late in the course of evolution; that is to say once all the sophisticated machinery required for the synthesis of proteins in eucaryotes had been developed. This structure, as well as the 3' polyadenylation of certain other viruses (Nepoviruses, Picornaviruses, for example) has never been found in the prokaryotic viruses.

Although its biological significance remains speculative, it presents a certain evolutionary character and certainly is in accordance with the proposal of Zimmern (1982) on the origin of RNA viruses: namely, that the 3'-tRNA like region represents a "signal RNA" attached to a coding fraction of an "antenna RNA."

When classifying plant viruses, one could take into account the nature of the 3' end (absence of special terminal sequences, poly A sequence, tRNA-like sequence). Mention should also be made of the observations of Salomon and Littauer (1974) and Lindley and Stebbing (1977) that the fixation of valine and serine by the RNA of the Mengovirus and the Encephalomyocarditis virus remains doubtful, and that RNA animal viruses have mainly poly A at their 3' end. It should also be stressed that the finding reported by Dreher *et al.* (1984) indicating that two different nucleotide regions at the 3'OH end of TYMV RNA are implicated in the recognition of replicase and valyl tRNA synthetase also contributes to making the tRNA-like structure a specific marker of evolution.

In conclusion, it can be stated that Tymoviruses, TYMV in particular, have served as initial models in attempts to solve the problem of replication of RNA viruses and the origin of the replicase.

Although considerable progress has been made in the case of Comoviruses and Picornaviruses (which are more or less related), a great deal remains to be done to clarify the problem posed by an enzyme present in healthy plants, which is also involved in the *in vitro* replication of viral RNA (Fraenkel-Conrat, 1983).

ACKNOWLEDGMENTS. We thank Dr. J. Witz for stimulating discussions and Mrs. J. Vonesch for her help in the preparation of this manuscript.

REFERENCES

Ahlquist, P., Strauss, E. G., Rice, C. M., Strauss, J. H., Haseloff, J., and Zimmern, D., 1985, Sindbis virus proteins nsP1 and nsP2 contain homology to nonstructural proteins from several RNA plant viruses, *J. Virol.* **53**:536.

Allen, T. C., Jr., and Fernald, E. K., 1971, Recovery and partial characterization of wild cucumber mosaic virus from *Marah oreganus*, *Plant Dis. Rep.* **55**:546.

Anderer, F. A., Schlumberger, H. D., and Frank, M., 1967, A serological screening method for the detection of free C-terminal amino acids in virus coat protein, *Biochim. Biophys. Acta* **140**:80.

Asselbergs, F. A. M., Meulenberg, E., Van Venrooij, W. J., and Bloemendal, H., 1980, Preferential translation of mRNAs in an mRNA-dependent reticulocyte lysate, *Eur. J. Biochem.* **109**:159.

Astier-Manifacier, S., and Cornuet, P., 1965, Isolation of turnip yellow mosaic virus RNA replicase and asymmetrical synthesis of polynucleotides identical to TYMV-RNA, *Biochem. Biophys. Res. Commun.* **18**:283.

Astier-Manifacier, S., and Cornuet, P., 1971, RNA-dependent RNA polymerase in Chinese cabbage, *Biochim. Biophys. Acta* **232**:484.

Atiri, G. I., 1984, The occurrence and importance of okra mosaic virus in Nigerian weeds, *Ann. Appl. Biol.* **104**:261.

Baltimore, D., Franklin, R. M., Eggers, H. J., and Tamm, I., 1963, Poliovirus-induced RNA polymerase and the effects of virus-specific inhibition on its production, *Proc. Natl. Acad. Sci. USA* **49**:863.

Bancroft, J. B., 1970, The self assembly of spherical plant viruses, *Adv. Virus Res.* **16**:99.

Barradas, M. M. 1983, Identification of the tomato white necrosis virus as a tymovirus, Thesis, Instituto de Biociencias da U.S.P. Sao Paulo.

Beachy, R. N., and Zaitlin, M., 1975, Replication of tobacco mosaic virus. VI. Replicative intermediate and TMV-RNA-related RNAs associated with polyribosomes, *Virology* **63**:84.

Beczner, L., Vassanyi, R., Salamon, P., and Dezgeri, M., 1976, Virus diseases of *Solanum dulcamara* in Hungary. I. Dulcamara mottle virus, *Acta Phytopathol. Acad. Sci. Hung.* **11**:245.

Bedbrook, J. R., and Matthews, R. E. F., 1973, Changes in the flow of early products of photosynthetic carbon fixation associated with the replication of TYMV, *Virology*, **53**:84.

Beer, S. V., and Kosuge, T., 1970, Spermidine and spermine–polyamine components of turnip yellow mosaic virus, *Virology* **40**:930.

Bénicourt, C., and Haenni, A. L., 1976, In vitro synthesis of turnip yellow mosaic virus coat protein in a wheat germ cell-free system, *J. Virol.* **20**:196.

Bercks, R., 1973, Scrophularia mottle virus, *CMI/AAB Descriptions of Plant Viruses*. No. 113.

Berna, A., Briand, J. P., Stussi-Garaud, C., Godefroy-Colburn, T., and Hirth, L., 1984, Immunodetection of a nonstructural protein of alfalfa mosaic virus (P2) in infected tobacco plants, *Ann. Virol. (Inst. Pasteur)* **135E**:285.

Bock, K. R., 1977, Clitoria yellow vein virus, *CMI/AAB Descriptions of Plant Viruses*. No. 171.

Bol, J. F., Clerx-Van Haaster, C. M., and Weening, C. J., 1976, Host and virus specific RNA polymerase in alfalfa mosaic virus infected tobacco, *Ann. Microbiol. (Inst. Pasteur)* **127A**:183.

Bouley, J. P., Briand, J. P., Jonard, G., Witz, J., and Hirth, L., 1975, Low pH RNA-protein interactions in turnip yellow mosaic virus. III. Reassociation experiments with other viral RNAs and chemically modified TYMV RNA, *Virology* **63**:312.

Bouley, J. P., Briand, J. P., Genevaux, M., Pinck, M., and Witz, J., 1976, The structure of eggplant mosaic virus: Evidence for the presence of low molecular weight RNA in top component, *Virology* **69**:775.

Bouley, J. P., Briand, J. P., and Witz, J., 1977, The stability of eggplant mosaic virus: action of urea and alkaline pH on top and bottom components, *Virology* **78**:425.

Bové, M. J., 1967, Virus de la mosaïque jaune du navet: Synthèse asymétrique in vitro d'un segment de RNA viral, Thèse de Doctorat d'Etat, Faculté des Sciences de Paris.

Bové, J. M., Bové, C., Rondot, M. J., and Morel, G., 1965, Chloroplasts and virus RNA synthesis, in: *The Biochemistry of Chloroplasts*, Volume 2 (T. W. Goodwin, ed.), pp. 329–339, Academic Press, London.

Briand, J. P., Bouley, J. P., Jonard, G., Witz, J., and Hirth, L., 1975, Low pH RNA–protein interactions in turnip yellow mosaic virus. II. Binding of synthetic polynucleotides to TYMV capsids and RNA, *Virology* **63**:304.

Briand, J. P., Jonard, G., Guilley, H., Richards, K. E., and Hirth, L., 1977, Nucleotide sequence (n = 159) of the amino acid accepting 3'OH extremity of turnip yellow mosaic virus RNA and the last portion of its coat protein cistron, *Eur. J. Biochem.* **72**:453.

Briand, J. P., Keith, G., and Guilley, H., 1978, Nucleotide sequence at the 5' extremity of TYMV genome RNA, *Proc. Natl. Acad. Sci. USA* **75**:3168.

Bricogne, G., 1984, Maximum entropy and the foundations of direct methods, *Acta Cryst.* **A40**:410.

Broadbent, L., and Heathcote, G. D., 1958, Properties and host range of turnip crinkle, rosette and yellow mosaic virus, *Ann. Appl. Biol.* **46**:585.

Brunt, A. A., 1970, Cacao yellow mosaic virus, *CMI/AAB Descriptions of Plant Viruses*. No. 11.

Brunt, A. A., and Kenten, R. H., 1965, Further studies on cocoa yellow mosaic virus, *J. Gen. Microbiol.* **38**:81.

Brunt, A. A., Barton, R. J., and Phillips, S., 1981, Miscellaneous, *Annu. Rep. Glasshouse Crops Res. Inst.* 152.

Bujarski, J. J., Dreher, T. W., and Hall, T. C., 1985, Deletions in the 3'-terminal tRNA like structure of brome mosaic virus RNA differentially affect aminoacylation and replication *in vitro*, *Proc. Natl. Acad. Sci. USA* **82**:1.

Butler, P. J. G., Finch, J. T., and Zimmern, D., 1977, Configuration of tobacco mosaic virus RNA during virus assembly, *Nature* **265**:217.

Caspar, D. L. D., 1956, Structure of bushy stunt virus, *Nature* **177**:475.

Caspar, D. L. D., and Klug, A., 1962, Physical principles in the construction of regular viruses, *Cold Spring Harbor Symp. Quant. Biol.* **27**:1.

Chauvin, C., Jaerot, B., Lebeurier, G., and Hirth, L., 1977, The structure of cauliflower mosaic virus. A neutron diffraction study. *Virology* **96**:640.

Chiko, A. W., 1983, Poinsettia mosaic virus in British Columbia, *Plant Dis.* **67**:427.

Chou, P. Y., and Fasman, G. D., 1978, Prediction of the secondary structure of proteins from their amino acid sequence, *Adv. Enzymol.* **47**:45.

Christie, S. R., and Crawford, W. E., 1978, Plant virus range of *Nicotiana benthamiana*, *Plant Dis. Rep.* **62**:20.

Cocker, F. M., and Cassells, A. C., 1983, Autoradiography of phosphorus-32 and carbon-14 incorporation into protoplasts as a means of determining the percentage of virus-infected protoplasts, *J. Virol. Methods* **6**:311.

Cohen, S. S., and Greenberg, M. L., 1981, Spermidine, an intrinsic component of turnip yellow mosaic virus, *Proc. Natl. Acad. Sci. USA* **78**:5470.

Cohen, S. S., Balint, R., and Sindhu, R. K., 1981, The synthesis of polyamines from methionine in intact and disrupted leaf protoplasts of virus-infected Chinese cabbage, *Plant Physiol.* **68**:1150.

Cosentino, V., Paigen, K., and Steere, R. I., 1956, Electron microscopy of turnip yellow mosaic virus and the associated abnormal protein, *Virology* **2**:139.

Crestani, O. A., Kitajima, E. W., Lin, M. T., Marinho, V. L. A., and Pimentel, J. P., 1984a, A new virus disease in passion-fruit yellow mosaic caused by a tymovirus, in Brazil, *Fitopatologia Brasileira* **9**:394.

Crestani, O. A., Astolfi, S., Azevedo, M. O., and Kitajima, E. W., 1984b, Composiçao quimica do virus do mosaico amarelo do Maracuja, *Fitopatologia Brasileira* **10**:175.

Crick, F. H. C., and Watson, J. D., 1956, Structure of small viruses, *Nature* **177**:473.

Crosbie, E. S., and Matthews, R. E. F., 1974a, Effects of TYMV infection on growth of *Brassica pekinensis* Rupr., *Physiol. Plant Pathol.* **4**:389.

Crosbie, E. S., and Matthews, R. E. F., 1974b, Effect on TYMV infection on growth of *Brassica pekinensis*, *Physiol. Plant Pathol.* **4**:339.

Croxall, H. E., Gwynne, D. C., and Broadbent, L., 1953, Turnip yellow mosaic in Broccoli, *Plant Pathology* **2**:122.

Dale, W. T., 1954, Sap-transmissible mosaic diseases of Solanaceous crops in Trinidad, *Ann. Appl. Biol.* **41**:240.

Debrot, E. A., Lastra, R., and De Uzcategui, R. C., 1977, *Solanum seaborthianum* a weed host of eggplant mosaic virus in Venezuela, *Plant Dis. Rep.* **61**:628.

DeRosier, D. J., and Haselkorn, R., 1966, Minor components associated with turnip yellow mosaic virus, *Virology* **30**:705.

Dorssers, L., Van der Krol, S., Van der Meer, J., Van Kammen, A., and Zabel, P., 1984, Purification of cowpea mosaic virus RNA replication complex: Identification of a virus-encoded 110,000-dalton polypeptide responsible for RNA chain elongation, *Proc. Natl. Acad. Sci. USA* **81**:1951.

Dougherty, W. G., Willis, L., and Johnson, R. E., 1985, Topographic analysis of tobacco etch virus capsid protein epitopes, *Virology* **144**:66.

Dreher, T. W., Bujarski, J. J., and Hall, T. C., 1984, Mutant viral RNAs synthesized *in vitro* show altered aminoacylation and replicase template activities, *Nature* **311**:171.

Duda, C. T., Zaitlin, M., and Siegel, A., 1973, *In vitro* synthesis of double-stranded RNA by an enzyme isolated from tobacco leaves, *Biochim. Biophys. Acta* **319**:62.

Dumas, P., Moras, D., Florentz, C., Giegé, R., Verlaan, P., Van Belkum, A., and Pleij, C. W. A., 1987, 3-D graphics modelling of the tRNA-like 3'-end of turnip yellow mosaic virus RNA: Structural and functional implications, *J. Biomol. Struct. Dyns.* **4**:707.

Dupin, A., Peter, R., Collot, D., Das, B. C., Peter, C., Bouillon, P., and Duranton, H., 1984, The primary structure of the eggplant mosaic virus (EMV) coat protein, *C. R. Acad. Sci. Paris* **298**:219.

Dupin, A., Collot, D., Peter, R., and Witz, J., 1985, Comparisons between the primary structure of the coat proteins of turnip yellow mosaic virus and eggplant mosaic virus, *J. Gen. Virol.* **66**:2571.

Durham, A. C. H., and Klug, A., 1971, Polymerization of tobacco mosaic virus protein and its control, *Nature New Biol.* **229**:42.

Ehresmann, B., Briand, J. P., Reinbolt, J., and Witz, J., 1980, Identification of binding sites of turnip yellow mosaic virus protein and RNA by *in situ* induced cross-links, *Eur. J. Biochem.* **108**:123.

Faed, E. M., Burns, D. J. W., and Matthews, R. E. F., 1972, Properties of minor nucleoprotein components found in TYMV preparations, *Virology* **48**:627.

Fauquet, C., Monsarrat, A., and Thouvenel, J. C., 1984, Voandzeia necrotic mosaic virus, *CMI/AAB Descriptions of Plant Viruses* No. 279.

Fernandez-Gonzalez, O., Renaudin, J., and Bové, J. M., 1980, Infection of chlorophyll-less protoplasts from etiolated Chinese cabbage hypocotyls by turnip yellow mosaic virus, *Virology* **104**:262.

Fernandez-Northcote, E. N., Vega, J. G., and Apablaza, G., 1982, Prevalence of potato virus Y relative to other potato viruses in Ecuador and Chile, *Phytopathology* **72**:169.

Finch, J. T., and Klug, A., 1960, X-ray "powder" diagrams of crystals of an artificial top component from turnip yellow mosaic virus, *J. Mol. Biol.* **2**:434.

Finch, J. T., and Klug, A., 1966, Arrangement of protein subunits and the distribution of nucleic acid in turnip yellow mosaic virus. II. Electron microscopic studies, *J. Mol. Biol.* **15**:344.

Florentz, C., 1983, Etude structurale de l'extrémité 3′-OH aminoacylable de l'ARN du virus de la mosaïque jaune du navet, Thesis, University of Strasbourg.

Florentz, C., Briand, J. P., Romby, P., Hirth, L., Ebel, J. P., and Giégé, R., 1982, The tRNA-like structure of turnip yellow mosaic virus RNA: Structural organization of the last 159 nucleotides from the 3′ OH terminus, *EMBO J.* **1**:269.

Florentz, C., Briand, J. P., and Giégé, R., 1984, Hypothesis. Possible functional role of viral tRNA-like structures, *FEBS Lett.* **176**:295.

Fraenkel-Conrat, H., 1976, RNA polymerase from tobacco necrosis virus infected and un-infected tobacco. Purification of the membrane-associated enzyme, *Virology* **72**:23.

Fraenkel-Conrat, H., 1983, RNA-dependent RNA polymerases of plants, *Proc. Natl. Acad. Sci. USA* **80**:422.

Francki, R. I. B., and Matthews, R. E. F., 1962, Some effects of 2-thiouracil on the multiplication of turnip yellow mosaic virus, *Virology* **17**:367.

Fraser, D., and Cosentino, V., 1957, The amino acid composition of turnip yellow mosaic virus and the associated abnormal protein, *Virology* **4**:126.

Fraser, L., and Matthews, R. E. F., 1979, Strain-specific pathways of cytological change in individual Chinese cabbage protoplasts infected with turnip yellow mosaic virus, *J. Gen. Virol.* **45**:623.

Fritsch, C., Stussi, C., Witz, J., and Hirth, L., 1973, Specificity of TMV RNA encapsidation *in vitro* coating of heterologous RNA by TMV protein, *Virology* **56**:33.

Fulton, R. W., and Fulton, J. L., 1980, Characterization of a Tymo-like virus common in Poinsettia, *Phytopathology* **70**:321.

Fulton, J. P., Scott, H. A., and Gamez, R., 1980, Beetles, in: *Vectors of Plant Pathogens* (K. F. Harris and K. Maramorosch, eds.), pp. 115–132, Academic Press, New York.

Garnier, M., Mamoun, R., and Bové, J. M., 1980, TYMV RNA replication in vivo. Replicative intermediate is mainly single stranded, *Virology* **104**:357.

Gibbs, A. J., 1978, Kennedya yellow mosaic virus, *CMI/AAB Descriptions of Plant Viruses* No. 193.

Gibbs, A., 1980a, A plant virus that partially protects its wild legume host against herbivores, *Intervirology* **13**:42.

Gibbs, A., 1980b, How ancient are the Tobamoviruses? *Intervirology* **14**:101.

Gibbs, A., and Harrison, B. D., 1969, Eggplant mosaic virus, and its relationships to Andean potato latent virus, *Ann. Appl. Biol.* **64**:225.

Gibbs, A. J., and Harrison, B. D., 1973, Eggplant mosaic virus, *CMI/AAB Descriptions of Plant Viruses* No. 124.

Gibbs, A., Hecht-Poinar, E., and Woods, R. D., 1966, Some properties of three related viruses: Andean potato latent, dulcamara mottle, and ononis yellow mosaic, *J. Gen. Microbiol.* **44**:177.

Gierer, A., and Schramm, G., 1956, Infectivity of ribonucleic acid from tobacco mosaic virus, *Nature* **177**:702.

Givord, L., 1978, Alternative hosts of okra mosaic virus near plantings of okra in southern Ivory Coast, *Plant Dis. Rep.* **62**:412.

Givord, L., 1979, La gamme d'hôtes du virus de la mosaïque du gombo (okra mosaic virus), *L'Agronomie Tropicale XXXIX* **1**:38.

Givord, L., 1980, Contribution à l'étude des tymovirus. Identification et propriétés d'un nouveau virus: Le virus de la mosaïque du Gombo. Essai de classification des tymovirus, Thesis, University of Strasbourg, France.

Givord, L., and Den Boer, L., 1980, Insect transmission of okra mosaic virus in the Ivory Coast, *Ann. Appl. Biol.* **94**:235.

Givord, L., and Hirth, L., 1973, Identification, purification and some properties of a mosaic virus of okra *(Hibiscus esculentus)*, *Ann. Appl. Biol.* **74**:359.

Givord, L., and Koenig, R., 1974, Okra mosaic virus, *CMI/AAB Descriptions of Plant Viruses* No. 128.

Givord, L., Pfeiffer, P., and Hirth, L., 1972, Un nouveau virus du groupe de la mosaïque jaune du navet: le virus de la mosaïque du gombo *(Hibiscus esculentus* L. Malvacée), *C. R. Acad. Sci. Paris* **D275**:1563.

Granett, A. L., 1973, Plantago mottle virus, a new member of the tymovirus group, *Phytopathology* **63**:1313.

Guilley, H., and Briand, J. P., 1978, Nucleotide sequence of turnip yellow mosaic virus coat protein messenger RNA, *Cell* **15**:113.

Guy, P., and Gibbs, A., 1981, A tymovirus of *Cardamine* sp. from Alpine Australia, *Aust. Plant Pathol.* **10**:12.

Guy, P., and Gibbs, A. J., 1985, Further studies on turnip yellow mosaic tymovirus isolates from an endemic Australian Cardamine, *Plant Pathol.* **34**:532.

Guy, P. L., Dale, J. L., Adena, M. A., and Gibbs, A. J., 1984, A taxonomic study of the host ranges of tymoviruses, *Plant Pathol.* **33**:337.

Haenni, A. L., and Chapeville, F., 1980, tRNA-like structures in viral RNA genomes, in: *Transfer RNA: Biological Aspects* (D. Soll, J. N. Abelson, and P. R. Schimmel, eds.), pp. 539–555, Cold Spring Harbor Laboratory, Cold Spring Harbor, N.Y.

Haenni, A. L., Bénicourt, C., Teixera, S., Prochiantz, A., and Chapeville, F., 1975, The specific role of tRNAs and tRNA-like structures in viral RNA genome replication and translation, *FEBS Proc. Meet.* **39**:121.

Haenni, A. L., Joshi, S., and Chapeville, F., 1982, tRNA-like structure in the genome of RNA viruses, *Prog. Nucleic Acid Res. Mol. Biol.* **27**:85.

Hall, T. C., 1979, Transfer RNA-like structures in viral genomes, *Int. Rev. Cytol.* **60**:1.

Hall, T. C., Shih, D. S., and Kaesberg, P., 1972, Enzyme-mediated binding of tyrosine to brome mosaic virus ribonucleic acid, *Biochem. J.* **129**:969.

Hanecak, B., Semler, B. L., Anderson, C. W., and Wimmer, E., 1982, Proteolytic processing of poliovirus polypeptide: Antibodies to polypeptide P3-7C inhibit cleavage at glutamine–glycine pairs, *Proc. Natl. Acad. Sci. USA* **79**:3973.

Harrison, S. C., 1983, Virus structure: High-resolution perspectives, *Adv. Virus Res.* **28**:175.

Harrison, S. C., 1985, First comparison of two animal viruses in their dimensions, *Nature* **317**:382.

Hatta, T., and Matthews, R. E. F., 1974, The sequence of early cytological changes in Chinese leaf cabbage following systemic infection with turnip yellow mosaic virus, *Virology* **59**:383.

Hein, A., 1959, Beitrage zur Kenntnis der Viruskrankheiten an Unkrautern. V. Ein Virus von *Scrophularia nodosa*, *Phytopath. Z.* **36**:290.

Hein, A., 1984, Ubertragung des TYMV durch Samen von *Camelina sativa* (Leindotter), *J. Plant Dis. Protect.* **91**:549.

Higgins, T. J. V., Whitfeld, P. R., and Matthews, R. E. F., 1978, Size distribution and *in vitro* translation of the RNAs isolated from turnip yellow mosaic virus, *Virology* **84**:153.

Hirth, L., Horn, P., and Strazielle, C., 1965, Structure et propriétés de l'acide ribonucléique extrait du virus de la mosaïque jaune du navet, *J. Mol. Biol.* **13**:735.

Hogle, J. M., Chow, M., and Filman, D. J., 1985, Three-dimensional structure of poliovirus at 2.9 Å resolution, *Science* **229**:1358.

Horvath, J., 1979, New artificial hosts and non-hosts of plant viruses and their role in the identification and separation of virus. XI. Tymovirus group: turnip yellow mosaic virus and belladonna mottle virus, *Acta Phytopathol. Acad. Sci. Hung.* **14**:297.

Horvath, J., Mamula, D., Juretic, N., and Besada, W. H., 1976, Natural occurrence of Belladonna mottle virus in Hungary, *Phytopath. Z.* **86**:193.

Hunter, T. R., Hunt, T., Knowland, J., and Zimmern, D., 1976, Messenger RNA for the coat protein of tobacco mosaic virus, *Nature* **260**:759.

Hunter, T., Jackson, R., and Zimmern, D., 1983, Multiple proteins and subgenomic RNAs may be derived from a single open reading frame on tobacco mosaic RNA, *Nucleic Acids Res.* **11**:801.

Igwegbe, E. C. K., 1983, New strain of okra mosaic virus in Nigeria, *Plant Dis.* **67**:320.

Jankulowa, M., Huth, W., Wittmann, H. G., and Paul, H. L., 1968, Untersuchungen über ein neues isometrisches Virus aus *Atropa belladonna* L. II. Serologische Reaktionen, Basenverhältnisse der RNS und Aminosäurenzusammensetzung des Proteins, *Phytopath. Z.* **63**:177.

Jeannel, R., 1949, Ordre des coléoptères, *Traité Zool.* **9**:892.

Johnson, M. W., 1964, The binding of metal ions by turnip yellow mosaic virus, *Virology* **24**:26.

Johnson, M. W., and Markham, R., 1962, Nature of the polyamine in plant viruses, *Virology* **17**:276.

Jonard, G., 1972, Contribution à l'étude des interactions RNA–protéine et protéine–protéine dans le virus de la mosaïque jaune du navet, Thesis, University of Strasbourg.

Jonard, G., Ralijaona, D., and Hirth, L., 1967, Action de l'urée sur le virus de la mosaïque jaune du navet. Propriétés du RNA obtenu lors de la formation de capsides artificielles, *C. R. Acad. Sci. Paris* **264**:2694.

Jonard, G., Witz, J., and Hirth, L., 1972, Formation of nucleoprotein complexes from dissociated turnip yellow mosaic virus RNA and capsids at low pH: Preliminary observations, *J. Mol. Biol.* **66**:165.

Jones, R. A. C., 1982, Tests for transmission of four potato viruses through potato true seed, *Ann. Appl. Biol.* **100**:315.

Jones, R. A. C., and Fribourg, C. E., 1977, Beetle, contact and potato true seed transmission of Andean potato latent virus, *Ann. Appl. Biol.* **86**:123.

Joshi, S., Haenni, A. L., Hubert, E., Huez, G., and Marbaix, G., 1978, *In vivo* aminoacylation and processing of turnip yellow mosaic virus RNA in *Xenopus leavis* oocytes, *Nature* **275**:339.

Joshi, S., Chapeville, F., and Haenni, A. L., 1982, Turnip yellow mosaic virus RNA is aminoacylated *in vivo* in Chinese cabbage leaves, *EMBO J.* **1**:935.

Joshi, R. L., Joshi, S., Chapeville, F., and Haenni, A. L., 1983, tRNA-like structures of plant viral RNAs: Conformational requirements for adenylation and aminoacylation, *EMBO J.* **2**:1123.

Joshi, R. L., Chapeville, F., and Haenni, A. L., 1985, Conformational requirements of tobacco mosaic virus RNA for aminoacylation and adenylation, *Nucleic Acids Res.* **13**:347.

Kaper, J. M., 1960a, Protein components of turnip yellow mosaic virus, *Nature* **186**:219.

Kaper, J. M., 1960b, Preparation and characterization of artificial top component from turnip yellow mosaic virus, *J. Mol. Biol.* **2**:425.

Kaper, J. M., 1969, Nucleic acid–protein interactions in turnip yellow mosaic virus, *Science* **166**:248.

Kaper, J. M., 1975, The chemical basis of virus structure, dissociation and reassembly, in: *Frontiers of Biology* (A. Neuberger and E. L. Tatum, eds.), North-Holland, Amsterdam.

Kaper, J. M., and Alting-Siberg, R., 1969a, Degradation of turnip yellow mosaic virus by freezing and thawing *in vitro*: A new method for studies on the internal organization of the viral components and for isolating native RNA, *Virology* **38**:407.

Kaper, J. M., and Alting-Siberg, R., 1969b, The effect of freezing on the structure of turnip yellow mosaic virus and a number of other simple plant viruses, *Cryobiology* **5**:366.

Katouzian-Safadi, M., and Berthet-Colominas, C., 1983, Evidence for the presence of a hole in the capsid of turnip yellow mosaic virus after RNA release by freezing and thawing. Decapsidation of turnip yellow mosaic virus *in vitro, Eur. J. Biochem.* **137**:47.

Katouzian-Safadi, M., Favre, A., and Haenni, A. L., 1980, Effect of freezing and thawing on the structure of turnip yellow mosaic virus, *Eur. J. Biochem.* **112**:479.

Keeling, J., and Matthews, R. E. F., 1982, Mechanism for release of RNA from turnip yellow mosaic virus at high pH, *Virology* **119**:214.

Kim, S. H., Sudath, P. L., Quigley, G. J., *et al.*, 1974, Tertiary structure of yeast phenyl RNA, *Science* **105**:435.

Klein, C., Fritsch, C., Briand, J. P., Richards, K., Jonard, G., and Hirth, L., 1976, Physical and functional heterogeneity in TYMV RNA: Evidence for the existence of an independent messenger coding for coat protein, *Nucleic Acids Res.* **3**:3043.

Klug, A., 1979, TMV reconstitution, *Harvey Lecture* **74**:141.

Klug, A., Finch, J. T., and Franklin, R. E., 1957, Structure of turnip yellow mosaic virus, *Nature* **179**:683.

Klug, A., Longley, W., and Leberman, R., 1966, Arrangement of protein subunits and the distribution of nucleic acid in turnip yellow mosaic virus, *J. Mol. Biol.* **15**:315.

Koenig, R., 1976, A loop-structure in the serological classification system of tymoviruses, *Virology* **72**:1.

Koenig, R., and Givord, L., 1974, Serological interrelationships in the turnip yellow mosaic virus group, *Virology* **58**:119.

Koenig, R., and Lesemann, D. E., 1979, Tymovirus group, *CMI/AAB Descriptions of Plant Viruses* No. 214.

Koenig, R., and Lesemann, D. E., 1980, Two isometric viruses in Poïnsettias, *Plant Dis.* **64**:782.

Koenig, R., and Lesemann, D. E., 1981, Tymoviruses, in: *Comparative Diagnosis of Viral Diseases*, Volume IV (E. Kurstak, ed.), pp. 33–60, Elsevier/North-Holland Biomedical Press, Amsterdam.

Koenig, R., Tremaine, J. H., and Shepard, J. F., 1978, *In situ* degradation of the protein chain of potato virus X at the N- and C-termini, *J. Gen. Virol.* **38**:329.

Koenig, R., Fribourg, C. E., and Jones, R. A. C., 1979, Symptomatological, serological, and electrophoretic diversity of isolates of Andean potato latent virus from different regions of the Andes, *Phytopathology* **69**:748.

Kohl, R. J., and Hall, T. C., 1974, Aminoacylation of RNA from several viruses: amino acid specificity and differential activity of plant, yeast and bacterial synthetases, *J. Gen. Virol.* **25**:257.

Kozak, M., 1978, How do eucaryotic ribosomes select initiation regions in messenger RNA? *Cell* **15**:1109.

Kurtz-Fritsch, C., and Hirth, L., 1972, Uncoating of two spherical plant viruses, *Virology* **47**:385.

Kuszala, M., Ziegler, V., Richards, K., Putz, C., Guilley, H., and Jonard, G., 1986, Beet necrotic yellow vein virus: Different isolates are serologically similar but differ in RNA composition, *Ann. Appl. Biol.* **109**:155.

Laflèche, D., and Bové, J. M., 1969, Development of double membrane vesicles in chloroplasts from turnip yellow mosaic virus infected cells, *Prog. Photosynth. Res.* **1**:74.

Laflèche, D., and Bové, J. M., 1971, Virus de la mosaïque jaune du navet: Site cellulaire de la réplication du RNA viral, *Physiol. Vég.* **9**:487.

Laflèche, D., Bové, C., Dupont, G., Mouchès, C., Astier, T., Garnier, M., and Bové, J. M., 1972, Site of viral RNA replication in the cells of higher plants: TYMV-RNA synthesis on the chloroplast outer membrane system, in: *Proceedings of the 8th FEBS Meeting, Amsterdam; RNA Viruses/Ribosomes*, Volume 27, pp. 43–71, North-Holland Publishing, Amsterdam.

Lana, A. F., 1980, Properties of a virus occurring in *Arachis hypogea* in Nigeria, *Phytopathology Z.* **97**:169.

Lana, A. O., and Bozarth, R. F., 1975, Studies on a virus induced mosaic disease of *Abelmoschus esculentus* in Nigeria, *Phytopath. Z.* **83**:77.

Lana, A. O., and Taylor, T. A., 1976, The insect transmission of an isolate of okra mosaic virus occurring in Nigeria, *Ann. Appl. Biol.* **82**:361.

Lana, A. F., Egunjobi, O. A., and Esuruoso, O. F., 1978, Studies of the soil transmission of the Nigerian okra mosaic virus, *Acta Phytopath. Acad. Scient. Hung.* **13**:307.

Leberman, R., 1966, The isolation of plant viruses by means of "simple" coacervates, *Virology* **30**:341.

Lebeurier, G., Nicolaieff, A., and Richards, K. E., 1977, Inside-out model for self-assembly of tobacco mosaic virus, *Proc. Natl. Acad. Sci. USA* **74**:149.

Lee, R. F., Niblett, C. L., Hubbard, J. D., and Johnson, L. B., 1979, Characterization of belladonna mottle virus isolates from Kansas and Iowa, *Phytopathology* **69**:985.

Legrand, M., 1983, Phenylpropanoid metabolism and its regulation in disease, in: *Biochemical Plant Pathology* (J. A. Callow, ed.), pp. 367–384, John Wiley and Sons, New York.

Lesemann, D. E., 1977, Virus group-specific and virus-specific cytological alterations induced by the members of the tymovirus group, *Phytopath. Z.* **90**:315.

Lesemann, D. E., Koenig, R., Huth, W., Brunt, A. A., Phillips, S., and Barton, R. J., 1983, Poïnsettia mosaic virus—a tymovirus? *Phytopath. Z.* **107**:250.

Lindley, I. J. D. and Stebbing, N., 1977, Aminoacylation of encephalomyocarditis virus RNA, *J. Gen. Virol.* **34**:177.

Longley, W., and Leberman, R., 1966, Structural similarity of artificial and natural top components of turnip yellow mosaic virus, *J. Mol. Biol.* **19**:223.

Mamula, D., 1976, New experimental hosts of Belladonna mottle virus (BMV), *Acta Bot. Croat.* **35**:41.

Markham, R., 1951, Physicochemical studies of the turnip yellow mosaic virus, *Discuss. Faraday Soc.* **11**:211.

Markham, R., 1959, *The Viruses*, Volume 2 (Burnett, F. M., and Stanley, W. M., eds.), Academic Press, New York.

Markham, R., and Smith, K. M., 1949, Studies on the virus of turnip yellow mosaic, *Parasitology* **39**:330.

Martini, C., 1958, The transmission of turnip viruses by biting insects and aphids, in: *Proceedings of the Third Conference on Potato Virus Diseases*, Wageningen-Lisse, 1958, pp. 106–113.

Matthews, R. E. F., 1955, Infectivity of turnip yellow mosaic virus containing 8-azaguanine, *Virology* **1**:165.

Matthews, R. E. F., 1960, Properties of nucleoprotein fractions isolated from turnip yellow mosaic virus preparations, *Virology* **12**:521.

Matthews, R. E. F., 1966, Reconstitution of turnip yellow mosaic virus RNA with TMV protein subunits, *Virology* **30**:82.

Matthews, R. E. F., 1970, Turnip yellow mosaic virus, *CMI/AAB Descriptions of Plant Viruses* No. 2.

Matthews, R. E. F., 1973, Induction of disease by viruses, with special reference to turnip yellow mosaic virus, *Annu. Rev. Phytopath.* **11**:147.

Matthews, R. E. F., 1974, Some properties of TYMV nucleoproteins isolated in cesium chloride density gradients, *Virology* **60**:54.

Matthews, R. E. F., 1981, *Plant Virology*, 2nd edition, Academic Press, New York.

Matthews, R. E. F., and Witz, J., 1985, Uncoating of TYMV *in vivo*, *Virology* **144**:318.

Mellema, J. R., Bénicourt, C., Haenni, A. L., Noort, A., Pleij, C. W. A., and Bosch, L., 1979, Translational studies with turnip yellow mosaic virus RNAs isolated from major and minor virus particles, *Virology* **96**:38.

Meshi, T., Ohno, T., Iba, H., and Okada, Y., 1981, Nucleotide sequence of a cloned cDNA copy of TMV (Cowpea strain) RNA, including the assembly origin, the coat protein cistron, and the 3' non-coding region, *Mol. Gen. Genet.* **184**:20.

Mills, D. R., Page, N. R., and Spiegelman, S., 1966, The *in vitro* synthesis of a noninfectious complex containing biologically active viral RNA, *Proc. Natl. Acad. Sci. USA* **56**:1778.

Milne, K. S., Grogan, R. G., and Kimble, K. A., 1969, Identification of viruses infecting cucurbits in California, *Phytopathology* **59**:819.

Moline, H. E., and Fries, R. E., 1974, A strain of belladonna mottle virus isolated from *Physalis heterophylla* in Iowa, *Phytopathology* **64**:44.

Montagnier, L., and Sanders, F. K., 1963, Replicative form of encephalomyocarditis virus ribonucleic acid, *Nature* **199**:664.

Morch, M. D., and Bénicourt, C., 1980, Polyamines stimulate suppression of amber termination codons *in vitro* by normal tRNAs, *Eur. J. Biochem.* **105**:445.

Morch, M. D., Zagorski, W., and Haenni, A. L., 1982, Proteolytic maturation of the turnip yellow mosaic virus polyprotein coded *in vitro* occurs by internal catalysis, *Eur. J. Biochem.* **127**:259.

Mouchès, C., 1984, Purification et propriétés de la RNA réplicase du virus de la mosaïque jaune du navet: Une sous-unité d'origine virale, une sous-unité d'origine cellulaire, Thesis, University of Bordeaux.

Mouchès, C., Candresse, T., and Bové, J. M., 1984, Turnip yellow mosaic virus RNA-replicase contains host and virus-encoded subunits, *Virology* **134**:78.

Nickerson, K. W., and Lane, L. C., 1977, Polyamine content of several RNA plant viruses, *Virology* **81**:455.

Nirenberg, M. W., and Matthaei, J. H., 1961, The dependence of cell-free protein synthesis in *E. coli* upon naturally occurring or synthetic polyribonucleotides, *Proc. Natl. Acad. Sci. USA* **47**:1588.

Noort, A., Van den Dries, C. L. A. M., Pleij, C. W. A., Jaspars, E. M. J., and Bosch, L., 1982, Properties of turnip yellow mosaic virus in cesium chloride solutions: The formation of high-density components, *Virology* **120**:412.

Oberg, B., and Philipson, L., 1972, Binding of histidine to tobacco mosaic virus RNA, *Biochem. Biophys. Res. Commun.* **48**:927.

Paul, H. L., 1971, Belladonna mottle virus, *CMI/AAB Descriptions of Plant Viruses* No. 52.

Paul, H. L., Gibbs, A., and Wittmann-Liebold, B., 1980, The relationships of certain tymoviruses assessed from the amino acid composition of their coat proteins, *Intervirology* **13**:99.

Pelham, H. R. B., and Jackson, R. J., 1976, An efficient mRNA-dependent translation system from reticulocyte lysates, *Eur. J. Biochem.* **67**:247.

Peng, X. X., and Shih, D. S., 1984, Proteolytic processing of the proteins translated from the bottom component RNA of cowpea mosaic virus, *J. Biol. Chem.* **259**:3197.

Peter, R., Stehelin, D., Reinbolt, J., Collot, D., and Duranton,, H., 1972, Primary structure of turnip yellow mosaic virus coat protein, *Virology* **49**:615.

Peters, D., and Derks, A. F. L. M., 1974, Host range and some properties of Physalis mosaic virus, a new virus of the turnip yellow mosaic virus group, *Neth. J. Pl. Pathol.* **80**:124.

Pfannenstiel, M. A., Mintz, K. P., and Fulton, R. W. 1982, Evaluation of heat therapy of Poïnsettia mosaic and characterization of the viral component, *Phytopathology* **72**:252.

Pinck, M., Yot, P., Chapeville, F., and Duranton, H., 1970, Enzymatic binding of valine to the 3' end of TYMV RNA, *Nature* **226**:954.

Pinck, M., Fritsch, C., Ravelonandro, M., Thivent, C., and Pinck, L., 1981, Binding of ribosomes to the 5' leader sequence N = 258 of RNA 3 from alfalfa mosaic virus, *Nucleic Acids Res.* **9**:1087.

Pleij, C. W. A., 1973, A study of TYMV-RNA. Fragmentation, aggregation, deaggregation, Ph.D. Thesis, University of Leiden.

Pleij, C. W. A., Neeleman, A., Van Vloten-Doting, L., and Bosch, L., 1976, Translation of turnip yellow mosaic virus RNA *in vitro*. A closed and an open coat protein cistron, *Proc. Natl. Acad. Sci. USA* **73**:4437.

Pleij, C. W. A., Mellema, J. R., Noort, A., and Bosch, L., 1977, The occurrence of the coat protein messenger RNA in the minor components of turnip yellow mosaic virus, *FEBS Lett.* **80**:19.

Pratt, D., Briand, J. P., and Van Regenmortel, M. H. V., 1980, Immunochemical studies of turnip yellow mosaic virus. I. Localization of four antigenic regions in the protein subunit, *Molec. Immunol.* **17**:1167.

Preil, W., Koenig, R., Engelhardt, M., and Meier-Dinkel, A., 1982, Eliminierung von Poïn-settia mosaic virus (PoïMV) and Poïnsettia cryptic virus (PoïCV) aus *Euphorbia pul-cherrima* Willd. durch Zellsuspensions Kultur, *Phytopath. Z.* **105**:193.

Proeseler, G., and Schmelzer, K., 1977, Laboratory and field tests about the trans-mission of Erysinum latent virus by *Phyllotreta* species (chrysomelidae), *Zbl. Bakt. II* **132**:716.

Provvidenti, R., 1979, Inheritance of resistance to plantago mottle virus in *Pisum sativum* L., *J. Heredity* **70**:350.

Provvidenti, R., and Granett, A. L., 1976, Occurrence of Plantago mottle virus in Pea, *Pisum sativum*, in New York State, *Ann. Appl. Biol.* **82**:85.

Quesniaux, V., Briand, J. P., and Van Regenmortel, M. H. V., 1983a, Immunochemical studies of turnip yellow mosaic virus. II. Localization of a viral epitope in the N-terminal residues of the coat protein, *Molec. Immunol.* **20**:179.

Quesniaux, V., Jaeglé, M., and Van Regenmortel, M. H. V., 1983b, Immunochemical studies of turnip yellow mosaic virus. III. Localization of two viral epitopes in residues 57-64 and 183-189 of the coat protein, *Biochim. Biophys. Acta* **743**:226.

Rayment, I., Baker, T. S., Caspar, D. L. D., and Murakami, W. T., 1982, Polyoma virus capsid structure at 22.5 Å resolution, *Nature* **295**:110.

Re, G. C., and Kaper, J. M., 1975, Chemical accessibility of tyrosyl and lysyl residues in turnip yellow mosaic capsids, *Biochemistry* **14**:4492.

Renaudin, J., Bové, J. M., Otsuki, Y., and Takebe, I., 1975a, Infection of Brassica leaf pro-toplasts by turnip yellow mosaic virus, *Mol. Gen. Genet.* **141**:59.

Renaudin, J., Saillard, C., Gandar, J., Mattern, P., and Bové, J. M., 1975b, Infection *in vitro* des protoplastes de chou de Chine par le virus de la mosaïque jaune du navet, *C. R. Acad. Sci. Paris* **t280**:2551.

Rietveld, K., Van Poelgeest, R., Pleij, C. W. A., Van Bloom, J. H., and Bosch, L., 1982, The tRNA-like structure at the 3' terminus of turnip yellow mosaic virus RNA. Differences and similarities with canonical tRNA, *Nucleic Acids Res.* **10**:1929.

Rietveld, K., Pleij, C. W. A., and Bosch, L., 1983, Three-dimensional models of the tRNA-like 3' termini of some plant viral RNAs, *EMBO J.* **2**:1079.

Rietveld, K., Linschooten, K., Pleij, C. W. A., and Bosch, L., 1984, The three-dimensional folding of the tRNA-like structure of tobacco mosaic virus RNA. A new building principle applied twice, *EMBO J.* **3**:2613.

Romaine, C. P., and Zaitlin, M., 1978, RNA-dependent RNA polymerases in uninfected and tobacco mosaic virus-infected tobacco leaves: viral-induced stimulation of a host polymerase activity, *Virology* **86**:241.

Rossmann, M. G., Arnold, E., Erickson, J. W., Frankenberger, E. A., Griffith, J. P., Hecht, H. J., Johnson, J. E., Kamer, G., Luo, M., Mosser, A. G., Rueckert, R. R., Sherry, B., and Vriend, G., 1985, Structure of a human common cold virus and functional relationship to other picornaviruses, *Nature* **317**:145.

Salomon, R., and Littauer, U. Z., 1974, Enzymatic acylation of histidine to mengovirus RNA, *Nature* **249**:32.

Schmelzer, K., 1976, Virus infestation at garden radish, *Zbl. Bakt. Abt. II. Bd.* **131**:703.

Scott, H. A., 1976, Desmodium yellow mottle virus, *CMI/AAB Descriptions of Plant Viruses* No. 168.

Shukla, D. D., and Gough, K. H., 1980, Erysimum latent virus, *CMI/AAB Descriptions of Plant viruses* No. 222.

Shukla, D. D., and Schmelzer, K., 1974, Studies on viruses and virus diseases of cruciferous plants. XVII. Serologically distinct strains of turnip yellow mosaic virus occurring in the GDR, *Acta Phytopath. Acad. Scient. Hung.* **9**:237.

Shukla, D. D., and Schmelzer, K., 1975, Studies on viruses and virus diseases of cruciferous plants. XIX. Analysis of the results obtained with ornamental and wild species, *Acta Phytopath. Acad. Scient. Hung.* **10**:217.

Shukla, D. D., Proeseler, G., and Schmelzer, K., 1975, Studies on viruses and virus diseases of cruciferous plants. XVIII. Beetle transmission and some new natural hosts of Ery-sinum latent virus, *Acta Phytopath. Acad. Scient. Hung.* **10**:211.

Siegel, A., Hari, V., Montgomery, I., and Kolacz, K., 1976, A messenger RNA for capsid protein isolated from tobacco mosaic virus-infected tissue, *Virology* **73**:363.

Silberklang, M., Prochiantz, A., Haenni, A. L., and RajBhandary, U. L., 1977, Studies on the sequence of the 3′ terminal region of turnip yellow mosaic virus RNA, *Eur. J. Biochem.* **72**:465.

Stefanac, Z., 1974, Belladonna mottle virus in Jugoslavia, *Acta Bot. Croat.* **33**:17.

Stols, A. L. H., 1965, Interaction of turnip yellow mosaic virus with quaternary ammonium salts, *Virology* **25**:508.

Strazielle, C., Benoit, H., and Hirth, L., 1965, Particularités structurales de l'acide ribonucléique extrait du virus de la mosaïque jaune du navet. II, *J. Mol. Biol.* **13**:735.

Stussi-Garaud, C., Lemius, J., and Fraenkel-Conrat, H., 1977, RNA-polymerase from tobacco necrosis virus-infected and uninfected tobacco. II. Properties of the bound and soluble polymerases and the nature of their products, *Virology* **81**:224.

Szczesna, E., and Filipowicz, W., 1980, Faithful and efficient translation of viral and cellular eukaryotic mRNAs in a cell-free S-27 extract of *Saccharomyces cerevisiae*, *Biochem. Biophys. Res. Commun.* **92**:563.

Szybiak, U., Bouley, J. P., and Fritsch, C., 1978, Evidence for the existence of a coat protein messenger RNA associated with the top component of each of three tymoviruses, *Nucleic Acids Res.* **5**:1821.

Ushiyama, R., and Matthews, R. E. F., 1970, The significance of chloroplast abnormalities associated with infection by turnip yellow mosaic virus, *Virology* **42**:293.

Van Belkum, A., Bingkun, J., Rietfeld, K., Pleij, C. W. A., and Bosch, L., 1987, Structural similarities among valine accepting tRNA-like structures in tymoviral RNAs and elongator tRNAs, *Biochemistry* **26**:1144.

Van Regenmortel, M. H. V., 1972, Wild cucumber mosaic virus, *CMI/AAB Descriptions of Plant Viruses* No. 105.

Van Regenmortel, M. H. V., 1982, List of virus abbreviations, in: *Serology and Immunochemistry of Plant Viruses* (M. H. V. Van Regenmortel, ed.), Academic Press, New York, p. xi.

Van Regenmortel, M. H. V., and von Wechmar, M. B., 1970, A reexamination of the serological relationship between tobacco mosaic virus and cucumber mosaic virus, *Virology* **41**:330.

Virudachalam, R., Sitaraman, K., Heuss, K. L., Markley, J. L., and Argos, P., 1983a, Evidence for pH-induced release of RNA from belladonna mottle virus and the stabilizing effect of polyamines and cations, *Virology* **130**:351.

Virudachalam, R., Sitaraman, K., Heuss, K. L., Argos, P., and Markley, J. L., 1983b, Carbon-13 and proton nuclear magnetic resonance spectroscopy of plant viruses: Evidence for protein–nucleic acid interactions in belladonna mottle virus and detection of polyamines in turnip yellow mosaic virus, *Virology* **130**:360.

Waterworth, H. E., Kaper, J. M., and Koenig, R., 1975, Purification and properties of a tymovirus from Abelia, *Phytopathology* **65**:891.

Weidemann, H. L., 1973, Zur Ubertragung des Scrophularia Mottle Virus, *Phytopath. Z.* **78**:278.

Weidemann, H. L., and Bode, O., 1973, Untersuchungen über ein neues isometrisches Virus aus Atropa belladonna L. IV. Versuche zur Ubertragberkeit des Belladonna mottle virus, *Phytopath. Z.* **76**:6.

Weissmann, C., 1965, Replication of viral RNA VII. Further studies on the enzymatic replication of MS 2 RNA, *Proc. Natl. Acad. Sci. USA* **54**:202.

Weissmann, C., and Feix, G., 1966, Replication of viral RNA. XI. Synthesis of viral "minus" strands *in vitro*, *Proc. Natl. Acad. Sci. USA* **55**:1264.

Wiegers, K. J., Gschwender, H. H., and Drzeniek, R., 1978, The effect of cesium salts on dense poliovirus particles, *Intervirology* **10**:329.

Wood, H. A., 1971, Buoyant density changes of cowpea mosaic virus components at different pH values, *Virology* **43**:511.

Xiong, C., Lebeurier, G., and Hirth, K., 1984, Detection *in vivo* of a new gene product (gene III) of cauliflower mosaic virus, *Proc. Natl. Acad. Sci. USA* **81**:6608.

Yot, P., Pinck, M., Haenni, A. L., Duranton, H., and Chapeville, F., 1970, Valine specific
    tRNA-like structure in turnip yellow mosaic virus RNA, *Proc. Natl. Acad. Sci. USA*
    67:1345.
Zagorski, W., Morch, M. D., and Haenni, A. L., 1983, Comparison of three different cell-
    free systems for turnip yellow mosaic virus RNA translation, *Biochimie* **65:**127.
Ziegler, V., Laquel, P., Guilley, H., Richards, K., and Jonard, G., 1985, Immunological
    detection of cauliflower mosaic virus gene V protein produced in engineered bacteria
    plants, *Gene* **36:**271.
Zimmern, D., 1982, Do viroids and RNA viruses derive from a system that exchanges
    genetic information between eucaryotic cells? *Trends Biochem. Sci.* **7:**205.

CHAPTER 7

Maize Rayado Fino and Related Viruses

R Gámez and P. León

I. INTRODUCTION

Rayado fino (Spanish for fine stippling), from which maize rayado fino virus (MRFV) derives its name, describes the characteristic fine dots and small chlorotic stripes associated with the disease induced in maize (*Zea mays* ssp. *mays* L.) by this virus (Gámez, 1969, 1980a,b). MRFV (Gámez, 1980a) is the type member of the maize rayado fino virus group, the establishment of which was approved by the International Committee on Taxonomy of Viruses in 1985 (R. I. Hamilton, personal communication). Other viruses included as possible members of the group are the serologically related oat blue dwarf virus (OBDV) (Banttari and Zeyen, 1973) and the recently described Bermuda grass etched-line virus (BELV) (Lockhart *et al.*, 1984, 1985).

MRFV was accidentally discovered in Costa Rica in field-collected maize leafhoppers of the species *Dalbulus maidis DeLong and Wolcott* (Gámez, 1969). This leafhopper had been long recognized as a vector of the corn (maize) stunt pathogen (Kunkel, 1946; Maramorosch, 1955). The etiology of corn stunt was poorly understood, and for a long time it was thought to be caused by a virus. Several different symptoms were associated with the disease. Two strains of the "virus," the Rio Grande and the Mesa Central, were first reported in southern United States and Mexico (Maramorosch, 1955). Rayado fino was later described in El Salvador as a third "strain" of corn stunt (Ancalmo and Davis, 1961), but its viral

R. GÁMEZ and P. LEÓN • Cellular and Molecular Biology, Research Center, University of Costa Rica, Costa Rica.

213

nature was not established until its discovery in Costa Rica (Gámez, 1969, 1973; Gámez et al., 1977). The Rio Grande and Mesa Central strains of corn stunt are now recognized as two different mollicutes, the corn stunt spiroplasma (CSS) (Davis and Worley, 1973) and the maize bushy stunt mycoplasma (MBSM), respectively (Bascope, 1977; Bradfute et al., 1977; Nault, 1980). Corn stunt is presently considered as a syndrome induced by a complex of leafhopper-borne viruses and mollicutes (Nault and Brad-fute, 1979).

OBDV was identified in 1962 as a virus transmitted by the aster leafhopper *Macrosteles fascifrons* Stal. (Banttari and Moore, 1962), al-though blue dwarf of oats (*Avena sativa* L.) was known in North America since 1952 (Moore, 1952; Goto and Moore, 1952). The aster leafhopper, the only recognized vector of OBDV, has long been considered the main vector of the aster yellows mycoplasma (Nielson, 1968).

BELV was first reported from Morocco in 1984, where it was found in naturally infected Bermuda grass, *Cynodon dactylon* (L.) Pers., and Johnson grass, *Sorghum halepense* L. It is transmitted by the leafhopper *Aconurella prolixa* Lethierry (Lockhart et al., 1984, 1985).

MRFV and OBDV are small ssRNA viruses that share a number of physical and biochemical characteristics. An important biological simi-larity is the ability of both viruses to multiply in their cicadellid leafhop-per vectors. These and other properties led Black (1979), and León and Gámez (1981), to propose that these viruses represent a new group. BELV is also transmitted by a cicadellid leafhopper, but its ability to multiply in its insect vector has not been yet investigated. The available infor-mation on this virus indicates that it shares with MRFV and OBDV several characteristics of the virion. In addition, the demonstration of serological relationships among these three viruses (Lockhart et al., 1984, 1985; E. Moreno, L. Taylor, C. Rivera, and R. Gámez, unpublished data), strongly supports their inclusion in the same taxonomical group.

II. GEOGRAPHICAL DISTRIBUTION, HOST RANGE, AND DISEASES

A. Geographical Distribution

The geographical distribution of MRFV is Neotropical, extending from the southern United States to northern Argentina and Uruguay (Gámez, 1969; Gámez et al., 1979; Kitajima et al., 1976; Lastra and de Uscategui, 1980; Martínez-López et al., 1974; Nault et al., 1980; Rocha-Peña, 1981). In the neotropics, MRFV and *D. maidis* are adapted to a wide variety of ecological conditions. Their successful adaptation to dissimilar environments in such a wide geographical and ecological range has been extensively analyzed elsewhere (Gámez, 1983a,b; Gámez and León, 1983,

1985; Gámez *et al.*, 1979). The distribution of MRFV appears to be circumscribed by that of its insect vector (Gámez and León, 1983, 1985). Several strains of MRFV have been characterized from Costa Rica (MRFV-CR), Colombia (MRFV-CO), Brazil (MRFV-BRA), Mexico (MRFV-MEX), Venezuela (MRFV-VEN), Florida (MRFV-FLA), and Texas (MRFV-TEX) (Falk and Tsai, 1986; Gámez, 1980a, Gingery *et al.*, 1982, Carballo and Lastra, 1987; E. Moreno, L. Taylor, C. Rivera, and R. Gámez, unpublished data).

OBDV has been reported to occur in the North American Great Plains, from Kansas through Minnesota and North Dakota in the United States to Manitoba in Canada (Banttari and Moore, 1962; Banttari and Zeyen, 1973; Creelman, 1965; Goto and Moore, 1952). Its presence in winter oats in Arkansas was recently reported (Timian, 1985). OBDV appears to be more restricted than MRFV in both its geographical and ecological distribution.

BELV is only known to occur in widely separated locations in northwestern and north-central Morocco (Lockhart *et al.*, 1985). It is possible that the virus and its insect vector exist in other ecologically similar areas of North Africa.

B. Host Range and Symptomatology

The host range of MRFV, as determined by several investigators, is summarized in a taxonomical analysis by Gámez (1983a) and Gámez and León (1985). These studies describe the reactions of 54 species and three subspecies in 30 genera that included the major Gramineae assemblages. Only Andropogonoid species within the genera *Zea*, *Tripsacum*, and *Rottboellia* are susceptible to the virus. *Zea* and *Tripsacum* are neotropical in origin; *Zea* includes most of the host plants: the widely cultivated maize, *Z. mays* ssp. *mays*, and the wild annual and perennial teosintes. The teosintes have a limited distribution in central and southern Mexico and northern Guatemala and Honduras. *T. australe* Cutter and Anderson, a South American relative of *Zea*, and *R. exaltata* L., an Asian weed, are the only other species known to be susceptible to MRFV (Gámez and León, 1983; 1985; Nault *et al.*, 1980).

D. maidis also has a very narrow host range, and is restricted mainly to the genus *Zea*, with maize as its principal host. The host range of MRFV is circumscribed by that of *D. maidis*. Taxonomic analysis indicates that both virus and vector have followed the evolutionary route of the genus *Zea*. Susceptibility to the virus and suitability to the insect vector appear to be phylogenetic traits within *Zea* that tend to disappear with taxonomic distance from *Z. mays* ssp. *mays* (Gámez and León, 1983, 1985; Nault, 1985).

Symptomatology of rayado fino disease, as observed by different investigators, has been summarized by Gámez (1980b). Disease symptoms appear in most cultivars 8–14 days after inoculation. Distinct small chlorotic dots develop at the base and along the veins of young leaves. These dots become increasingly numerous in subsequent new leaves and some of them coalesce to form short stripes. These symptoms are usually more severe in the younger basal half of the foliar lamina. A few long continuous stripes may be observed in tolerant genotypes, but numerous long stripes and severe chlorosis develop in the more susceptible maize cultivars. The formation of holes in the leaf blade due to the collapse of the cells within the chlorotic dots occurs in highly susceptible cultivars, as well as wilting and death of the whole plant. In tolerant varieties, disease symptoms are more conspicuous in leaves formed three to four weeks after inoculation, but become gradually milder and eventually absent in leaves that develop when the plants approach maturity. Variable degrees of stunting may be observed according to plant genotype and time of infection. However, leaf reddening symptoms, commonly associated with the diseases caused by the maize stunting mollicutes and transmitted by *D. maidis*, are never produced by MRFV. Other symptoms include reduction in size of the root system and the ear, which may bear fewer or no grains in the more susceptible varieties. In general, disease symptoms are more severe in plants infected at the early stages of development (Gámez, 1969, 1973, 1980b; Martínez-López, 1977; Martínez-López *et al.*, 1974; Rocha-Peña, 1981; Toler *et al.*, 1985).

An extensive study of the experimental host range of OBDV included 67 species in 14 different and phylogenetically distant families of plants. Eighteen species representing seven families of monocotyledonous and dicotyledonous plants are susceptible to the virus. Development of enations is the most consistent and characteristic symptom of the disease in nine of these species, whereas in nine others the infections are symptomless. The host range of OBDV is narrower than that of its leafhopper vector (Bantarri and Zeyen, 1973; Westdal, 1968).

OBDV naturally infects oats, barley (*Hordeum vulgare* L.), and flax (*Linum usitatissimum* L.). The reaction of these and other species to the virus has been documented by several investigators (Banttari and Moore, 1962; Banttari and Zeyen, 1971; Westdal, 1968). Disease symptoms generally appear approximately two to three weeks after inoculation. The first evidence of infection in oats is a slight stunting and a general blue-green appearance of the plant. Leaves grow shorter and stiffer and project at abnormally wider angles, giving the infected plant a sturdy appearance. Varieties differ in the degree of stunting. More tillers are formed, frequently appearing at nodes above the crown. Spikes are shorter and smaller on plants that tassel, and these produce many sterile florets and few seeds. In the field, infected plants remain green longer than normal. Associated with these symptoms is a thickening of the lateral veins of leaves, followed by the formation of ridgelike enations of the abaxial surface of leaves and sometimes on the leaf sheath. Enations appearing in older leaves may be the only symptom in some infected plants. Recovery from the disease has been observed under greenhouse conditions (Westdal, 1968).

Symptoms in infected barley plants are similar to those induced in oats. Enations and the typical blue-green colorations of leaves are not as prominent in barley, which does not recover from the disease (Banttari and Zeyen, 1973; Westdal, 1968).

Disease symptoms in flax are mainly characterized by crinkle of leaves, resulting from the swelling of lateral veins on the margins, small indentations along the adaxial surfaces, and enations on the abaxial surfaces of the leaves. Stunting and reduced tillering and seed set are also characteristic symptoms in flax (Banttari and Zeyen, 1971; Westdal, 1968). Simultaneous infection by OBDV and aster yellows mycoplasma in flax results in more severe symptoms than those induced by single infections of either pathogen. Dually infected plants are severely stunted and show general chlorosis. The stem apex is swollen, deformed, and chlorotic and the leaves show veinal enations (Banttari and Zeyen, 1972, 1979).

BELV is only known to infect Bermuda grass and Johnson grass, several cultivars of maize, durum wheat (*Triticum durum* L.), and oats (Lockhart *et al.*, 1984, 1985). Symptoms on leaves of Bermuda grass and Johnson grass consist of white etched lines and chlorotic spots, and plants are noticeably stunted. Infected maize cultivars are also stunted, and leaves show narrow broken lines and short streaks identical to symptoms induced by MRFV. Durum wheat plants are also markedly stunted, and leaf veins are thickened and slightly deformed. No symptoms are observed in oats (Lockhart *et al.*, 1985).

III. TRANSMISSION, EPIDEMIOLOGY, AND CONTROL

A. Transmission

MRFV, OBDV, and BELV are all persistently transmitted by cicadellid leafhoppers, but not mechanically or through seeds of infected plants (Banttari and Zeyen, 1973; Gámez, 1980a; Lockhart *et al.*, 1985). Transmission of MRFV and OBDV by their leafhopper vector can occur simultaneously with transmission of mollicutes (Banttari and Zeyen, 1979; Castro-Caicedo, 1985; Gámez, 1973; Wolanski and Maramorosch, 1979).

The natural vectors of MRFV are Deltocephaline leafhoppers of the genus *Dalbulus*. Their biological characteristics and role as virus vectors have been extensively analyzed (Gámez, 1983a,b; Gámez and León, 1983, 1985; Nault, 1985). At present, *D. maidis* is recognized as the main vector of MRFV (Gámez, 1983a,b; Gámez and León, 1983, 1985; Gámez *et al.*, 1979). Other leafhoppers are experimental vectors, and include several species in the American genera *Dalbulus*, *Graminella*, *Baldulus*, and *Stirellus* (Nault *et al.*, 1980; Rocha-Peña, 1981; Wolanski and Maramorosch, 1979). Except for *D. maidis*, all other species of leafhoppers capable of transmitting the virus probably have limited importance as vectors due to their feeding preferences for hosts other than corn, their low rates of transmission, low population densities, and restricted distribution (Delgadillo-Sánchez, 1984; Gámez, 1983a,b; Gámez and León, 1983, 1985; Nault et al., 1980).

The characteristics of MRFV transmission by D. maidis have been summarized in detail (Gámez, 1983a,b; Gámez and León, 1983, 1985). Leafhoppers transmit the virus after long incubation periods of nine to 32 days. Transmission is intermittent and leafhoppers eventually lose the ability to transmit before they die. Only a small proportion (10–34%) of insects in a wild or experimental population are capable of transmitting the virus (Gámez, 1973; González and Gámez, 1974; Martínez-López et al., 1974; Nault et al., 1980), and yet nearly 80% of the insects are infected as demonstrated by enzyme-linked immunosorbent assay (ELISA) (Rivera and Gámez, 1986). The ability of D. maidis to transmit MRFV appears to be under genetic control as the number of transmitters can be rapidly increased by selective breeding. The transmission rate in these selected populations drops to normal levels after a few generations of outcrossing by random matings (Martínez-López, 1977; Nault et al., 1980; Paniagua and Gámez, 1976). Active transmitters could be recessive homozygotes for rare alleles that are rapidly diluted in the outcrosses (Gámez and León, 1983).

The existence of a propagative cycle of the virus in this leafhopper was originally indicated by the main features of the virus–vector relationship (Gámez, 1973), and later confirmed by direct evidence provided by ELISA and electron microscopy (Gingery et al., 1982; Rivera and Gámez, 1986; Rivera et al., 1981). Leafhoppers appear to vary in their response to MRFV infection as substantial differences in antigen concentration are observed in individual insects (Rivera and Gámez 1986). Virus replication occurs in the insect without any detectable cytopathological effect (Kitajima and Gámez, 1983; Rivera et al., 1981). These observations are congruent with the absence of deleterious effects of MRFV infection on the longevity and reproductive biology of D. maidis (Godoy, 1985; González and Gámez, 1974).

The characteristics of the transmission of OBDV by its insect vector (Banttari and Zeyen, 1970, 1973) are very similar to those described for MRFV. A limited number of individuals (25–30%) in a wild population are capable of transmitting the virus. This ability is under genetic control and may be substantially increased (up to 69%) by selective inbreeding for three generations (Timian and Alm, 1973). It is not known if the ability to transmit OBDV is restricted to the aster leafhopper, as no other species of leafhoppers has been tested as a virus vector.

OBDV was the first small ssRNA virus shown to multiply in its leafhopper vector. The dual ability of the virus to replicate in both its insect and plant hosts was indicated by the characteristics of the transmission described by Banttari and Zeyen (1970, 1973). Direct evidence of multiplication was provided by the serial passage of the virus through eight successive insect populations, and visualization of the viral particles in the insect's body (Banttari and Zeyen, 1970, 1976).

The characteristics of the transmission of BELV by *A. prolixa* have not been established (Lockhart *et al.*, 1985).

B. Epidemiology

The epidemiology of MRFV has been examined in areas of Costa Rica and Mexico where the virus is prevalent. Rayado fino disease appears to have the main characteristics of tropical endemic pathosystems. In Costa Rica there is an alternation between endemic and epidemic phases of the disease coinciding with the tropical rainy and dry seasons (Gámez, 1983b; Gámez and Saavedra, 1986). According to the proposed epidemiological model, MRFV overseasons in *D. maidis*. Immigrant leafhoppers that survive the dry season invade the first maize plantings during the early rainy season in April–May. A new insect generation emerges and becomes inoculative when plants near the tasseling stage and the virus spread in the field approaches its maximum. This new vector generation disperses in search of new hosts and invades neighboring fields that have overlapping crops, normally found during the rainy season. The immigrant leafhoppers are responsible for the spread of MRFV within fields. Infected plants are seldom distributed uniformly, and frequently show distinct aggregations in particular areas of the field, which suggests that leafhoppers feed on several adjacent plants. Total insect population, viruliferous vector population, and disease incidence build up steadily during the rainy season but drop abruptly with the onset of the dry season in November. Fields dry out rapidly and both virus and leafhopper remain endemic in the sporadic fields grown under irrigation during the dry season. Alternatively, insects could migrate to other regions or aestivate in the same area, particularly where no maize is grown during the prolonged dry season. No alternate hosts for either virus or vector are known in Latin America (Gámez, 1983a; Gámez and Saavedra, 1986). Although important environmental differences exist between the highlands of Costa Rica and central Mexico, the epidemiological model of MRFV appears to be similar in both areas (Delgadillo-Sánchez, 1984; Rodríguez and Palacios, 1987).

Observation on virus incidence and disease losses, carried out by different investigators in Brazil, Costa Rica, Colombia, Mexico, and Venezuela, have been recently summarized (Gámez, 1983a,b; Gámez and León, 1983, 1985; Gámez and Saavedra, 1986). Disease incidence may range from 0 to 100%: the high incidences have been observed in some areas of Mexico, El Salvador, and Colombia. Yield losses are variable and may reach 100% in some highly susceptible and newly developed or introduced cultivars.

The epidemiology of OBDV has not been comprehensively studied. Observations and reports by various authors indicate that, although generally present, the disease occurs at very low incidence throughout the north-central United States and southern Canada, ranging from a few individuals to a maximum of 5% infected plants, which appeared randomly distributed in oat fields during surveys in 1957–1959 (Banttari and Moore, 1962; Banttari and Zeyen, 1970; Timian, 1985; Westdal, 1986). The wide host range of this virus and the difficulties in the diagnosis of the disease in the field suggested that OBDV could be of importance in Manitoba (Westdal, 1968). However, recent epidemiological studies using sensitive serological techniques show that this virus is of no significance in North Dakota (Timian, 1985). Preliminary results indicate that OBDV incidence in North Dakota depends on migrating populations of viruliferous leafhoppers from overwintering hosts in the southern United States (Timian, 1985).

C. Control

There are no satisfactory control measures for MRFV (Gámez, 1980b), and no immune genotypes have been detected among representative races of maize and several hundred cultivars and accessions from Latin America and the United States (Espinoza and Gámez, 1980; Gámez, 1973, 1980b, 1983a; Martínez-López, 1977; Martínez-López et al., 1974; Nault et al., 1980; Rocha-Peña, 1981; Toler et al., 1985). Furthermore, no commercially acceptable level of resistance was found in North American germplasm (Toler et al., 1985). Locally adapted Central American or Mexican materials appear in general to be more tolerant to the virus and to show lower degrees of disease incidence within a population (Gámez, 1983a,b). It may be possible to increase tolerance to the virus in highly susceptible varieties by breeding. From the epidemiological model described above it may be predicted that crop rotation, mixed plantings, and selection of planting dates should result in significant decreases in virus infection as observed in Colombia (Martínez-López, 1977). The use of systematic insecticides did not reduce virus incidence (Martínez-López, 1977).

No information is available on the control of OBDV and BELV, which are presumably of no economic significance.

IV. PURIFICATION

A. Purification of the Viruses

MRFV has been purified from infected leaves by polyethylene glycol (PEG) precipitation cycles followed by sucrose gradient centrifugation (Gámez et al., 1977; León and Gámez, 1981), or by chloroform extraction and concentration by high speed centrifugation (Gingery et al., 1982). Extraction with organic solvents yields significantly higher quantities of viral capsids, but leads to an extensive degradation of the RNA. Two components, full particles and empty shells, are always present. Yields are around 20–40 µg of full particles per gram of fresh leaves. MRFV is stable in cesium salts, and in density and zonal gradients the two components are easily resolved (Gámez *et al.*, 1977, León and Gámez, 1981).

BELV has been purified by differential centrifugation after acid precipitation and separation in a linear sucrose density gradient (Lockhart *et al.*, 1985). Purification of OBDV was accomplished in a cellulose column by elution with PEG and NaCl-containing solutions as eluants, followed by sucrose density gradient centrifugation (Banttari and Zeyen, 1969).

B. Purification of the RNA

Undegraded MRFV-RNA can be obtained from virus purified through PEG precipitation cycles of leaf extracts followed by centrifugation onto a 40% sucrose cushion. The opaque wide band over the sucrose, which contains the virus particles, is reprecipitated with PEG, resuspended and layered over a 10–40% sucrose gradient. After centrifugation the bottom component is immediately extracted with phenol three times, and then several times with ether before ethanol precipitation (A. M. Espinoza, P. Ramírez, and P. León, unpublished data).

OBDV-RNA has been purified by dissociating the purified virus in bentonite at pH 9.0 followed by sucrose density gradient centrifugation (Pring *et al.*, 1973).

V. PARTICLE PROPERTIES

A. Physicochemical Properties

Properties of MRFV particles have been determined using zonal and isopycnic centrifugation of purified preparations (Table I). MRFV contains a positive ssRNA (Section VII) encapsidated within an isometric protein shell with icosahedral symmetry. Ultraviolet absorption and isopycnic properties (Table I) indicate that 25–30% of the MRFV mass is due to its 2×10^6 molecular weight RNA, so that the full particle should have an approximate mass of $6-7 \times 10^6$. The total mass of the top and bottom components, deduced from the molecular weight of the individual capsid proteins and RNA, is estimated to be 4.5 and 6.5×10^6, respectively. Sedimentation properties are consistent with these estimates (León and Gámez, 1981; Gingery *et al.*, 1982).

The physicochemical properties of several different strains of MRFV, as well as those of OBDV and BELV, appear in Table I. It should be noted

TABLE I. Physicochemical Properties of MRFV and Related Viruses

| Properties | MRFV[a] | | | MRFV-FLA[e] | OBDV[f] | BELV[g] |
	MRFV-CR[b]	MRFV-VEN[c]	MRFV-TEX[d]			
Number of sedimenting components	2	2	2	2	1	2
Density in CsCl						
Bottom comp.	1.46 g/ml	1.42 g/ml	1.43 g/ml	ND[h]	ND	ND
Top comp.	1.28 g/ml	1.30 g/ml	1.27 g/ml	ND	ND	ND
S value						
Bottom comp.	120	ND	125	ND	119	119
Top comp.	54	ND	47	ND	—	55
A_{260}/A_{280}						
Bottom comp.	1.58	1.35	1.72	ND	1.63	ND
Top comp.	0.87	0.83	0.58	ND	—	ND
Particle size (nm)	30–33	33	22–27	ND	28–30	28
RNA content (%)	33–36	33	25.5	ND	ND	ND
Mol. wt. RNA ($\times 10^{-6}$)	2.0	2.0	2.4	ND	2.1	ND
Mol. wt. proteins ($\times 10^{-3}$)						
Main band	22	22	22.4	22	ND	22[i]
Minor band	28	27	25.6	29	ND	26

[a] Strains of MRFV from Costa Rica (MRFV-CR), Venezuela (MRFV-VEN), Texas (MRFV-TEX), and Florida (MRFV-FLA).
[b] León and Gámez (1981).
[c] Carballo and Lastra (1987).
[d] Gingery et al. (1982).
[e] Falk and Tsai (1986).
[f] Banttari and Zeyen (1973)
[g] Lockhart et al. (1985).
[h] ND, property not determined.
[i] Lockhart et al. (1985) observed only one protein band in 7.5% acrylamide gels but two bands were observed by E. Moreno, L. Taylor, C. Rivera, and R. Gámez (unpublished data).

that the single component of OBDV is similar to the bottom component of MRFV.

B. Capsid Protein Studies

Two main capsid protein subunits, along with multimeric forms of these proteins, are detected by means of SDS–PAGE in both the top and bottom components in several isolates of MRFV. Various molecular weights have been reported for the capsid proteins (Table I) (Falk and Tsai, 1986; Gingery et al., 1982; P. León, E. Moreno, L. Taylor, and R. Gámez, unpublished data). We have estimated molecular weights of 22,000 (p22) and 28,000 (p28) for the two capsid proteins in denaturing acrylamide gels (SDS–PAGE). In two-dimensional gels (O'Farrell, 1975), two spots are detected with a clear anodal displacement of p28 with respect to p22 (P. León, E. Moreno, L. Taylor, and R. Gámez, unpublished data). The p22 and p28 fail to stain with periodic acid-Schiff or to bind lectins, which indicates that sugars are not covalently attached to these proteins (Ramírez et al., 1983). Mercaptoethanol and dithiotreitol produce no detectable changes in band mobility in SDS–PAGE (P. León, E. Moreno, L. Taylor, and R. Gámez, unpublished data). The two proteins are never found in equimolar amounts; rather, the p22 is always more abundant than p28 (Falk and Tsai, 1986; Gingery et al., 1982; León and Gámez, 1981; P. León, E. Moreno, L. Taylor, and R. Gámez, unpublished data). Densitometric analysis of our stained gels with freshly isolated bottom and top component revealed a 7 : 1 mass ratio for p22 : p28, which corresponds to a 9 : 1 molar ratio. In aged or heated preparations a substantial proportion of p22 hydrolyzes to generate smaller peptides, lowering the estimated ratio to the 3 : 1 values reported by Falk and Tsai (1986). Our analyses are based on the assumption that both proteins bind Coomassie blue in proportion to their masses (León and Gámez, 1986).

Electrophoretic and immunological studies of the two proteins have resolved some interesting features. First, it is clear that p28 does not generate p22 on protein reruns (Falk and Tsai, 1986; P. León, E. Moreno, L. Taylor, and R. Gámez, unpublished data). The two proteins are immunologically distinguishable so it appears unlikely that p28 is a precursor molecule to p22. Immunologic studies with monospecific polyclonal antibodies indicate that two minor intermediate size peptides (p26 and p24), between p22 and p28, are related to p22 rather than to p28. In addition, p22 regenerates dimer, trimer, and other multimeric forms best detected in overloaded gels. The p28 is more stable, forms multimers to a very limited extent, and in blot transfers reacts more strongly than p22. Several polyclonal sera prepared against both proteins produce Western blotting patterns indistinguishable from those in the Coomassie brilliant blue-stained gels. Unexpectedly, in acrylamide gels, limited proteolysis of the two capsid proteins with trypsin and V8 protease generate similar peptide patterns (Falk and Tsai, 1986; P. León, E. Moreno, L. Taylor, and R. Gámez, unpublished data). It seems unlikely at the present time that both proteins are coded by extensively overlapping RNA sequences. Probably similar peptide patterns reveal an ancestral duplication event in the evolution of these two viral proteins.

Nothing is known about the proteins of OBDV. Regarding BELV, some conflicting results have been obtained (Section VI). Lockhart et al. (1985) reported a single 26,500–27,000 molecular weight band, but using the same BELV isolate, two immunoreactive bands with molecular weights similar to those of MRFV proteins have been detected (E. Moreno, L. Taylor, C. Rivera, and R. Gámez, unpublished data).

C. Electron Microscopy Studies of Isolated Particles

Negatively stained top and bottom components from MRFV and BELV stain differentially with dye penetrating the top but not the bottom component (Gámez *et al.*, 1977; León and Gámez, 1981; Lockhart *et al.*, 1985). Morphological subunits have been resolved in MRFV particles with views down through a twofold axis of symmetry, with diagonal pentamers and hexamers, as well as other configurations. A triangulation number T = 3 has been proposed, with 32 morphological subunits and 180 protein subunits (Gámez *et al.*, 1978). All members of the group have particles with diameters of approximately 30 nm (Table I).

D. Capsid Structure of MRFV

An icosahedral symmetry is not an unusual organization among small plant viruses. For a T = 3 icosahedral shell, 180 protein subunits are expected, with 20 hexamers and 12 pentamers. Quasiequivalent contacts (Caspar and Klug, 1962) between identical proteins subunits presumably hold together the closed shell with no major contribution to stability by the RNA. *In vitro* capsid stability studies of MRFV combining electron microscopy and enzymatic hydrolysis of the RNA, show that particles release their RNA while maintaining their overall appearance (Pereira and León, 1986). In sedimentation studies at different pHs, the two components dissociate below pH 3.0 (León and Gámez, 1981). Detergents are necessary to solubilize the capsid proteins, but even in the presence of SDS the p22 protein reaggregates to form dimers, trimers, and other multimers (P. León, E. Moreno, L. Taylor, and R. Gámez, unpublished data).

A model for the capsid structure of MRFV requires the accommodation of the two proteins present in nonequimolar ratio. From our estimates (León and Gámez, 1986), it appears that the molar proportion fitting the observed mass ratio is one in which 180 copies of p22 and 20 copies of p28 form the capsid. Other possible molar combinations are widely different from the ratio observed. The available data are consistent with a model for the MRFV capsid in which the p28 polypeptide projects externally from a closed shell formed exclusively by p22. The model also avoids the problem of forming quasiequivalent contacts between different protein types (León and Gámez, 1986).

VI. SEROLOGY AND IMMUNOCHEMISTRY

A. Gel Double-Immunodiffusion Tests

All three members of the MRFV group are good immunogens in rabbits, producing sera with dilution titers for immunodiffusion assays of up to 1 : 128. No cross-reactivity of MRFV with many other small isometric ssRNA viruses belonging to several groups has been found (Gámez, 1980a; Lockhart et al., 1985).

Double-immunodiffusion tests indicated clear immunologic affinities among the three members of this virus group, with BELV occupying an intermediate position between MRFV and OBDV but being more closely related to MRFV (Lockhart et al., 1985; E. Moreno, L. Taylor, C. Rivera, and R. Gámez, unpublished data). OBDV and MRFV are more distantly related to each other. Spur formation is detected in most heterologous reactions among MRFV-CR, MRFV-BRA, MRFV-MEX, and MRFV-VEN, except for MRFV-CO and MRFV-BRA, which appear to be identical (E. Moreno, L. Taylor, C. Rivera, and R. Gámez, unpublished data).

B. Blot Transfers and Immunological Detection of Viral Proteins

In Western blot transfers, antisera against MRFV-CR reacted with the two proteins of the MRFV-CR, MRFV-CO, MRFV-MEX, MRFV-BRA, and MRFV-VEN strains, as well as with all other protein bands separated by SDS–PAGE, including the high-molecular-weight multimeric band. The antiserum against MRFV-BRA (kindly provided by Dr. M. T. Lin, Universidad de Brasilia) is monospecific for p28, reacting strongly with this protein, but neither with p22, nor with any of the multimeric forms (E. Moreno, L. Taylor, C. Rivera, and R. Gámez, unpublished data).

Antisera against BELV reacted only with its cognate viral antigens. The two proteins labeled by the immunochemical reaction have molecular weights estimated at 22,000 and 26,000. Thus the heavier protein band of BELV is smaller than the corresponding p28 of MRFV. An antiserum against MRFV-VEN (kindly provided by Dr. R. Lastra, IVIC, Venezuela), reacted with p22, but neither with p26 from BELV, nor with p28 from MRFV-CR and MRFV-BRA (E. Moreno, L. Taylor, C. Rivera, and R. Gámez, unpublished data), suggesting that p22 is more conserved in the evolution of this virus group. The results of several blot transfers indicate that p28 is more antigenic than p22 when homologous reactions are performed (Ramírez et al., 1983), since it is more strongly labeled despite its lower concentration. This also seems to be the case with BELV, in which p26 is more intensely labeled with the cognate antiserum. Available information indicates that p28 is responsible for spur formation among the different strains of MRFV tested by agar double-diffusion (E. Moreno, L. Taylor, C. Rivera, and R. Gámez, unpublished data).

The immunological studies are consistent with the biological and physicochemical investigations of this virus group, and confirm the relatedness of MRFV, OBDV, and BELV, and the existence of distinct strains of MRFV.

VII. GENOME PROPERTIES

A. Properties of the RNA

The nucleic acid purified from MRFV-TEX reacts positively with orcinol but not with diphenylamine (Gingery *et al.*, 1982). Molecules isolated from MRFV-CR produce a strong hyperchromic shift of absorbance in the ultraviolet when treated with RNAse but not the DNAse, and behave in a manner typical of ssRNA in hydroxyapatite columns (León and Gámez, 1981). Likewise, OBDV nucleic acid is sensitive to RNAse but not to DNAse (Pring *et al.*, 1973). A single RNA species with a molecular weight of approximately 2.1×10^6 is observed in denaturing agarose gels (S. Silva and P. León, unpublished data). With surface-spreading techniques, typical RNA molecules slightly longer than 1 μm and equivalent to 2.0×10^6 are visualized (León and Gámez, 1981). The RNAs from MRFV and OBDV sediment as single species in sucrose density gradients. Other properties of the RNA from different strains and members of this virus group are compared in Table I.

The structure of the 5′ and 3′ ends has not been investigated in detail. The viral RNA does not bind to oligo (dT)-cellulose columns. It is not labeled by ^{125}I in the presence of chloramine T, under conditions that label the genome-bound peptide of a comovirus (Ramírez *et al.*, 1987). Thus, we conclude that it does not contain a poly-A at its 3′ end or a peptide covalently linked to its 5′ end.

B. *In vitro* Expression of the MRFV-RNA

MRFV-RNA strongly stimulates the incorporation of labeled amino acids into polypeptides in the nuclease-treated reticulocyte lysate system. Fluorographic analysis of the labeled products reveals a large diversity of newly synthesized polypeptides, with progressive increases in the high-molecular-weight products with time of incubation; the largest polypeptide has a molecular weight of approximately 160K. The capsid proteins were not detected in blot transfers and immunoprecipitates of the reaction products. The protease inhibitor PMSF allows the accumulation of a 110K product, but otherwise few changes in the fluorographic pattern are produced by this inhibitor and by amino acid analogs (Espinoza, 1984; A. M. Espinoza, P. Ramírez, and P. León, unpublished data). Therefore, proteolytic processing does not appear to be the major pathway for the generation of the diverse patterns observed. Notably, the presence of total RNA from infected maize tissues or exogenous tRNAs in the reaction mixture results in the synthesis of only the 110K and 160K proteins, and the virtual disappearance of all the other lower-molecular-weight bands. It may be that the diversity of products synthesized in the reticulocyte lysates is the result of premature termination in specific regions of the MRFV-RNA. "Supressor" tRNAs present in maize could also avoid termination of stop codons and allow the accumulation of the longest polypeptides in the reticulocyte lysate (P. Ramírez, A. M. Espinoza, and P. León, unpublished data).

Nearly 90% of the coding capacity of the RNA is required by the two capsid proteins and the large translation product, assuming that only one reading frame is utilized. A smaller proportion of the coding capacity would be utilized if p22 and p28 share coding sequences (Falk and Tsai, 1986). A 5' leader sequence with appropriate information is necessary for ribosome binding in the plant and animal cells. We suggest that the MRFV genome is polygenic but functionally monocistronic, presumably with subgenomic messenger RNAs that are not encapsidated, directing the synthesis of coat proteins in infected cells.

VIII. RELATIONS WITH CELLS AND TISSUES

A. In Plant Hosts

The ultrastructural alterations induced by MRFV infection in the foliar lamina of infected maize plants appear to vary with the plant genotype, as demonstrated by scanning electron microscopy studies (Espinoza and Gámez, 1980). The chlorotic spots or short stripes typical of the macroscopic disease symptoms did not necessarily correlate with the number and size of altered areas in six cultivars examined. The nature and degree of the morphological distortions of epidermal cells also differ with the genetic constitution of the plant. These alterations include changes in the size, shape, and arrangement of the cells, depression of some areas of the lamina, loss of definition of cell walls, and structural modifications of the stomata.

Electron microscopic examination of thin sections of maize leaf tissues infected with MRFV-BRA and MRFV-CR consistently revealed the presence of loose aggregates of the isometric virus particles, approximately 25 nm in diameter, in vacuoles of parenchyma and epidermal cells, commonly interspersed with an electron-dense amorphous material. In a few cases, such particles were also found in the lumen of the sieve tubes. Leaf parenchyma cells appeared shrunken, but cellular organelles such as nucleus, chloroplasts, and mitochondria were not visibly altered (Kitajima and Gámez, 1977; Kitajima et al., 1976). Similar observations, although restricted to phloem cells, have been reported for maize tissues infected with MRFV-MEX (Wolanski and Maramorosch, 1979). Large crystalline arrays of virus particles may occur within membrane-bound vesicles in the cytoplasm of phloem sieve cells or parenchyma cells (E. Kitajima, R. Pereira, and R. Gámez, unpublished; Wolanski and Maramorosch, 1979).

In plants dually infected with MRFV and CSS, both pathogens were consistently observed in phloem cells but never occurred in the same cell (Wolanski and Maramorosch, 1979).

Detailed light and electron microscope studies of OBDV-infected oats indicated that the virus is restricted to phloem tissues (Banttari and Zeyen, 1973; Zeyen and Banttari, 1972). Marked abnormalities occur in phloem development, including hyperplasia and hypertrophy of young phloem and adjacent parenchyma. Most phloem elements in hyperplastic areas are parenchymatous, have truncated end walls, and lack sieve plates. In infected flax leaves, the typical enlarged lateral veins on which enations develop are associated with hyperplasia of phloem elements and hyperplasia and hypertrophy of cells that normally develop into fibers and chloroplast-containing parenchyma. Chloroplasts in these cells were either small and abnormal or absent. Vascular bundles of stems showed mild hyperplasia of phloem, and hyperplasia and hypertrophy of fibers. Phloem elements, fibers, and intercellular spaces were occasionally occluded (Banttari and Zeyen, 1971). In both flax and oats, OBDV appears to be restricted to the phloem, where it normally occurs in membrane-bounded inclusions. Numerous virus particles and abundant crystalline and paracrystalline formations were observed in oat cells, in the region between mature and immature phloem. These crystalline inclusions were large enough to be seen by light microscopy. OBDV is not associated with cell organelles, but the virus may induce substantial alterations of mitochondria (Banttari and Zeyen, 1971, 1973; Zeyen and Banttari, 1972).

In flax dually infected with OBDV and the aster yellows mycoplasma, there is extensive hyperplasia of phloem elements and hyperplasia and hypertrophy of fibers and cortical parenchyma of stems. Phloem destruction and disorganization is more pronounced in doubly than in singly infected plants. Although both virus and mycoplasma were easily located in these plants, only rarely could both pathogens be observed in the same phloem element (Banttari and Zeyen, 1972, 1979).

The relations of BELV with cells and tissues of its plant hosts have not been investigated.

B. In Insect Hosts

Examination of the internal organs of MRFV-infected leafhoppers revealed the presence of discrete groups of 20–25 nm isometric particles in the ectoderm, fatbody, midgut, follicular cells of the ovary, muscles, and salivary glands. Viral particles commonly occurred within cytoplasmic bodies bound by membranes, some of which resembled lysosomes. There is no evidence of cytopathological effects of the viral infection in

D. maidis (Kitajima and Gámez, 1983; Kitajima *et al.*, 1976; Rivera *et al.*, 1981).

OBDV particles have been detected in membrane-bounded crystalline and paracrystalline aggregates in cells of the neural lamella surrounding the supraesophaegeal ganglia, or in similar type of aggregates in fat body cells of infected leafhoppers (Banttari and Zeyen, 1976).

The interaction of BELV with cells and tissues of its vector has not been studied.

IX. EVOLUTIONARY CONSIDERATIONS

Biological and physicochemical considerations indicate that the MRFV group is a homogeneous group of evolutionarily related viruses, which infect mainly grasses and replicate in their leafhopper vector. The success of the Gramineae, particularly in recent times with the development of agriculture, has opened new possibilities for this group of pathogens, and has probably favored the appearance of viruses specifically adapted to different grass species. Coevolutionary theory predicts that reciprocal adaptations may arise between interacting organisms resulting in highly specialized associations, as in the case of MRFV, *Dalbulus* and maize (Gámez and León, 1983, 1985).

Speculating on the origin of MRFV, we find no conclusive evidence for a plant virus origin as opposed to an insect origin, but tentatively assume that the physicochemical similarities to many small ssRNA plant viruses are indicative of such an origin (Gámez and León, 1983, 1986). This implies that the MRFV group arose from a plant virus secondarily adapted to insect cell replication. Such an adaptation would necessarily incorporate new and essential regulatory information for maturation and/or translation in animal cells. If this event occurred early in the evolutionary history of the group, the dual capacity of the virus to replicate in plants and insects was already incorporated before the emergence of some of the extant viruses, such as MRFV.

We hypothesize that the ancestral virus entered a preestablished leafhopper, plant interaction, perhaps as grasses emerged as dominant plant forms. Presumably a circulative, nonpropagative manner of transmission gave way to propagative infection in the vector, with resultant long range dispersability in space and time. This also provided specificity, which depended on the feeding habits of the insect. It is very likely that other members of this virus group will be encountered in different grasses and deltocephaline vectors in Africa, Asia, Europe, and the Americas.

ACKNOWLEDGMENTS. The research on MRFV was supported in part by grants from the Vicerrectoría de Investigación, Universidad de Costa Rica, and the Consejo Nacional de Investigaciones Científicas y Tecnológicas de Costa Rica (CONICIT). P. L. is a research fellow of CONICIT. The authors are grateful to Drs. L. M. Black, E. Hiebert, L. R. Nault, and I. Herskowitz for their constructive criticism of this manuscript, and to the Organization for Tropical Studies and Francisco González for the word-processing of the text.

REFERENCES

Ancalmo, O., and Davis, W. C., 1961, Achaparramiento (corn stunt), *Plant Dis. Rep.* **45**:281.

Banttari, E. E., and Moore, M. B., 1962, Virus cause of blue dwarf of oats and its transmission to barley and flax, *Phytopathology* **52**:897.

Banttari, E. E., and Zeyen, R. J., 1969, Chromatographic purification of oat blue dwarf virus, *Phytopathology* **59**:183.

Banttari, E. E., and Zeyen, R. J., 1970, Transmission of oat blue dwarf by the aster leafhopper following natural acquisition or inoculation, *Phytopathology* **60**:399.

Banttari, E. E., and Zeyen, R. J., 1971, Histology and ultrastructure of flax crinkle, *Phytopathology* **61**:2149.

Banttari, E. E., and Zeyen, R. J., 1972, Ultrastructure of flax with a simultaneous virus and mycoplasma infection, *Virology* **49**:305.

Banttari, E. E., and Zeyen, R. J., 1973, Oat blue dwarf virus, *C. M. I./A. A. B. Descriptions of Plant Viruses* No. 123, Commonwealth Mycological Institute, Kew, Surrey, England.

Banttari, E. E., and Zeyen, R. J., 1976, Multiplication of oat blue dwarf virus in the aster leafhopper, *Phytopathology* **66**:896.

Banttari, E. E., and Zeyen, R.J., 1979, Interactions of mycoplasma-like organisms and viruses in dually infected leafhoppers, planthoppers and plants, in: *Leafhopper Vectors and Plant Disease Agents* (K. Maramorosch and K. Harris, eds.), pp. 327–347, Academic Press, New York.

Bascope, B., 1977, El agente causal de la llamada "raza Mesa Central" del achaparramiento del maíz, Masters thesis, Colegio de Posgraduados, Chapingo, Mexico.

Black, L. M., 1979, Vector cell monolayers and plant viruses, in: *Advances in Virus Research*, Volume 25 (M. A. Lauffer, F. B. Bang, K. Maramorosch and K. M. Smith, eds.), pp. 191–271, Academic Press, New York.

Bradfute, O., Nault, L. R., Robertson, D. C., and Toler, R. W., 1977, Maize bushy stunt—a disease associated with a non-helical mycoplasma organism (abstr.), *Proc. Am. Phytopathol. Soc.* **4**:171.

Carballo, O., and Lastra, R., 1987, El virus rayado fino del maíz en Venezuela: Sintomatología, purificación y características de las partículas virales, *Agron. Trop.* (in press).

Caspar, D. L. D., and Klug, A., 1962, Physical principles in the construction of regular viruses, *Cold Spring Harbor Symp. Quant. Biol.* **27**:1.

Castro-Caicedo, B. L., 1985, Transmisión del virus rayado fino del maíz y espiroplasma del achaparramiento por *Dalbulus maidis* (D. L. & W.), y evaluación de tres insecticidas en el control de la enfermedad viral. Tesis M. C., Colegio de Posgraduados, Chapingo, Mexico.

Creelman, D.W., 1965, Summary of the prevalence of plant diseases in Canada in 1964, *Can. Plant Dis. Rep.* **53**:37.

Davis, R. E., and Worley, J. F., 1973, Spiroplasma: Motile, helical microorganism associated with corn stunt disease, *Phytopathology* **63**:403.

Delgadillo-Sánchez, F., 1984, Supervivencia del virus del rayado fino del maíz en México, Tesis M. C., Colegio de Posgraduados, Chapingo, Mexico.

Espinoza, A. M., 1984, Traducción *in vitro* del ARN del virus del rayado fino del maíz en el lisado de reticulocitos de conejo, Masters thesis, Universidad de Costa Rica.

Espinoza, A. M., and Gámez, R., 1980, La ultraestructura de la superficie foliar de cultivares de maíz infectados con el virus del rayado fino, *Turrialba* **30**:413.

Falk, B. W., and Tsai, J. H., 1986, The two capsid proteins of maize rayado fino virus contain common peptide sequences, *Intervirology* **25**:111.

Gámez, R., 1969, A new leafhopper-borne virus of corn in Central America, *Plant Dis. Rep.* **53**:929.

Gámez, R., 1973, Transmission of rayado fino virus of maize (*Zea mays*) by *Dalbulus maidis*. *Ann. Appl. Biol.* **73**:285.

Gámez, R., 1980a, Maize rayado fino virus, *C. M. I./A. A. B., Descriptions of Plant Viruses* No. 220, Commonwealth Mycological Institute, Kew Surrey, England.

Gámez, R., 1980b, Rayado fino disease of maize in the American tropics, *Trop. Pest Management* **26**:26.

Gámez, R., 1983a, The ecology of maize rayado fino virus in the American tropics, in: *Plant Virus Epidemiology* (R. T. Plumb and J. M. Thresh, eds.), pp. 267–275, Blackwell Scientific, Oxford, England.

Gámez, R., 1983b, Maize rayado fino disease: The virus–host–vector interaction in Neotropical environments, in: *Proceedings, International Maize Virus Disease Colloquium and Workshop* (August 2–6, 1982) (D. T. Gordon, J. K. Knoke, L. R. Nault, and R. M. Ritter, eds.), pp. 62–68, Ohio Agricultural Research and Development Center, Wooster, Ohio.

Gámez, R., and León, P., 1983, Maize rayado fino virus: Evolution with plant host and insect vector, in: *Current Topics in Vector Research* (K. F. Harris, ed.), pp. 149–168, Praeger Scientific, New York.

Gámez, R., and León, P., 1985, Ecology and evolution of a neotropical leafhopper-virus-maize association, in: *The Leafhoppers and Planthoppers* (L. R. Nault and J. G. Rodríguez, eds.), pp. 331–350, J. Wiley & Sons, Inc., New York.

Gámez, R., and Saavedra, R., 1986, Maize rayado fino: A model of a leafhopper-borne virus disease in the Neotropics, in: *Plant Virus Epidemics* (G. MacLean, R. Garret, and W. Ruesink, eds.), pp. 315–325, Academic Press, Melbourne.

Gámez, R., Kozuka, Y., and Fukuoka, T., 1977, Purification of isometric particles associated with maize rayado fino virus, *Rev. Biol. Trop.* **25**:151.

Gámez, R., Fukuoka, T., and León, P., 1978, Properties and morphology of the leafhopper-borne maize rayado fino virus, abstracts, *IVth International Congress for Virology*, (August 30–September 6, 1978) p. 281, The Hague, The Netherlands.

Gámez, R., and Kitajima, E. W., and Lin, M. T., 1979, The geographical distribution of maize rayado fino virus, *Plant Dis. Reptr.* **63**:830.

Gingery, R. E., Gordon, D. T., and Nault, L. R., 1982, Purification and properties of an isolate of maize rayado fino virus from the United States, *Phytopathology* **72**:1313.

Godoy, C., 1985, Efecto de la infección del virus del rayado fino del maíz en la biología reproductiva de su insecto vector *Dalbulus maidis* De Long and Wolcott, Tesis, Lic. Biol., Universidad de Costa Rica.

González, V., and Gámez, R., 1974, Algunos factores que afectan la transmisión del virus del rayado fino del maíz por *Dalbulus maidis* (DeLong & Wolcott), *Turrialba* **24**:51.

Goto, S., and Moore, M. B., 1952, Some oat diseases in Minnesota in 1951, *Plant Dis. Reptr.* **36**:69.

Kitajima, E. W., and Gámez, R., 1977, Histological observations on maize leaf tissues infected with maize rayado fino virus, *Turrialba* **27**:71.

Kitajima, E. W., and Gámez, R., 1983, Electron microscopy of maize rayado fino virus in the internal organs of its leafhopper vector, *Intervirology* **19**:129.

Kitajima, E. W., Yano, T., and Costa, A. S., 1976, Purificacion and intracellular location of isometric virus-like particles associated with Brazilian corn streak virus infection, *Cienc. Cult.* **28**:427.

Kunkel, L. O., 1946, Leafhopper transmission of corn stunt, *Proc. Natl. Acad. Sci. USA* **32**:246.

Lastra, R., and de Uscategui, R. C., 1980, El virus rayado fino del maíz en Venezuela, *Turrialba* **30**:405.

León, P., and Gámez, R., 1981, Some physicochemical properties of maize rayado fino virus, *J. Gen. Virol.* **56**:67.

León, P., and Gámez, R., 1986, Biología molecular del virus del rayado fino del maíz, *Rev. Biol. Trop.* **34**:111.

Lockhart, B. E., Lin, M. T., and Kitajima, E. W., 1984, A virus serologically related to maize rayado fino virus found in *Cynodon dactylon* at Morocco, *Fitopat. Bras.* **9**:396.

Lockhart, B. E., Khaless, N., Lennon, A. M., and el Maatauoi, M., 1985, Properties of Bermuda grass etched-line virus, a new leafhopper transmitted virus related to maize rayado fino and oat blue dwarf viruses, *Phytopathology* **75**:1258.

Maramorosch, K., 1955, The occurrence of two distinct types of corn stunt in Mexico, *Plant Dis. Reptr.* **39**:896.

Martínez-López, G., 1977, New maize virus diseases in Colombia, in: *Proceedings International Maize Virus Disease Colloquium and Workshop* (16–19 August, 1976) (L. E. Williams, D. T. Gordon, and L. R. Nault, eds.), pp. 20–29, Ohio Agricultural Research and Development Center, Wooster, Ohio.

Martínez-López, G., Rico de Cujía, L. R., and de Luque, S. C., 1974, Una nueva enfermedad del maíz en Colombia transmitida por el saltahojas *Dalbulus maidis* (DeLong & Wolcott), *Fitopatología* **9**:93.

Moore, M. B., 1952, The cause and transmission of blue dwarf and red leaf of oats, *Phytopathology* **42**:471.

Nault, L. R., 1980, Maize bushy stunt and corn stunt: A comparison of disease symptoms, pathogen host ranges and vectors, *Phytopathology* **70**:659.

Nault, L. R., 1985, Evolutionary relationships between maize leafhoppers and their host plants, in: *The Leafhoppers and Planthoppers* (L. R. Nault and J. G. Rodríguez, eds.), pp. 309–330, J. Wiley and Sons, New York.

Nault, L. R., and Bradfute, O. E., 1979, Corn stunt: involvement of a complex of leafhopper-borne pathogens, in: *Leafhopper Vectors and Plant Disease Agents* (K. Maramorosch and K. Harris, eds.), pp. 561–586, Academic Press, New York.

Nault, L. R., Gingery, R. E., and Gordon, D. T., 1980, Leafhopper transmission and host range of maize rayado fino virus, *Phytopathology* **70**:709.

Nielson, M. W., 1968, The leafhopper vectors of phytopathogenic viruses (Homoptera, Cicadellidae), *Taxonomy, Biology and Virus Transmission*, United States Department of Agriculture, Technical Bulletin No. 1382.

O'Farrell, P. H., 1975, High resolution two-dimensional electrophoresis of proteins, *J. Biol. Chem.* **250**:4000.

Paniagua, R., and Gámez, R., 1976, El virus del rayado fino del maíz: Estudios adicionales sobre la relación del virus y su insecto vector, *Turrialba* **26**:39.

Pereira, R., and León, P., 1986, Estabilidad del virus del rayado fino del maíz determinada correlacionando degradación enzimática del ARN viral y microscopía electrónica de las partículas, *Rev. Biol. Trop.* **34**:163.

Pring, D. R., Zeyen, R.J., and Banttari, E. E., 1973, Isolation and characterization of oat blue dwarf virus ribonucleic acid, *Phytopathology* **63**:393.

Ramírez, P., Bonilla, J. A., Moreno, E., and León, P., 1983, Electrophoretic transfer of viral proteins to nitrocellulose sheets and detection with peroxidase-bound lectins and protein A, *J. Immunol. Meth.* **62**:15.

Ramírez, P., Espinoza, A. M., Fuentes, A. L., and León, P., 1987, Physicochemical properties of bean rugose mosaic virus, *Phytopathology* **77** (in press).

Rico de Cujía, L. A., and Martínez-López, G., 1977, Efecto de la temperatura en la incubación del virus del rayado colombiano del maíz, *Rev. Inst. Colomb. Agropec.* **12**:13.

Rivera, C., and Gámez, R., 1986, Multiplication of maize rayado fino virus in its leafhopper vector *Dalbulus maidis*, *Intervirology* **25**:76.

Rivera, C., Kozuka, Y., and Gámez, R., 1981, Rayado fino virus: Detection in salivary glands and evidence of increase in virus titer, *Turrialba* **31**:78.

Rocha-Peña, O., 1981, Algunos aspectos relacionados con el virus del rayado fino del maíz en México, Tesis, M. C., Colegio de Posgraduados, Chapingo, México.

Rodríguez, R., and Palacios, E., 1987, Estudios patométricos de la virosis causada por el virus rayado fino del maíz. 1. Aplicación del modelo logístico a la descripción y caracterización de la epifitia, *Agrociencia* (in press).

Timian, R. G., 1985, Oat blue dwarf virus in its plant host and insect vector, *Plant Dis.* **69:**706.

Timian, R. G., and Alm, K., 1973, Selective inbreeding of *Macrosteles fascifrons* for increased efficiency in virus transmission, *Phytopathology* **63:**109.

Toler, R. W., Skinner, G., Bockholt, A. J., and Harris, K., 1985, Reactions of maize (*Zea mays*) accessions to maize rayado fino virus, *Plant Dis.* **69:**56.

Westdal, P. H., 1968, Host range studies of oat blue dwarf virus, *Can. J. Bot.* **46:**1431.

Wolanski, B. S., and Maramorosch, K., 1979, Rayado fino and corn stunt spiroplasma: Phloem restriction and transmission by *Dalbulus elimatus* and *Dalbulus maidis, Fitopatol. Bras.* **4:**47.

Zeyen, R. J., and Banttari,E. E., 1972, Histology and ultrastructure of oat blue dwarf virus infected oats, *Can. J. Bot.* **50:**2511.

CHAPTER 8

Luteoviruses

R. CASPER

I. INTRODUCTION

The Luteoviruses were recognized as a plant virus group by the International Committee on Taxonomy of Viruses in 1975 (Shepherd *et al.*, 1976). Only three viruses were considered to be definite members of the group at that time, but 13 others were listed as possible members. The list of definite members has been expanding over the years, but Luteovirus taxonomy is still in a "chaotic state," as Matthews (1983) pointed out.

The group name is derived from Latin "luteus," which means yellow. Barley yellow dwarf virus, the type member of the Luteovirus group, causes yellowing, as do other group members in their respective hosts. There are, however, also members that can be latent in many hosts. The host range varies with different members. Some, e.g., beet western yellows virus (BWYV), infect a very wide range of dicotyledonous and monocotyledonous plants. Others, as potato leaf roll virus (PLRV), are restricted to some hosts in a few families. All Luteoviruses are more or less limited to the phloem (Jensen, 1969) and neighboring cells of the host plant. Virus particles occur in very low concentration in plant sap and are not normally transmitted by mechanical inoculation. Luteoviruses are transmitted by many different aphid species in a persistent manner and apparently do not replicate in the vector (Eskandari *et al.*, 1979). They occur worldwide in many important crops and cause heavy economic losses. In addition, many weeds and grasses are natural hosts for Luteoviruses and are often the source for field infections. Reviews on general characteristics or special aspects of the Luteoviruses have been published by Ashby and John-

R. CASPER ● Plant Virus Institute, Federal Biological Research Center for Agriculture and Forestry, D-3300 Braunschweig, Federal Republic of Germany.

stone (1985), Johnstone *et al.* (1984), Duffus (1977a), Francki *et al.* (1985), Plumb (1983), and Rochow and Duffus (1981). To avoid unnecessary repetition, much of the information given in these reviews will not be discussed again here.

Comprehensive information on particular Luteoviruses is given in the CMI/AAB descriptions of barley yellow dwarf virus (BYDV) (Rochow, 1970), bean leafroll virus (BLRV) (Ashby, 1984), BWYV Duffus, (1972), carrot red leaf virus (CRLV) (Waterhouse and Murant, 1982), PLRV (Harrison, 1984), soybean dwarf virus (SDV) (Tamada and Kojima, 1977), and tobacco necrotic dwarf virus (TNDV) (Kubo, 1981).

II. TAXONOMY AND PROPERTIES OF INDIVIDUAL VIRUSES

Since the recognition of the Luteovirus group in 1975 (Shepherd *et al.*, 1976) a growing number of viruses has been included in this group. While in 1975 only RMV, RPV, and SGV vector-specific viruses allied to barley yellow dwarf virus, beet western yellows virus, and soybean dwarf virus were considered as members, by 1982 15 viruses were listed as definite, 3 as probable, and 16 as possible members (Matthews, 1982). Since then, some of the latter two groups have been established as members. All the Luteoviruses are difficult to work with, because they are mainly restricted to the phloem of the host plant and are transmitted by aphids only. No mechanical transmission of a Luteovirus has been achieved so far, except in one specific case (Falk *et al.*, 1979). Vector and/or host specificity of some strains and isolates additionally complicate transmission studies. For example, *Myzus persicae* Sulz. is an efficient vector for several Luteoviruses but some strains or isolates of barley yellow dwarf virus need specific vectors like *Macrosiphum avenae F.* or *Schizaphis graminum* Rond. The occurrence of many Luteovirus strains or isolates, the necessity of transmitting them by vectors, which may differ for different viruses or virus strains, the restriction to phloem tissues, low virus concentration in host plants, the difficulty of obtaining purified preparations for physical, chemical, or serological studies—all these problems make work with Luteoviruses a painstaking and sometimes frustrating effort.

Many of the first Luteovirus descriptions were based on symptomatology only, but virus relationships or identities could not be investigated because of the difficulties mentioned above. Symptom expression may vary depending on strain or isolate and host plant, and therefore should not be used for classification of Luteoviruses. Vector specificity or the ability of the aphid to infect certain host plants may vary because of different feeding preferences of biotypes of the aphid species. This may be true especially when Luteoviruses from different geographical regions are to be compared. The best known vectors of BYDV (*Metopolophium dirhodum* Walk., *Rhopalosiphum maidis* Fitch, *R. padi* L., and *Schizaphis graminum, Sitobion avenae* F.), used for diagnosis of BYDV strains in the United States and Canada, are not endemic to all countries where BYDV has been detected. Transmission ability and efficiency may vary within an aphid species depending upon the environmental temperatures and upon the host plant on which the aphid has been reared. This all could probably account for discrepancies in reports on host ranges of Luteoviruses with otherwise similar properties, e.g., in serological tests (Ashby and Johnstone, 1985).

We have tried a grouping of Luteoviruses based only on serological relationships (Table I). At least a distant serological relationship can be expected between all or most members of the group, as we have learned from studies with other virus groups. But strong serological reactions may occur only between identical viruses or closely related strains of a particular virus. There is some confusion about the use of the terms "isolate" and "strain" in the Luteovirus group. In this chapter we use the term

TABLE I. Members of the Luteovirus Group[a]

Member	Reference
Beet western yellows virus (BWYV)	Rochow and Duffus (1981)
Turnip yellows virus	Duffus (1972)
Malva yellows virus	Duffus (1972)
Radish yellows virus	Duffus (1960, 1972)
Beet mild yellowing virus	Govier (1985)
Barley yellow dwarf virus	
Isolates RPV and RMV	Rochow (1984)
Rice giallume virus	Osler (1984)
Carrot red leaf virus	Waterhouse and Murant (1982)
Barley yellow dwarf virus (BYDV)	Burnett (1984)
Isolates PAV, MAV and SGV	Rochow (1984)
Bean leafroll virus (BLRV)	Ashby (1984)
Pea leafroll virus	Ashby and Johnstone (1985)
Legume yellows virus	Ashby (1984)
Soybean dwarf virus (SDV)	Tamada and Kojima (1977)
Subterranean clover red leaf virus	Ashby and Johnstone (1985)
Potato leafroll virus (PLRV)	Harrison (1984)
Solanum yellows virus	Milbrath and Duffus (1978)
Tobacco necrotic dwarf virus	Kubo (1981)
Tomato yellow top virus	Thomas (1985)
Groundnut rosette assistor virus (GRAV)	Casper *et al.* (1983)
Possible members	
Banana bunchy top virus	Dale *et al.* (1986)
Filaree red-leaf virus	Sylvester and Osler (1977)
Raspberry leaf curl virus	Stace-Smith and Converse (1970)
Strawberry mild yellow-edge virus	Martin and Converse (1985)
Tobacco yellow vein assistor virus	Adams and Hull (1972)

[a] Includes synonyms and closely related strains. A recent reference on the taxonomy of the particular virus is given. Viruses without references in the past twenty years are not included.

"isolate" for viruses showing no distinct differences to the type strain. We use "strain" to describe viruses showing a strong serological relationship, but differing in some properties (e.g., host range, vector specificity). We consider, for example, BYDV isolates RPV and RMV to be strains of BWYV because of their strong serological relationships to strains of the latter virus.

Different serological methods can be used for virus identification and determination of relationships. Precipitin tests in agar or other media are not very suitable for investigating Luteoviruses, since they require rather high virus concentrations. Infectivity neutralization by antiviral antibodies has been employed successfully for serological comparisons (Duffus, 1977b; Duffus and Gold, 1969; Duffus and Milbrath, 1977; Duffus and Rochow, 1978; Duffus and Russel, 1972, 1975; Gold and Duffus, 1967; Rochow and Ball, 1967). Infectivity neutralization is assayed by tests for virus activity following serological reactions. Infectivity is inactivated in the presence of antiserum. Depending on the degree of relationship, an antiserum against a related virus can achieve total or partial infectivity neutralization. The test is specific and has potential for diagnosis especially with viruses that are difficult to isolate in substantial quantities, but it requires time-consuming aphid transmissions and specific antisera.

Immunoelectron microscopical methods have also been used for investigating relationships in the Luteovirus group (Casper et al., 1983; Diaco et al., 1986; Govier, 1985; Paliwal, 1977, 1982) but these methods are rather laborious when quantitative results are needed (Francki et al., 1985). The decoration method is very useful in detecting mixed infections of two Luteoviruses (Fig. 1).

It was not before the introduction of the enzyme-linked immunosorbent assay (ELISA) into plant virology (Clark and Adams, 1977) that Luteoviruses could be detected serologically in crude sap of host plants (Casper, 1977). An antiserum prepared by M. Kojima in Japan seemingly failed to react in agar gel diffusion tests, but it proved its excellent reactivity when prepared for ELISA and used for the detection of PLRV in crude sap from leaves, stems, roots, and tubers of field-grown potatoes in Germany (Casper, 1977). ELISA not only possesses the sensitivity necessary to detect the low concentration of Luteoviruses in plant tissue and even vectors (Clarke et al., 1980; Denechere et al., 1979; Govier, 1985; Jedlinski, 1981; Kastirr et al., 1985; Paliwal, 1982; Tamada and Harrison, 1981), but it also reveals the degree of serological relationship when tests are done with defined amounts of purified viruses, or when viruses in crude sap are compared using antisera to different viruses. Virus antigens can be determined more or less quantitatively by ELISA. We found that the direct double antibody sandwich method (DAS–ELISA), as developed by Clark and Adams (1977) without any modification, was an excellent tool to demonstrate different degrees of relationship between Luteoviruses. Because of its sensitivity, reliability, and simplicity, we think ELISA

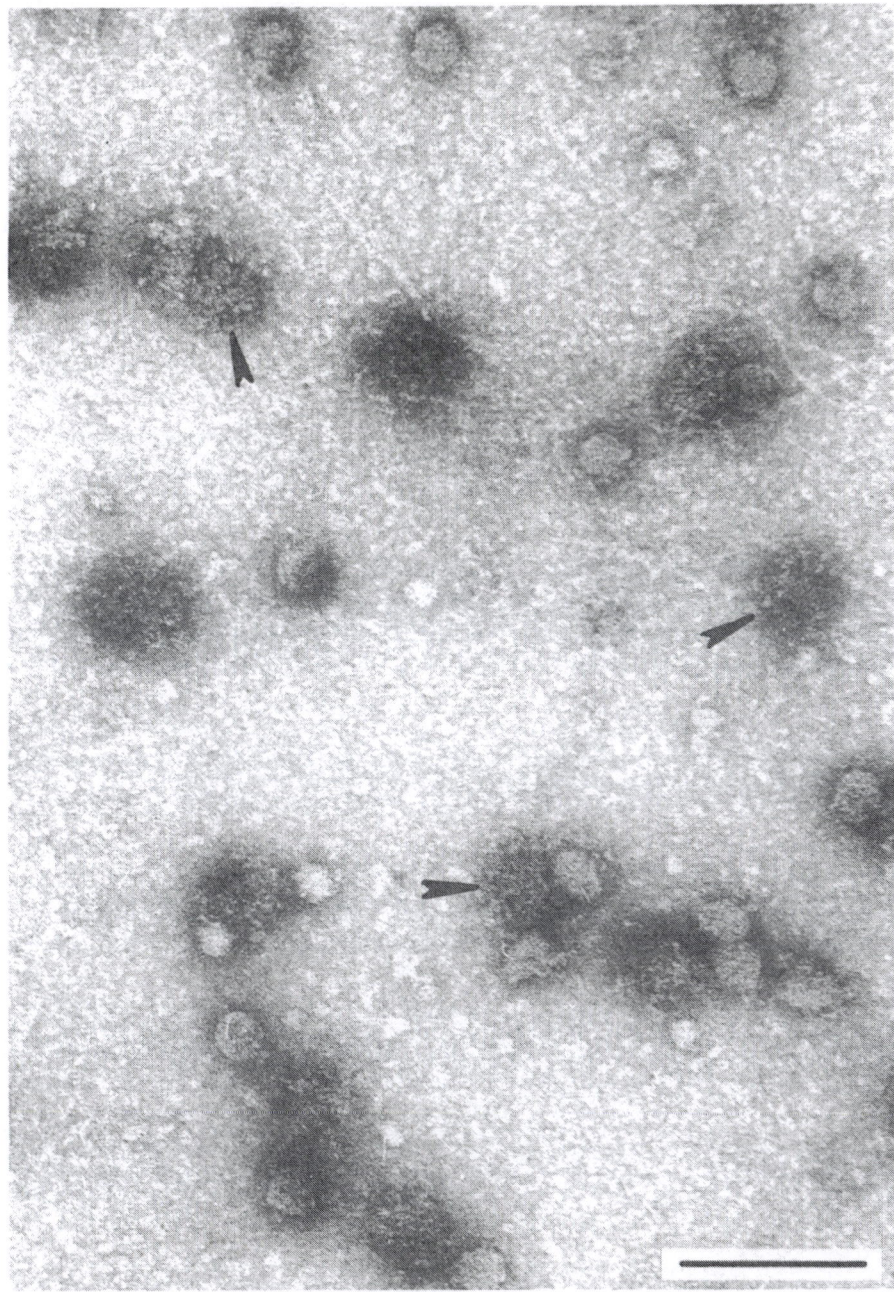

FIGURE 1. Partially purified virus preparation from *Physalis floridana* infected with potato leafroll and beet western yellows viruses. Decoration test with antiserum to BWYV reveals antibody coating of approximately half of the particles (arrows). Scale bar, 100 nm. Courtesy D.-E. Lesemann.

will be the method of choice for the investigation of serological relationships in the Luteovirus group.

In some cases it may be difficult to decide on the degree of viral relatedness on the basis of reported results obtained with only one antiserum. Moreover, results obtained in different laboratories with different antisera or virus isolates are not always comparable.

Our grouping of the Luteoviruses (Table I) is based on reports from the literature (e.g., Rochow and Duffus, 1981) and our own results (N. Chansilpa and R. Casper, unpublished data). Some of the arrangements may be preliminary, but they should be helpful in bringing some order to Luteovirus taxonomy. As more antisera become available, relationships can be defined more precisely and mistakes that are unavoidable at present can be eliminated.

Recently, cloned cDNAs to the RNAs of BYDV (Waterhouse *et al.*, 1986), PLRV (Prill *et al.*, 1986), and SCRLV (Jayasena *et al.*, 1984) have been produced. Nucleic acid hybridization tests have potential for the general diagnosis of infection by any Luteovirus as well as detection of specific strains. So far this method does not seem to be more sensitive than ELISA, but relationships can possibly be more reliably investigated, since the whole viral genome is open for comparative tests.

Monoclonal antibodies have been prepared for Luteovirus detection and may be used in the near future for large scale application (Diaco *et al.*, 1986; Hewish *et al.*, 1983; Hsu *et al.*, 1983, 1984; Martin and Stace-Smith, 1984; Rabenstein *et al.*, 1984). They are an additional tool for the investigation of viral relationships.

A. Beet Western Yellows Virus

Beet western yellows virus (BWYV) seems to be the most common Luteovirus, but it is divided into many strains (Table I). It was first described by Duffus in 1960, and in a general description, its properties were summarized by Duffus in 1972. Synonyms of BWYV include: turnip yellows virus, Malva yellows virus, radish yellows virus, and turnip mild yellows virus (Duffus, 1972). The name beet western yellows virus does not have priority, but has been widely used in the literature and should therefore be accepted to avoid any additional confusion in Luteovirus taxonomy.

Close serological relationships between BWYV from the United States and BMYV from England have been reported (Duffus and Russel, 1975). Govier (1985) again showed a close serological relationships between BMYV from sugar beet in England and BWYV infecting sugar beets and lettuce in the United States. These isolates were indistinguishable in ELISA and immunospecific electron microscopy (ISEM). In our own experience with ELISA, a large number of serologically identical or closely related isolates of BMYV can be found in sugar beets in Germany. They all show a very close relationship or are identical to our German BWYV isolates from rape (*Brassica napus L.*), radish (*Raphanus sativus L.*), and lettuce (*Lactuca sativa L.*). Based on the reports mentioned above and by Rochow and Duffus (Rochow and Duffus, 1981; H. Kleinhempel, personal communication) and our own findings (N. Chansilpa and R. Casper, unpublished data), we consider BWYV and BMYV as one virus with strains differing in symptom expression and host range.

The RPV isolate of barley yellow dwarf virus (BYDV) shows a close serological relationship to BWYV. This relationship has been well established in a careful investigation (Rochow and Duffus, 1978). In another publication, the same authors (Rochow and Duffus, 1981) reported that most Luteoviruses are serologically related; but as quantitative data become available on the extent of the reactions, the degree of relationships among the different Luteoviruses will become clearer. For example, although BWYV is related to three strains of BYDV (RPV, MAV, PAV), it is more closely related to the RPV strain than to MAV and PAV. Discussing the identification and naming of the Luteoviruses that cause barley yellow dwarf, Rochow suggests "since RPV is especially closely related to some isolates of beet western yellows virus, a more logical argument is that RPV-like isolates should be called beet western yellows virus" (Rochow, 1984). Other suggestions, such as Catherall's (1974) "ryegrass chlorotic streak virus" for RPV-like isolates, or Rochow's (1984) designation of his five isolates as NY (New York) isolates, do not seem as logical as using the name beet western yellows virus for the isolates reacting strongly with BWYV antisera. It is well established, especially through the extensive work of Rochow, that isolates of BYDV fall into two groups on the basis of serological reactions. This has been supported by investigations of their cytopathogenic effects (Gill and Chong, 1979a,b). We therefore believe it is not premature to group RPV- and RMV-like isolates of BYDV as strains of BWYV.

B. Rice Giallume Virus

Rice giallume virus (RGV) should also be considered a strain of BWYV because of its close serological relationship to RPV-like isolates of BYDV (Osler, 1984). RGV has been mainly described from Italy, where in some years it causes nearly total losses of rice crops depending on area, year, and cultivar (Osler, 1984).

C. Carrot Red Leaf Virus

Carrot red leaf virus (CLRV) (Waterhouse and Murant, 1982) is another probable strain of BWYV. This virus seems to infect only species of the Umbelliferae, and so far is the only definitive Luteovirus known to infect this plant family. It was first reported from Australia, later from Japan, Great Britain, Germany, and Canada, and probably is present in many more countries. CRLV is serologically related to several Luteoviruses but "most closely to barley yellow dwarf virus (RPV strain) and perhaps beet western yellows virus, more distantly to tobacco necrotic dwarf, potato leaf roll and bean leaf roll viruses, and very distantly to barley yellow dwarf virus (MAV strain) and soybean dwarf virus" (Waterhouse and Murant, 1982). It therefore seems to be justified to include CRLV in the list of strains of BWYV.

D. Barley Yellow Dwarf Virus

Barley Yellow Dwarf Virus (BYDV) was the topic of a workshop at the Centro Internacional de Mejoramiento de Maiz y Trigo in Mexico in 1983. The proceedings of this workshop (Burnett, 1984) provide a comprehensive review of the present status of all aspects of research on barley yellow dwarf diseases. The causal agent BYDV can be divided into several strains, which have been grouped according to their vector specificity (Rochow, 1969, 1970, 1984):

1. MAV, transmitted specifically by *Sitobion avenae*.
2. PAV, transmitted nonspecifically by *Rhopalosiphum padi* and *Sitobion avenae*.
3. RPV, transmitted specifically by *Rhopalosiphum padi*.
4. RMV, transmitted specifically by *Rhopalosiphum maidis*.
5. SGV, transmitted specifically by *Sitobion graminum*.

Because of close serological relationships to BWYV (Rochow, 1984), we have assigned RPV and RMV to BWYV. This leaves only the MAV, SGV, and PAV strains as "true" BYDV.

E. Bean Leaf Roll Virus

The properties of bean leaf roll virus (BLRV) have recently been summarized (Ashby, 1984). Synonyms include: pea leaf roll virus, pea top yellows virus, pea tip yellowing virus, and legume yellows virus. Ashby (1984) suggests that "on the basis of priority and descriptiveness bean leaf roll is the most appropriate name for the virus," and "the name leaf roll is descriptive of the symptoms caused in beans, whereas in peas the main symptom is apical yellowing." BLRV reacted strongly in agar gel diffusion tests with an antiserum to SDV and weakly with BWYV antiserum (Ashby and Huttinga, 1979). BLRV antiserum reacted weakly with purified PLRV and with purified CRLV (Waterhouse and Murant, 1981). With antiserum to BLRV a distant serological relationship between subterranean clover red leaf virus and BLRV was demonstrated (Ashby and Kyriakou, 1982). More serological tests are needed to obtain a clear picture of the degree of relationships and the place of BLRV in the continuum of serologically related Luteoviruses.

F. Soybean Dwarf Virus

Soybean dwarf virus (SDV) was first described in Japan (Tamada and Kojima, 1977). It infects mainly legumes. Two strains have been reported: a dwarfing strain and a yellowing strain, which are very similar to sub-terranean clover red leaf virus (SCRLV) based on their serological inter-actions, vector specificity, and host ranges (Ashby and Johnstone, 1985; Ashby and Kyriakou, 1982). According to Johnstone and McLean (1986), SCRLV and SDV should not be considered as separate viruses. Duffus (1977b) detected a serological relationship between SDV and BWYV using infectivity neutralization tests, but the degree of relationship of SDV to other Luteoviruses has not been reported.

G. Potato Leaf Roll Virus

Potato leaf roll virus (PLRV) has a limited host range, but is found all over the world where potatoes are cultivated (Harrison, 1984). So far, no serologically distinct strains have been reported. An antiserum from M. Kojima in Japan has been used in ELISA successfully for the detection of PLRV isolates in field grown potatoes in Germany (Casper, 1977). Another antiserum from M. Kojima was used successfully for PLRV de-tection in the state of Oregon (Clarke et al., 1980). This already gives an indication that PLRV isolates from different geographical regions must not necessarily be serologically distinct. ELISA test kits commercially available from Germany, Switzerland, and the United States for PLRV detection in seed potato certification programs have been used in many potato-growing countries, and so far no PLRV strains undetectable with such kits have been reported.

H. Tomato Yellow Top Virus

Tomato yellow top virus (TYTV) has been reported from Austra-lia, New Zealand, Brazil, and the United States (Thomas, 1984). A com-parison using PLRV antisera from Scotland, Canada and the Nether-lands demonstrated a very close relationship between TYTV and PLRV. TYTV can be considered as identical with PLRV (J.E. Thomas, personal communication).

I. Tobacco Necrotic Dwarf Virus

Tobacco necrotic dwarf virus (TNDV) (Kubo, 1981) is recorded only from Japan. It is serologically closely related to PLRV and distantly related to SDV and CLRV (Kubo, 1981). It seems justifiable to consider TNDV as a serological strain of PLRV that may be distinguished from PLRV by host reactions.

J. Solanum Yellows Virus

Solanum yellows virus (Milbrath and Duffus, 1978) originally appeared to be a distinct member of the Luteovirus group, but recent serological investigations showed a close relationship to PLRV (J. Duffus, personal communication; J.E. Thomas, personal communication). We therefore consider Solanum yellows virus a strain of PLRV.

K. Groundnut Rosette Assistor Virus

Groundnut rosette assistor virus (GRAV) (Hull and Adams, 1968) has been listed as a possible member of the Luteovirus group (Matthews, 1982). Casper *et al.* (1983) showed that GRAV is a Luteovirus by demonstrating distant serological relationships of GRAV to BWYV, PLRV, pea leaf roll virus (same as BLRV), and BYDV. No strain specification for the used BYDV antiserum is given. Reddy *et al.* (1985) confirmed the serological relationship reported by Casper *et al.* (1983). GRAV is definitely a Luteovirus, but it may be too early to decide on its position in the continuum of serological relationships within the group.

L. Strawberry Mild Yellow-Edge Virus

Strawberry mild yellow-edge virus (SMYEV) is considered a possible member of the Luteovirus group based on symptomatology and virus-vector relationship. Martin and Converse (1985) supplied further evidence to include SMYEV in the Luteovirus group: it is transmitted in a persistent manner, is circulative in its vector, and 23-nm particles are present in very low concentration in strawberry plant tissue. However, serological affinities with other Luteoviruses were not detected using indirect (ab')$_2$-ELISA. Investigations on SMYEV are hindered by tannins and viscous materials in the host sap, which is especially disadvantageous in serological tests (Martin and Converse, 1985). Spiegel *et al.* (1985) reported trapping but no decoration of SMYEV particles in serologically specific electronmicroscopy (SSEM) by antiserum against BWYV. Florance *et al.* (1986) found particles with diameters ranging from 25 to 34 nm in cross-sections of phloem parenchyma cells of the youngest leaf from plants infected for four weeks. This diameter appears larger than the 23-nm diameter published for purified particles (Martin and Converse, 1985). Evidence to include SMYEV in the Luteovirus group certainly is growing, but since final positive evidence has not yet been furnished we consider SMYEV still a probable member of the Luteovirus group.

M. Banana Bunchy Top Virus

New evidence to include banana bunchy top virus (BBTV) in the Luteovirus group has recently been presented by the isolation of disease-associated dsRNA (Dale *et al.*, 1986). Only small amounts of dsRNA were found in diseased banana plants even when relatively large amounts (100 g) of host tissue were used. The amount of isolated dsRNA was highly influenced by growth temperatures and time after inoculation. The largest amount of dsRNA was obtained from plants grown at 30°C and harvested 23 days after infection by the banana aphid *Pentalonia nigronervosa* Coq. By 29 days after aphid transmission only very faint bands were detected in electrophoresis, indicating that much less dsRNA had been extracted. Even less dsRNA was obtained from plants grown at 25°C. The authors consider this to be additional evidence that banana bunchy top disease is caused by a virus. The dsRNA pattern in electrophoresis from bunchy top infected banana is similar to that of BYDV (Gildow *et al.*, 1983) and also a strain of BWYV (Falk and Duffus, 1984). But so far, no particles resembling a Luteovirus have been observed by electron microscopy.

Among the possible members known from the literature, only those few with references within the last 20 years are listed in this review (Table I). We did not include viruses of rather uncertain identity, especially if no new information on them has been published.

III. TRANSMISSION AND CONTROL

Luteoviruses are transmitted by aphids, but usually not mechanically, by dodder, or any other vector. Falk *et al.* (1979) reported a specific case of mechanical transmission of BWYV in complex with lettuce speckles mottle virus. In nature, aphids provide the only means for virus spread during the growing season. *Myzus persicae* Sulz. seems to be the most efficient and important vector for most, but not all, Luteoviruses. The viruses are circulative and persistent in the aphid vector. The minimum acquisition time needed by the aphid to acquire the virus has been reported to be 5 min for BWYV (Duffus, 1972), but may be several hours for other Luteoviruses. The acquisition time is followed by a latent period of at least 12 hr, and the virus can be transmitted with an inoculation access period of 10 to 30 min. The given times can vary depending on the efficiency of the vector, the virus concentration in the plant host, the virus strain, temperature, and other environmental factors (Damsteegt and Hewings, 1986, 1987).

No evidence for virus replication in the vector has been found (Eskandari *et al.*, 1979) except for one report with PLRV. Weidemann (1982) found virus antigen accumulating in the nuclei of cells of the midgut and principal salivary gland of *M. persicae* shortly after virus acquisition, and therefore suggested that PLRV multiplies in these organs. This suggestion may need further confirmation.

Seed transmission of a Luteovirus has been reported as early as 1948. Clinch *et al.* (1948) found in sugar beet (*Beta vulgaris* L.) a seed transmission rate between 25 and 55% of "virus yellows," probably BWYV.

There were several more reports on seed transmission of BWYV in sugar beets. In a recent study seed transmission of beet mild yellows virus (BWYV) has been reinvestigated (Fritzsche et al., 1983). They used seed from fields with 50–90% yellows virus infection. One hundred and forty beet plants grown from this seed were used as virus source for aphid transmissions to *Sinapis alba* L. Nine successful transmissions were indicated by BMYV symptom development on *S. alba*. The authors deduct from their results a seed transmission rate of 6.4% in the investigated commercial seed. No serological or electronmicroscopical confirmation of the transmission was given in this publication.

Since no thoroughly convincing evidence for seed transmission of Luteoviruses has yet been published, it is generally accepted that these viruses are not seed-transmissible.

Luteoviruses are found in most places where the host plants are grown. Some, like BWYV, PLRV, and BYDV, are probably common throughout the world. The damage caused by these and other Luteoviruses results in enormous economic losses in agriculture (Duffus, 1977a; Gill 1980; Grafton et al., 1982; Hampton, 1983).

For the control of Luteoviruses in crops, four major methods have been reported: elimination of virus source plants and supply of virus-free planting material; application of insecticides to minimize vector population; adjustment of crop planting time to avoid maximum aphid populations; and planting cultivars resistant or tolerant to field infection (Duffus, 1977a; Rochow and Duffus, 1981).

For the elimination of virus source plants, reliable diagnosis is essential. One of the most reliable methods is currently ELISA (Clark and Adams, 1977). Since 1977 (Casper, 1977) it has been applied in many countries for all Luteoviruses for which specific antisera are available. ELISA proved its usefulness for the reliable diagnosis in field inspection, elimination of virus sources, and investigation of host ranges (Chevallier and Putz, 1982; Clarke, 1981; Clarke et al., 1980; Clement et al., 1986; D'Arcy, 1984b; Doupnik et al., 1982; Ehlers et al., 1983; Eweida and Ryden, 1984; Govier, 1985; Gugerli, 1980; Gugerli and Gehriger, 1980; Haase et al., 1985; Hewings and D'Arcy, 1984; Hill and Jackson, 1984; Holmes, 1985; Kastirr et al., 1985; Kojima et al., 1982; Lister et al., 1983, 1985; Lister and Rochow, 1979; Proeseler et al., 1985; Reichenbaecher et al., 1985; Richter et al., 1985; Rochow et al., 1986; Singh and Somerville, 1983; Tamada and Harrison, 1980a,b; Torrance and Jones, 1982; Torrance et al., 1986).

Selecting tubers free of PLRV by ELISA for production of seed potatoes and in certification programs requires great effort, but has been successfully performed in some European countries (Germany, the Netherlands, Switzerland).

Luteovirus infection can be reduced by spraying insecticides (Dubey et al., 1981; Gibson et al., 1982; Jenkyn and Plumb, 1983; Johnstone, 1984; Johnstone and Rapley, 1981; Stoltz and Forster, 1984), but the impact on the environment may not be justified by the rather limited success. Successful spraying of pesticides for vector control has to be based on sound knowledge of virus epidemiology (Plumb, 1984).

BYDV infection often can be reduced by planting barley and other susceptible small grains late in the fall or early in the spring (Duffus, 1977a; Plumb, 1984). Johnstone and Rapley (1979) found that the incidence of SCRLV in *Vicia faba* can be reduced by favorable planting dates. However, finding the best sowing date each year to avoid infection may be very difficult (Plumb, 1984).

BYDV may be effectively controlled by genetic resistance in barley (Catherall and Hayes, 1967; Schaller, 1984), but in wheat, (Tola and Kronstad, 1984) reports on resistance to BYDV are very limited, and no resistance has been found in oats (Jedlinski, 1984; Jedlinski *et al.*, 1977). Of all control measures, resistant cultivars offer the only positive and economical method for reducing losses in barley (Schaller, 1977, 1984; Schaller and Qualset, 1980). In the weed *Agropyron spec.*, high levels of resistance to BYDV have been found (Sharma *et al.*, 1984). In potatoes, resistance to PLRV has been detected in *Solanum brevidens* and successfully transferred to potato (Austin *et al.*, 1985). Mndolwa *et al.* (1984) report on resistance of potato to aphid vectors and PLRV.

IV. PURIFICATION

Luteoviruses are relatively stable, but occur in very low concentrations in their hosts. Their restriction mainly to the phloem makes it especially difficult to extract the small amounts of virus the host plants contain. Conventional procedures are therefore not applicable. Yield of purified virus is low, but has been remarkably increased by enzymatic maceration of tissues to release cell-bound virus (Takanami and Kubo, 1979a). Most protocols published in recent years include enzyme-assisted tissue maceration (Ashby and Huttinga, 1979; Ashby and Kyriakou, 1982; Clarke, 1981; D'Arcy, 1984a; D'Arcy *et al.*, 1983; Eweida and Oxelfelt, 1985; Hammond *et al.*, 1983; Hewish and Shukla, 1983; Johnstone *et al.*, 1982; Kühne *et al.*, 1985; Paliwal, 1978; Proll *et al.*, 1984; Rochow *et al.*, 1971; Rowhani and Stace-Smith, 1979; Waterhouse and Helms, 1984; Waterhouse and Murant, 1981).

We used with good success the following method (N. Chansilpa and R. Casper, unpublished data, based on Takanami and Kubo, 1979a): Disrupt fresh leaves and preferably stems in a mechanical blender with 0.1 M sodium citrate buffer, pH 6, containing 0.5% mercaptoethanol (2 ml/g plant tissue). Add 2% industrial-grade pectin glycosidase (Rohament-P, Roehm GmbH, Darmstadt, West Germany). Incubate 3 to 7 hr at 24–37°C with shaking. Clarify by filtration through cheesecloth. Add 0.67 volumes of a 1 : 1 (v/v) chloroform/butanol mixture and stir for 15–30 min at room temperature, then centrifuge at 6000 rpm for 10 min. Filtrate supernatant through filter paper. Add polyethylene glycol, M_r 6000, to 8% (w/v) and NaCl to 2% (w/v). Incubate for 1 hr at 4°C. Centrifuge at 8000 rpm for 20 min. Resuspend the precipitate carefully in a small amount of 0.02 M phosphate buffer, pH 7.5, containing 1% Triton X-100. Centrifuge at 30,000 rpm for 4 hr using a 20% sucrose cushion. Resuspend pellet in phosphate buffer as above and further purify the virus by rate zonal density centrifugation in 10–40% sucrose gradients.

This protocol proved to be effective for the purification of PLRV and BWYV from *Physalis floridana* and BLRV from *Vicia faba*. It need not be followed in all details, but may be modified depending on virus properties or host plant. Since Luteoviruses may precipitate at low temperatures (4°C), we always work at room temperature. However, careful slow re-suspension of centrifugation pellets by stirring for several hours is essential for obtaining maximum virus yield.

V. PARTICLE PROPERTIES

Purified preparations of Luteoviruses usually contain one sediment-ing component. However, Proll *et al.* (1985) reported on a top component of an isolate of BYDV. The particles of the top component with a sedi-mentation coefficient of 55 S seemed to be nucleic acid-free protein shells. They showed serological relationship with other BYDV-PAV isolates from Germany, and gave positive heterologous reactions with antiserum against BYDV-PAV obtained from R.M. Lister, in the United States. Hewings and D'Arcy (1986) found a top component associated with BWYV from Cal-ifornia. This component had a maximum absorbance of 275 nm and a $S_{20,w}$ of 62 S. No nucleic acid was detected in the top component particles, and they were not infectious. N. Chansilpa, R. Casper, and D. E. Lesemann (unpublished data) found with preparations of one isolate of BWYV from sugar beet in density gradients a top component that was absent in other BWYV preparations. Viruslike particles with a diameter of 17 nm were detected in one density gradient fraction with all PLRV, BWYV, and BYDV isolates investigated so far.

The diameter of Luteovirus particles reported in recent publications ranges from 24 nm for PLRV (Takanami and Kubo, 1979b) to 27 nm for BLRV (Ashby and Huttinga, 1979). Particle diameters of other Luteovi-ruses are within this range. The Luteovirus particles are isometric, many appearing hexagonal in outline in preparations negatively stained with uranyl acetate (Fig. 2).

Reported sedimentation coefficients $(S_{20,w})$ are 114 S for BWYV (Hew-ings and D'Arcy, 1986), 106 to 120 S for BYDV depending on strain and authors (Hammond *et al.*, 1983; Hewings and D'Arcy, 1986; Proll *et al.*, 1985; Rochow and Brakke, 1964), 108 and 114 S for SDV strain D and Y, respectively (Hewings and D'Arcy, 1986), 112 to 127 S for PLRV (Kubo, 1981; Rowhani and Stace-Smith, 1979; Takanami and Kubo, 1979b;

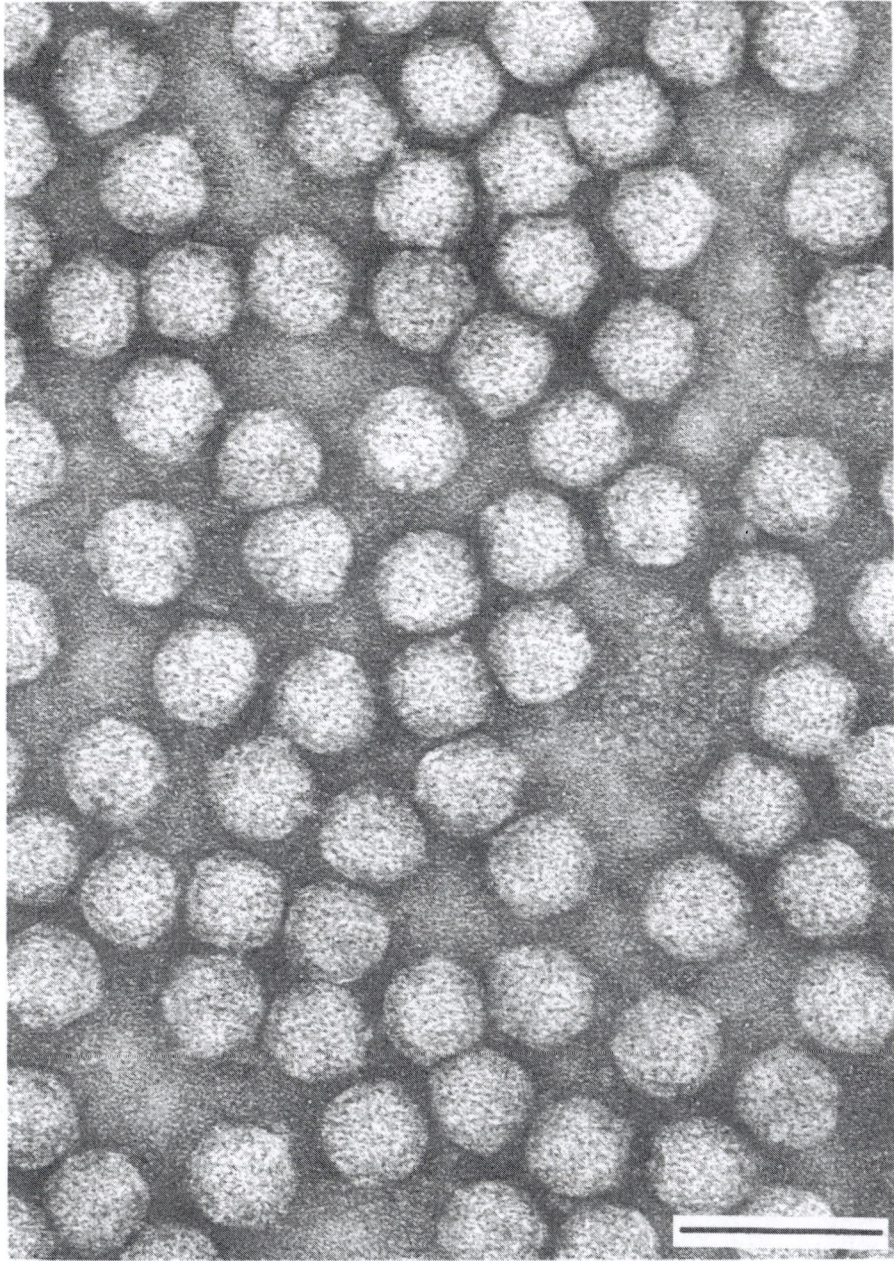

FIGURE 2. Particles of potato leaf roll virus in a purified preparation negatively stained with uranyl acetate. Scale bar, 50 nm. Courtesy D.-E. Lesemann.

Waterhouse and Murant, 1981), and 104 S for CLRV (Waterhouse and Murant, 1982).

The particles contain single-stranded RNA with M_r between 1.8×10^6 for PLRV (Waterhouse and Murant, 1982) and 2.4×10^6 for BLRV (Ashby, 1984). About 30% of the particle weight is RNA (Hewings et al., 1986; Rowhani and Stace-Smith, 1979; Takanami and Kubo, 1979b).

The coat protein of PLRV (Rowhani and Stace-Smith, 1979), CRLV (Waterhouse and Murant, 1982), and TNDV (Kubo and Takanami, 1979) consists of one protein species of an M_r of about 26,000 estimated by polyacrylamide gel electrophoresis. More extreme values have been reported for SDV (M_r 22,000) (Tamada and Kojima, 1977) and for BLRV (M_r 32,500) (Ashby and Huttinga, 1979).

VI. SEROLOGY

Methods useful for serological studies on Luteoviruses and their applications for diagnosis and taxonomy have been described in detail in Sections II and III of this chapter.

VII. GENOME PROPERTIES AND REPLICATION

Little information on the genomic properties of Luteoviruses has been published so far. The genomic RNA of PLRV is positive-stranded, being both infective and a messenger RNA. A small protein (M_r 7000) is bound to its RNA, which remains infective when this protein is degraded by protease. Unlike the RNAs of cowpea mosaic virus and of the Nepoviruses, PLRV RNA does not contain a polyadenylated sequence. In messenger-dependent rabbit reticulocyte lysates, PLRV RNA induced specific polypeptide synthesis. The major product had a M_r of 71,000, but none of the products of the cell-free system had the size of the coat protein subunit (M_r 26,000). With the wheat germ system, an additional translation product with a molecular weight close to that of PLRV coat protein has been obtained, but it did not react with PLRV-specific antiserum (Mayo and Barker, 1984). Perhaps these features are characteristic for all Luteoviruses. The role of low-molecular-weight, minor RNA species found with RNA from PLRV (Barker et al., 1984), BYDV (Brakke and Rochow, 1974) or TNDV (Takanami and Kubo, 1979b) remains unknown. They may represent encapsidated subgenomic messenger RNA or just degradation products. RNA replication and expression in plant tissue still is not thoroughly investigated. Tobacco protoplasts can be infected by PLRV and TNDV virions and by PLRV RNA (Mayo et al., 1982; Kubo and Takanami, 1979).

VIII. CYTOPATHOLOGY

Luteoviruses appear to be restricted to and replicated in the phloem tissue of their respective host plant. This makes determining the sequence of cytological events in the plant cell following infection difficult (Francki et al., 1985). Esau and Hoefert (1972 a,b) found double-membraned vesicles in the cytoplasm of BWYV-infected sugarbeet leaves. Similar vesicles have also been observed in cells infected with BYDV, CRLV, PLRV, and SCRLV (Francki et al., 1985; Jayasena et al., 1981; Shepardson et al., 1980). The possible role of these vesicles in Luteovirus replication is discussed by Francki et al. (1985). Virus particles that were first seen close to the nucleolus, later became scattered throughout the nucleoplasm (Francki et al., 1985).

In PLRV-infected cells, Francki et al. (1985) observed dilation of the cristae of the mitochondria and the appearance of vesicles containing either fibrillar or opaque material in the cytoplasm. In cells infected with CRLV, mitochondria undergo morphological changes. Fibrillar material contained in vesicles at the mitochondrial periphery seems likely to be dsRNA, possibly replicative form RNA, as suggested by Francki et al. (1985).

Evidence from cytopathology indicates that BYDV strains may be categorized into two groups in accordance with results from serological and cross-protection studies. One of the groups—RPV and RMV isolates—produced cytopathological effects similar to those induced by BWYV (Aapola and Rochow, 1971). Oat cells infected with RPV and RMV isolates (=BWYV) show a nucleus more or less normal in outline and virus progeny first appears around the nucleolus. With all other BYDV isolates, a distorted nucleus and occurrence of first virus progeny in the cytoplasm was observed (Gill and Chong, 1979a). Infection of protoplasts from tobacco mesophyll cells with TNDV and replication of the virus in the protoplasts has been reported (Kubo and Takanami, 1979).

IX. CONCLUSIONS

The enormous damage of Luteoviruses to agricultural crops has been recognized for more than 50 years, but despite strong research efforts in the interim much of our knowledge dates only from the past 10 years, when purification of Luteoviruses was achieved in several countries. This became possible through the introduction of enzyme-assisted purification methods and the exchange of antisera between laboratories. Subsequently the taxonomic grouping of many Luteoviruses, which until then were differentiated mainly on the basis of symptom expression in certain host plants, became much easier. In our laboratory, e.g., with BWYV, a large number of isolates or strains can be differentiated based on serological behavior. With some other, so far poorly investigated isolates, a new name may have to be taken into consideration. Once the degree of serological relationship has been determined, such isolates should be grouped under a generally accepted name.

The possibility of purifying Luteoviruses and obtaining substantial amounts of virus for biochemical and gene technological experiments will supply more knowledge in the years ahead, and may eventually lead to an effective control of Luteovirus infections with the aim of avoiding the losses they cause to agriculture.

REFERENCES

Aapola, A. I. F., and Rochow, W. F., 1971, Relationship among three isolates of barley yellow dwarf virus, *Virology* **46**:127.

Adams, A. N., and Hull, R., 1972, Tobacco yellow vein, a virus dependent on assistor viruses for its transmission by aphids, *Ann. Appl. Biol.* **71**:135.

Ashby, J. W., 1984, Bean leaf roll virus, *CMI/AAB Descriptions of Plant Viruses* No. 286.

Ashby, J. W., and Huttinga, H., 1979, Purification and some properties of pea leafroll virus, *Neth. J. Pl. Path.* **85**:113.

Ashby, J. W., and Johnstone, G. R., 1985, Legume luteovirus taxonomy and current research, *Australas. Plant Path.* **14**:2.

Ashby, J. W., and Kyriakou, A., 1982, Purification and properties of subterranean clover red leaf virus, *N. Z. J. Agric. Res.* **25**:607.

Austin, S., Bear, M. A., and Helgeson, J. P., 1985, Transfer of resistance to potato leaf roll virus from *Solanum brevidens* into *Solanum tuberosum* by somatic fusion, *Plant Sci.* **39**:75.

Barker, H., Mayo, M. A., and Robinson, D. J., 1984, Polygenic resistance to potato leafroll virus (PLRV), *Rep. Scott. Crop Res. Inst.* **1983**:194.

Brakke, M. K., and Rochow, W. F., 1974, Ribonucleic acid of barley yellow dwarf virus, *Virology* **61**:240.

Burnett, P. A., 1984, Preface, *Barley Yellow Dwarf, A Proceedings of the Workshop, CIMMYT, Mexico*, 6.

Casper, R., 1977, Detection of potato leafroll virus in potato and *Physalis floridana* by enzyme-linked immunosorbent assay (ELISA), *Phytopathol. Z.* **90**:364.

Casper, R., Meyer, S., Lesemann, D.-E., Reddy, D. V. D., Rajeshwari, R., Misari, S. M., and Subbarayudu, S. S., 1983, Detection of a luteovirus in groundnut rosette diseased groundnuts (*Arachis hypogaea*) by enzyme-linked immunosorbent assay and immunoelectron microscopy, *Phytopath. Z.* **108**:12.

Catherall, P. L., 1974, Chlorotic streaking in Italian ryegrass, *Plant Pathol.* **23**:116.

Catherall, P. L., and Hayes, J. D., 1967, Assessment of varietal reaction and breeding for resistance to the yellow dwarf virus in barley, *Euphytica* **15**:39.

Chevallier, D., and Putz, C., 1982, Detection of sugarbeet yellowing viruses in leaf extracts by enzyme-linked immunosorbent assay (ELISA), *Ann. Virol. (Inst. Pasteur)* **133E**:473.

Clark, M. F., and Adams, A. N., 1977, Characteristics of the microplate method of enzyme-linked immunosorbent assay for the detection of plant viruses, *J. Gen. Virol.* **34**:475.

Clarke, R. G., 1981, Potato leafroll virus purification and antiserum preparation for enzyme-linked immunosorbent assays, *Am. Potato J.* **58**:291.

Clarke, R. G., Converse, R. H., and Kojima, M., 1980, Enzyme-linked immunosorbent assay to detect potato leafroll virus in potato tubers and viruliferous aphids, *Plant Dis.* **64**:43.

Clement, D. L., Lister, R. M., and Foster, J. E., 1986, ELISA-based studies on the ecology and epidemiology of barley yellow dwarf virus in Indiana, *Phytopathology* **76**:86.

Clinch, P. E. M., Loughnane, I. B., and McKay, R., 1948, Transmission of a disease resembling virus yellows through the seed of sugar beet, *Nature* **161**:28.

Dale, J. L., Phillips, D. A., and Parry, J. N., 1986, Double-stranded RNA in banana plants with bunchy top disease. *J. Gen. Virol.* **67**:371.

Damsteegt, V. D., and Hewings, A. D., 1986, Comparative transmission of soybean dwarf virus by three geographically diverse populations of *Aulacorthum* (= *Acyrthosiphon*) *solani, Ann. Appl. Biol.* **109**:453.

Damsteegt, V. D., and Hewings, A. D., 1987, Relationships between *Aulacorthum solani* and soybean dwarf virus: Effect of temperature on transmission, *Phytopathology* **77**:515.

D'Arcy, C. J., 1984a, Purification of barley yellow dwarf viruses, *Barley Yellow Dwarf, A Proceedings of the Workshop, CIMMYT, Mexico,* 36.

D'Arcy, C. J., 1984b, Surveying for barley yellow dwarf, *Barley Yellow Dwarf, A Proceedings of the Workshop, CIMMYT, Mexico,* 40.

D'Arcy, C. J., Hewings, A. D., Burnett, P. A. and Jedlinski, H., 1983, Comparative purification of two luteoviruses, *Phytopathology* **73**:755.

Denechere, M., Cante, F., and Lapierre, H., 1979, ELISA detection of barley yellow dwarf virus in its aphid vector *Rhopalosiphum padi* (L.), *Ann. Phytopathol.* **11**:507.

Diaco, R., Lister, R. M., Hill, J. H., and Durand, D. P., 1986, Detection of homologous and heterologous barley yellow dwarf virus isolates with monoclonal antibodies in serologically specific electron microscopy, *Phytopathology* **76**:225.

Doupnik, B., Jr., Stuckey, R. E., Bryant, G. R., and Pirone, T. P., 1982, Enzyme-linked immunosorbent assay for barley yellow dwarf virus using antiserum produced to virus from field-infected plants, *Plant Dis.* **66**:812.

Dubey, G. S., Singh, R. S., and Chaudhary, R. G., 1981, Effect of roguing and application of systemic insecticides on incidence of aphids and virus diseases of potato, *Indian J. Agric. Res.* **15**:87.

Duffus, J. E., 1960, Radish yellows, a disease of radish, sugar beet, and other crops, *Phytopathology* **50**:389.

Duffus, J. E., 1972, Beet western yellows virus, *CMI/AAB Descriptions of Plant Viruses* No. 89.

Duffus, J. E., 1977a, Aphids, viruses, and the yellow plague, in: *Aphids as Virus Vectors* (K. F. Harris and K. Maramorosch, eds.), pp. 361–383, Academic Press, New York.

Duffus, J. E., 1977b, Serological relationships among beet western yellows, barley yellow dwarf, and soybean dwarf viruses, *Phytopathology* **67**:1197.

Duffus, J. E., and Gold, A. H., 1969, Membrane feeding and infectivity neutralisation used in a serological comparison of potato leaf roll and beet western yellows viruses, *Virology* **37**:150.

Duffus, J. E., and Milbrath, G. M., 1977, Susceptibility and immunity in soybean to beet western yellows virus, *Phytopathology* **67**:269.

Duffus, J. E., and Rochow, W. F., 1978, Neutralisation of beet western yellows virus by antisera against barley yellow dwarf virus, *Phytopathology* **68**:45.

Duffus, J. E., and Russell, G. E., 1972, Serological relationship between beet western yellows and turnip yellows viruses, *Phytopathology* **62**:1274.

Duffus, J. E., and Russel, G. E., 1975, Serological relationship between beet western yellows and beet mild yellowing viruses, *Phytopathology* **65**:811.

Ehlers, U., Vetten, H. J., and Paul, H. L., 1983, Detection of potato leafroll virus in primarily infected tubers by enzyme-linked immunosorbent assay, *Phytopathol. Z.* **107**:37.

Esau, K., and Hoefert, L. L., 1972a, Development of infection with beet western yellows virus in the sugarbeet, *Virology* **48**:724.

Esau, K., and Hoefert, L. L., 1972b, Ultrastructure of sugarbeet leaves infected with beet western yellows virus, *J. Ultrastruct. Res.* **40**:556.

Eskandari, F., Sylvester, E. S., and Richardson, J., 1979, Evidence for lack of propagation of potato leafroll virus in its aphid vector, *Myzus persicae, Phytopathology* **69**:45.

Eweida, M., and Oxelfelt, P., 1985, Purification and properties of two Swedish isolates of barley yellow dwarf virus and their serological relationships with American isolates, *Ann. Appl. Biol.* **106**:475.

Eweida, M., and Ryden, K., 1984, Nachweis des Virus der Gersten-Gelbverzwergung (BYDV) in Getreide und Gräsern durch ELISA in Schweden, *Z. Pflanzenkr. Pflanzenschutz* **91**:131.

Falk, B. W., and Duffus, J. E., 1984, Identification of small single- and double-stranded RNAs associated with severe symptoms in beet western yellows virus-infected *Capsella bursa-pastoris*, *Phytopathology* **74**:1224.

Falk, B. W., Duffus, J. E., and Morris, T. J., 1979, Transmission, host range, and serological properties of the viruses that cause lettuce speckles disease, *Phytopathology* **69**:612.

Florance, E. R., Allen, T. C., and Converse, R. H., 1986, The ultrastructure of strawberry mild yellow-edge virus, a luteovirus, in cells of *Fragaria vesca*, cultivar UC-4 (abstract), *Phytopathology* **76**:843.

Francki, R. I. B., Milne, R. G., and Hatta, T. (eds.), 1985, Luteovirus group, in: *Atlas of Plant Viruses*, Volume I, pp. 137–152, CRC Press, Inc., Boca Raton, Florida.

Fritzsche, R., Proeseler, G., Karl, E., and Kleinhempel, H., 1983, Nachweis der Samen-übertragbarkeit des Milden Rübenvergilbungs-Virus (BMYV), *Phytopathol. Z.* **106**:360.

Gibson, R. W., Rice, A. D., and Sawicki, R. M., 1982, Effects of the pyrethroid deltamethrin on the acquisition and inoculation of viruses by *Myzus persicae*, *Ann. Appl. Biol.* **100**:49.

Gildow, F. E., Ballinger, M. E., and Rochow, W. F., 1983, Identification of double-stranded RNAs associated with barley yellow dwarf virus infection of oats, *Phytopathology* **73**:1570.

Gill, C. C., 1980, Assessment of losses on spring wheat naturally infected with barley yellow dwarf virus, *Plant Dis.* **64**:197.

Gill, C. C., and Chong, J., 1979a, Cytopathological evidence for the division of barley yellow dwarf virus isolates into two subgroups, *Virology* **95**:59.

Gill, C. C., and Chong, J., 1979b, Cytological alterations in cells infected with corn leaf aphid-specific isolates of barley yellow dwarf virus, *Phytopathology* **69**:363.

Gold, A. H., and Duffus, J. E., 1967, Infectivity neutralization: A serological method as applied to persistent viruses of beet, *Virology* **31**:308.

Govier, D. A., 1985, Purification and partial characterisation of beet mild yellowing virus and its serological detection in plants and aphids, *Ann. Appl. Biol.* **107**:439.

Grafton, K. F., Poehlmann, J. M., Sechler, D. T., and Sehgal, O. P., 1982, Effect of barley yellow dwarf virus infection on winter survival and other agronomic traits in barley, *Crop Sci.* **22**:596.

Gugerli, P., 1980, Potato leafroll virus concentration in the vascular region of potato tubers examined by enzyme-linked immunosorbent assay (ELISA), *Potato Res.* **23**:137.

Gugerli, P., and Gehriger, W., 1980, Enzyme-linked immunosorbent assay (ELISA) for the detection of potato leafroll virus and potato virus Y in potato tubers after artificial break of dormancy, *Potato Res.* **23**:353.

Haase, D., Proeseler, G., Richter, J., and Eisbein, K., 1985, Möglichkeiten der Diagnose des Gerstengelbverzwergungs-Virus, *Nachrichtenbl. Pflanzenschutz DDR* **3**:52.

Hammond, J., Lister, R. M., and Foster, J. E., 1983, Purification, identity and some properties of an isolate of barley yellow dwarf virus from Indiana, *J. Gen. Virol.* **64**:667.

Hampton, R. O., 1983, Pea leafroll in Northwestern U.S. pea seed production areas, *Plant Dis.* **67**:1306.

Harrison, B. D., 1984, Potato leafroll virus, *CMI/AAB Descriptions of Plant Viruses* No. 291.

Hewings, A. D., Damsteegt, V. D., and Tolin, S. A., 1986, Purification and some properties of two strains of soybean dwarf virus, *Phytopathology* **76**:759.

Hewings, A. D., and D'Arcy, C. J., 1984, Maximizing the detection capability of a beet western yellows virus ELISA system, *J. Virol. Methods* **9**:131.

Hewings, A. D., and D'Arcy, C. J., 1986, Comparative characterization of two luteoviruses: beet western yellows virus and barley yellow dwarf virus, *Phytopathology* **76**:1270.

Hewish, D. R., and Shukla, D. D., 1983, Purification of barley yellow dwarf virus by gel filtration on Sephacryl S-1000 Superfine, *J. Virol. Methods* **7**:223.

Hewish, D. R., Shukla, D. D., Johnstone, G. R., and Sward, R. J., 1983, Monoclonal antibodies to luteoviruses (abstract), *Fourth International Congress of Plant Pathology, Melbourne, Australia* (17-24 August) No. 463.

Hill, S. A., and Jackson, E. A., 1984, An investigation of the reliability of ELISA as a practical test for detecting potato leafroll virus and potato virus Y in tubers, *Plant Pathol.* **33**:21.

Holmes, S. J. I., 1985, Barley yellow dwarf virus in ryegrass and its detection by ELISA, *Plant Pathol.* **34**:214.

Hsu, H. T., Aebig, J., and Rochow, W. F., 1984, Differences among monoclonal antibodies to barley yellow dwarf viruses, *Phytopathology* **74**:600.

Hsu, H. T., Aebig, J., Rochow, W. F., and Lawson, R. H., 1983, Isolations of hybridomas secreting antibodies reactive to RPV and MAV isolates of barley yellow dwarf virus and carnation etched ring virus (abstract), *Phytopathology* **73**:790.

Hull, R., and Adams, A. N., 1968, Groundnut rosette and its assistor virus, *Ann. Appl. Biol.* **62**:139.

Jayasena, K. W., Hatta, T., Francki, R. I. B., and Randles, J. W., 1981, Luteovirus-like particles associated with subterranean clover red leaf virus infection, *J. Gen. Virol.* **57**:205.

Jayasena, K. W., Randles, J. W., and Barnett, O. W., 1984, Synthesis of a complementary DNA probe specific for detecting subterranean clover red leaf virus in plants and aphids, *J. Gen. Virol.* **65**:109.

Jedlinski, H., 1981, The rice root aphid, *Rhopalosiphum rufiabdominalis*, a vector of barley yellow dwarf virus in Illinois, and the disease complex, *Plant Dis.* **65**:975.

Jedlinski, H., 1984, The genetics of resistance to barley yellow dwarf virus in oats, *Barley Yellow Dwarf, A Proceedings of the Workshop, CIMMYT, Mexico,* 101.

Jedlinski, H., Rochow, W. F., and Brown, C. M., 1977, Tolerance to barley yellow dwarf virus in oats, *Phytopathology* **67**:1408.

Jenkyn, J. F., and Plumb, R. T., 1983, Effects of fungicides and insecticides applied to spring barley sown on different dates in 1976-79, *Ann. Appl. Biol.* **102**:421.

Jensen, S. G., 1969, Occurrence of virus particles in the phloem tissue of BYDV-infected barley, *Virology* **38**:83.

Johnstone, G. R., 1984, Control of primary infections of subterranean clover red leaf virus, a luteovirus, in a broad bean crop with the synthetic pyrethroid Deltamethrin, *Australas. Plant Pathol.* **13**:55.

Johnstone, G. R., Ashby, J.W., Gibbs, A. J., Duffus, J. E., Thottappilly, G., and Fletcher, J. D., 1984, The host ranges, classification and identification of eight persistent aphid-transmitted viruses causing diseases in legumes, *Neth. J. Pl. Path.* **90**:225.

Johnstone, G. R., Duffus, J. E., Munro, D., and Ashby, J. W., 1982, Purification of a Tasmanian isolate of subterranean clover red leaf virus, and its serological interactions with a New Zealand isolate and other luteoviruses, *Aust. J. Agric. Res.* **33**:697.

Johnstone, G. R., and McLean, G. D., 1986, Virus diseases of subterranean clover—a review, *Aust. J. Exp. Agric.* (in press).

Johnstone, G. R., and Rapley, P. E. L., 1979, The effect of time of sowing on the incidence of subterranean clover red leaf virus infection in broad bean (Vicia faba), *Ann. Appl. Biol.* **91**:345.

Johnstone, G. R., and Rapley, P. E. L., 1981, Control of subterranean clover red leaf virus in broad bean crops with aphicides, *Ann. Appl. Biol.* **99**:135.

Kastirr, R., Reichenbächer, D., and Haase, D., 1985, ELISA-Nachweis des milden Rübenvergilbungs-Virus (beet mild yellowing virus) und des Gerstengelbverzwergungs-Virus (barley yellow dwarf virus) im Vektor, *Arch. Phytopathol. u. Pflanzenschutz* **21**:331.

Kojima, M., Takizawa, T., Uyeda, I., and Shikata, E., 1982, Application of enzyme-linked immunosorbent assay to diagnosis of potato leafroll disease, *Ann. Phytopat. Soc. Japan* **48**:458.

Kubo, S., 1981, Tobacco necrotic dwarf virus, *CMI/AAB Descriptions of Plant Viruses* No. 234.

Kubo, S., and Takanami, Y., 1979, Infection of tobacco mesophyll protoplasts with tobacco necrotic dwarf virus, a phloem-limited virus, *J. Gen. Virol.* **42**:387.

Kühne, T., Proeseler, G., Richter, J., Stanarius, A., and Proll, E., 1985, Mildes Rübenvergilbungs-Virus (beet mild yellowing virus): Vermehrung, Reinigung und Herstellung von Antiseren, *Arch. Phytopathol. u. Pflanzenschutz* **21**:3.

Lister, R. M., Clement, D., and Skaria, M., 1985, Stability of ELISA activity of barley yellow dwarf virus in leaf samples and extracts, *Plant Dis.* **69**:854.

Lister, R. M., Hammond, J., and Clement, D. L., 1983, Comparison of intradermal and intramuscular injection for raising plant virus antisera for use in ELISA, *J. Virol. Methods* **6**:179.

Lister, R. M., and Rochow, W. F., 1979, Detection of barley yellow dwarf virus by enzyme-linked immunosorbent assay, *Phytopathology* **69**:649.

Martin, R. R., and Converse, R. H., 1985, Purification, properties and serology of strawberry mild yellow-edge virus, *Phytopathol. Z.* **114**:21.

Martin, R. R., and Stace-Smith, R., 1984, Production and characterization of monoclonal antibodies specific to potato leaf roll virus, *Can. J. Plant Pathol.* **6**:206.

Matthews, R. E. F., 1982, Classification and nomenclature of viruses, fourth report of the International Committee on Taxonomy of Viruses, *Intervirology* **17**:3.

Matthews, R. E. F. (ed.), 1983, *A Critical Appraisal of Viral Taxonomy*, CRC Press, Boca Raton, Fla.

Mayo, M. A., and Barker, H., 1984, Translation products and RNA species of potato leafroll virus (PLRV), *Rep. Scott. Crop Res. Inst.* **1983**:186.

Mayo, M. A., Barker, H., Robinson, D. J., Tamada, T., and Harrison, B. D., 1982, Evidence that potato leafroll virus RNA is positive-stranded, is linked to a small protein and does not contain polyadenylate, *J. Gen. Virol.* **59**:163.

Milbrath, G. M., and Duffus, J. E., 1978, Solanum yellows virus, an apparently distinct member of the luteovirus group, *Phytopathol. News* **12**:170.

Mndolwa, D., Bishop, G., Corsini, D., and Pavek, J., 1984, Resistance of potato clones to the green peach aphid and potato leafroll virus, *Am. Potato J.* **61**:713.

Osler, R., 1984, Rice giallume, a disease related to barley yellow dwarf in Italy, *Barley Yellow Dwarf, A Proceedings of the Workshop*, CIMMYT, Mexico, 125.

Paliwal, Y. C., 1977, Rapid diagnosis of barley yellow dwarf virus in plants using serologically specific electron microscopy, *Phytopathol. Z.* **89**:25.

Paliwal, R. C., 1978, Purification and some properties of barley yellow dwarf virus, *Phytopathol. Z.* **92**:240.

Paliwal, Y. C., 1982, Detection of barley yellow dwarf virus in aphids by serologically specific electron microscopy, *Can. J. Bot.* **60**:179.

Plumb, R. T., 1983, Barley yellow dwarf virus—a global problem, in: *Plant Virus Epidemiology* (R. T. Plumb and J. M. Thresh, eds.), p. 185, Blackwell Scientific Publications, Oxford.

Plumb, R. T., 1984, Chemical and cultural control of barley yellow dwarf, *Barley Yellow Dwarf, A Proceedings of the Workshop*, CIMMYT, Mexico, 52.

Prill, B., Breyel, E., and Casper, R., 1986, Klonierung von Luteoviren (Kartoffelblattrollvirus und beet western yellows virus), *Mitt. Biol. Bundesanst. Land- und Forstwirtschaft, Berlin-Dahlem* **232**:388.

Proeseler, G., Richter, K., Kalinina, I., and Kühne, T., 1985, Weitere Ergebnisse bei der Anwendung des ELISA zum Nachweis des Milden Rübenvergilbungs-Virus (beet mild yellowing virus), *Arch. Phytopathol. u. Pflanzenschutz* **21**:437.

Proll, E., Eisbein, K., Haase, D., and Richter, J., 1985, Ein Isolat des Gerstengelbverzwergungs-Virus (barley yellow dwarf virus) mit einer top-Komponente, *Arch. Phytopathol. u. Pflanzenschutz* **21**:243.

Proll, E., Richter, J., Hamann, U., Stanarius, A., Eisenbrandt, K., and Spaar, D., 1984, Kartoffelblattroll-Virus (potato leafroll virus): Reinigung und Herstellung von Antiseren, *Arch. Phytopathol. u. Pflanzenschutz* **20**:1.

Rabenstein, F., Kühne, T., Richter, J., and Kleinhempel, H., 1984, Herstellung monoklonaler Antikörper gegen das Milde Rübenvergilbungs-Virus (beet mild yellowing virus), *Arch. Phytopathol. u. Pflanzenschutz* **20**:517.

Reddy, D. V. R., Murant, A. F., Raschke, J. H., Mayo, M. A., and Ansa, O. A., 1985, Properties and partial purification of infective material from plants containing groundnut rosette virus, *Ann. Appl. Biol.* **107**:65.

Reichenbächer, D., Ulbricht, G., Golke, U., Kilian, J., Schulze, M., and Horn, A., 1985, Zur Frage der visuellen Auswertung eines Ultramikro-ELISA bei Kartoffelviren, *Arch. Phytopathol. u. Pflanzenschutz* **21**:95.

Richter, J., Reichenbächer, D., and Ulbricht, G., 1985, Einsatz des ELISA zum Nachweis von Kartoffelviren bei der Pflanzgutkontrolle, *Saat u. Pflanzgut* **26**:141.

Rochow, W. F., 1969, Biological properties of four isolates of barley yellow dwarf virus, *Phytopathology* **59**:1580.

Rochow, W. F., 1970, Barley yellow dwarf, *CMI/AAB Descriptions of Plant Viruses* No. 32.

Rochow, W. F., 1984, Appendix I, barley yellow dwarf, A Proceedings of the Workshop, *CIMMYT, Mexico* (December 6-8, 1983), 204.

Rochow, W. F., Aapola, A. I. E., Brakke, M. K., and Carmichael, L. E., 1971, Purification and antigenicity of three isolates of barley yellow dwarf virus, *Virology* **46**:117.

Rochow, W. F., and Ball, E. M., 1967, Serological blocking of aphid transmission of barley yellow dwarf virus, *Virology* **33**:359.

Rochow, W. F., and Brakke, M. K., 1964, Purification of barley yellow dwarf virus, *Virology* **24**:310.

Rochow, W. F., and Duffus, J. E., 1978, Relationships between barley yellow dwarf and beet western yellows viruses, *Phytopathology* **68**:51.

Rochow, W. F., and Duffus, J. E., 1981, Luteoviruses and yellow diseases, in: *Handbook of Plant Virus Infections and Comparative Diagnosis* (E. Kurstak, ed.), pp. 147–170, Elsevier/North-Holland Biomedical Press, Amsterdam.

Rochow, W. F., Muller, I., Tufford, L. A., and Smith, D. M., 1986, Identification of luteoviruses of small grains from 1981 through 1984 by two methods, *Plant Dis.* **70**:461.

Rowhani, A., and Stace-Smith, R., 1979, Purification and characterization of potato leafroll virus, *Virology* **98**:45.

Schaller, C. W., Rasmusson, D. C., Qualset, C. O., 1963, Sources of resistance to the yellow-dwarf virus in barley, *Crop Sci.* **3**:342–344.

Schaller, C. W., 1984, The genetics of resistance to barley yellow dwarf virus in barley, *Barley Yellow Dwarf*, A Proceedings of the Workshop, CIMMYT, Mexico, 93.

Schaller, C. W., and Qualset, C. O., 1980, Breeding for resistance to the barley yellow dwarf virus, in: *Proceedings, Third International Wheat Conference, Madrid, Spain*, pp. 528–541, University of Nebraska Agricultural Experiment Station publication MP41.

Sharma, H. C., Gill, B. S., and Uyemoto, J. K., 1984, High levels of resistance in *Agropyron* species to barley yellow dwarf and wheat streak mosaic viruses, *Phytopathol. Z.* **110**:143.

Shepardson, S., Esau, K., and McCrum, R., 1980, Ultrastructure of potato leaf phloem infected with potato leafroll virus, *Virology* **105**:379.

Shepherd, R. J., Francki, R. I. B., Hirth, L., Hollings, M., Inouye, T., MacLeod, R., Purcifull, D. E., Sinha, R. C., Tremaine, J. H., Valenta, V., and Wetter, C., 1976, New groups of plant viruses approved by the International Committee on Taxonomy of Viruses (September 1975), *Intervirology* **6**:181.

Singh, R. P., and Somerville, T. H., 1983, Effect of storage temperatures on potato virus infectivity levels and serological detection by enzyme-linked immunosorbent assay, *Plant Dis.* **67**:1133.

Spiegel, S., Cohen, J., and Converse, R. H., 1985, Detection of strawberry mild yellow-edge virus by serologically specific electron microscopy, *Acta Horticulturae* **186**:95.

Stace-Smith, R., and Converse, R. H., 1970, Raspberry leaf curl, in: *Virus Diseases of Small Fruits and Grapevines (A Handbook)* (N. W. Frazier, ed.), pp. 120–122, University of California, Div. of Agric. Sciences, Berkeley, California.

Stoltz, R. L., and Forster, R. L., 1984, Reduction of pea leaf roll of peas (*Pisum sativum*) with systemic insecticides to control the pea aphid (Homoptera: Aphididae) vector, *J. Econ. Entomol.* **77**:1537.

Sylvester, E. S., and Osler, R., 1977, Further studies on the transmission of the filaree red-leaf virus by the aphid *Acyrthosiphon pelargonii zerozalphum, Environ. Entomol.* **6**:39.

Takanami, Y., and Kubo, S., 1979a, Enzyme-assisted purification of two phloem-limited plant viruses: Tobacco necrotic dwarf and potato leafroll, *J. Gen. Virol.* **44**:153.

Takanami, Y., and Kubo, S., 1979b, Nucleic acids of two phloem-limited viruses: Tobacco necrotic dwarf and potato leafroll, *J. Gen. Virol.* **44**:853.

Tamada, T., and Harrison, B. D., 1980a, Factors affecting the detection of potato leafroll virus in potato foliage by enzyme-linked immunosorbent assay, *Ann. Appl. Biol.* **95**:209.

Tamada, T., and Harrison, B. D., 1980b, Application of enzyme-linked immunosorbent assay to the detection of potato leafroll virus in potato tubers, *Ann. Appl. Biol.* **96**:67.

Tamada, T., and Harrison, B. D., 1981, Quantitative studies on the uptake and retention of potato leafroll virus by aphids in laboratory and field conditions, *Ann. Appl. Biol.* **98**:261.

Tamada, T., and Kojima, M., 1977, Soybean dwarf virus, *CMI/AAB Descriptions of Plant Viruses* No. 179.

Thomas J. E., 1984, Characterization of an Australian isolate of tomato yellow top virus, *Ann. Appl. Biol.* **104**:79.

Thomas, J. E., 1985, Recent studies on virus diseases of tomato in Queensland, Australia, *Fifth Conference of the ISHS-Vegetable Virus Working Group, Bet Dagan, Israel* (1-4 September) No. 29.

Tola, J. E., and Kronstad, W. E., 1984, The genetics of resistance to barley yellow dwarf virus in wheat, *Barley Yellow Dwarf, A Proceedings of the Workshop, CIMMYT, Mexico,* 83.

Torrance, L., and Jones, R. A. C., 1982, Increased sensitivity of detection of plant viruses obtained by using a fluorogenic substrate in enzyme-linked immunosorbent assay, *Ann. Appl. Biol.* **101**:501.

Torrance, L., Plumb, R. T., Lennon, E. A., and Gutteridge, R. A., 1986, A comparison of ELISA with transmission tests to detect barley yellow dwarf virus-carrying aphids, in: *Developments and Applications in Virus Testing* (R. A. C. Jones, and L. Torrance, eds.), pp. 165–176, Publ. Association of Applied Biologists, Wellesbourne.

Waterhouse, P. M., Gerlach, W. L., and Miller, W. A., 1986, Serotype-specific and general luteovirus probes from cloned cDNA sequences of barley yellow dwarf virus, *J. Gen. Virol.* **67**:1273.

Waterhouse, P. M., and Helms, K., 1984, Purification of particles of subterranean clover red leaf virus using an industrial-grade cellulase, *J. Virol. Methods* **8**:321.

Waterhouse, P. M., and Murant, A. F., 1981, Purification of carrot red leaf virus and evidence from four serological tests for its relationship to luteoviruses, *Ann. Appl. Biol.* **97**:191.

Waterhouse, P. M., and Murant, A. F., 1982, Carrot red leaf virus, *CMI/AAB Descriptions of Plant Viruses* No. 249.

Weidemann, H. L., 1982, Zur Vermehrung des Kartoffelblattrollvirus in der Blattlaus *Myzus persicae* (Sulz.), *Z. Angew. Entomol.* **94**:321.

CHAPTER 9

Maize Chlorotic Dwarf and Related Viruses

R. E. GINGERY

I. INTRODUCTION AND HISTORY

Maize chlorotic dwarf virus (MCDV), which has a polyhedral particle
about 30 nm in diameter, was designated as the type member of the Maize
Chlorotic Dwarf Virus group in 1981 by the International Committee on
Taxonomy of Viruses (Matthews, 1982). The rice tungro spherical virus
(RTSV) has similar characteristics and is listed as a probable member.
RTSV is one of two morphologically distinct particles involved in rice
tungro and related diseases of rice in the Far East. The second is that of
the rice tungro bacilliform virus (RTBV), which is 35 nm in diameter and
150–350 nm long (Hibino *et al.*, 1978; Saito, 1977; Saito *et al.*, 1981). In
many rice tungro studies the contribution of RTSV cannot be separated
from that of RTBV. Therefore, this discussion will focus on MCDV and
the disease it causes, maize chlorotic dwarf (MCD), but will include
information on rice tungro and related diseases if the spherical particle
was clearly identified. Diseases considered to be related to the rice tungro
disease found in the Philippines are: rice waika in Japan, leaf yellowing
in India, penyakit habang and mentek in Indonesia, penyakit merah in
Malaysia, and yellow-orange leaf in Thailand (Ling and Tiongco, 1979).
Several reviews summarize information on this group of diseases (Ling,
1972; Ling and Tiongco, 1979; Shikata, 1979).

R. E. GINGERY • United States Department of Agriculture, Agricultural Research Service, Department of Plant Pathology, The Ohio State University, Ohio Agricultural Research
and Development Center, Wooster, Ohio 44691.

MCDV was discovered in 1969 by Rosenkranz, who described a pathogen transmitted by the blackfaced leafhopper, *Graminella nigrifrons* (Forbes), from a stunted maize plant in southern Ohio. He named the new pathogen the corn stunt agent-Ohio strain (CSA-OH), but did not demonstrate its viral nature. Shortly thereafter, a 30-nm isometric viruslike particle was identified from such plants (Bradfute *et al.*, 1972a,b; Pirone *et al.*, 1972) and was named MCDV (Nault *et al.*, 1973). To assess MCDV's importance, Gordon and Nault (1977) conducted an extensive survey of maize-stunting diseases in the United States and found MCDV infection associated with 76% of stunted maize plants from 16 states. These results and others indicate that MCD is currently either the most important virus disease affecting maize in the United States or the second most important after maize dwarf mosaic, which is caused by the maize dwarf mosaic virus (Gordon *et al.*, 1981).

Properties distinguishing MCDV and RTSV from other viruses of similar morphology are rapid sedimentation rates and semipersistent relationships with their leafhopper vectors; i.e., the insects can transmit the virus for only a few days following virus acquisition (Nault *et al.*, 1973; Ling, 1966). All other known virus-leafhopper relationships are of the persistent type in which the virus is transmitted for weeks or for the life of the vector.

II. GEOGRAPHICAL DISTRIBUTION, DISEASES, AND HOST RANGES

A. Geographical Distribution

MCDV occurs in 19 states throughout the southeastern U.S.A., from the Gulf of Mexico on the south to states bordering the Ohio River plus Pennsylvania on the north, and from the Atlantic coast westward to eastern Texas (Ayers *et al.*, 1978; Damsteegt, 1976; Gordon and Nault, 1977; Gordon *et al.*, 1981). This is essentially the area defined by the overlapping distributions of the overwintering host, johnsongrass (*Sorghum halepense* (L.) Pers.), and the vector, *G. nigrifrons*. MCDV probably occurs in countries south of the United States where johnsongrass is present because *Graminella* species, including *G. nigrifrons*, are known to have neotropical distributions (Kramer, 1967; Stoner and Gustin, 1967). In fact, there is one report of MCDV in Mexico (Gordon *et al.*, 1981).

B. Main Diseases

MCDV causes a variety of symptoms including plant stunting, leaf reddening and yellowing (Bradfute *et al.*, 1972a,b; Gordon and Nault,

1977; Rosenkranz, 1969), and a subtle chlorosis of the smallest or tertiary leaf veins that is diagnostic for MCDV infection (Gordon and Nault, 1977; Louie *et al.*, 1974). This veinal chlorosis has been referred to as chlorotic striping of tertiary veins (Gordon and Nault, 1977), vein banding (Louie *et al.*, 1974), and vein clearing (Nault and Bradfute, 1979). Leaf yellowing, leaf reddening, and plant stunting are correlated with MCDV infection, but only vein clearing is unique (Gordon and Nault, 1977). In most maize varieties, the first symptom of MCDV infection in the greenhouse is a chlorotic mottle at the base of the whorl 5–8 days after inoculation (Nault and Bradfute, 1979). Vein clearing follows, usually on all subsequently developing leaves, but sometimes not on the youngest leaves of tolerant varieties. Other symptoms, mainly in field-infected plants, are "miniature" plants in which leaves and internodes are proportionately shortened (Rosenkranz, 1969); necrosis at the base of the stalk resulting in early death of the plant; high incidence of tassel seed in some lines (Choudhury and Rosenkranz, 1973); leaves with dull and rough upper surfaces that are less easily torn from leaf sheaths than are healthy leaves (Louie *et al.*, 1974); and chlorosis and tearing of leaf margins in severely infected plants (Nault and Bradfute, 1979). It is puzzling that usually only vein clearing is observed in maize plants infected with MCDV in the greenhouse. Perhaps selecting for vein clearing during serial transfers attenuates the virus (Gingery *et al.*, 1981), or maybe a second particle, analogous to the situation in rice tungro, is associated with MCDV in the field but is lost in greenhouse transfers.

Rice plants infected with RTSV alone show only moderate stunting. RTBV causes the more severe tungrolike symptoms (leaf discoloration, severe stunting) that are intensified if RTSV is also present (Hibino *et al.*, 1978).

C. Host Range

Only graminous species are susceptible to MCDV (Nault *et al.*, 1976). In addition to maize, susceptible cultivated species include sorghum (*Sorghum bicolor* (L.) Moench), Sudan grass (*S. sudanense* (Piper) Stapf.), Proso millet (*Panicum miliaceum* L.), pearl millet (*Pennisetum americanum* (L.) Leeke), and wheat (*Triticum aestivum* L.). Wheat and sorghum are symptomless, and symptoms on the others are mild. Although johnsongrass is the only perennial grass known to be susceptible to MCDV, there are several susceptible annual grasses including crabgrass (*Digitaria sanguinalis* (L.) Scop.), *Coix lacryma-jobi* L., *Echinochloa crusgalli* (L.) Beauv., *Eleusine indica* (L.)Gaertn., *Panicum capillare* L., *Pennisetum glaucum* (L.) R. Br., the foxtails *Seteria faberii* Herrm., *S. lutescens* (Weigel) F. T. Hubb, *S. magma* Griseb., and *S. veridis* (L.) Beauv., and the teosinte, *Z. luxurians* (Durieu and Ascherson). Of the 18 susceptible species, ten are Panicoids, six are Andropogonoids, one is a Chloridoid, and one is a Festucoid. Forty-two other graminous species are not susceptible (Nault *et al.*, 1976).

D. Losses

Although MCDV presumably causes significant losses annually in the United States, few systematic studies have been done. Yield losses have not been documented satisfactorily, but it seems clear from experimental studies that they can be quite high (Gordon *et al.*, 1981). The amount of loss appears to depend on the age of the plant at the time of infection, the susceptibility of the maize genotype, and probably other unknown factors. Main *et al.* (1984) reported a 5% loss in North Carolina, and Scott *et al.* (1977) observed that the younger the plant at the time of infection, the greater the yield loss.

III. TRANSMISSION, EPIDEMIOLOGY, AND CONTROL

A. Transmission

So far, MCDV has been transmitted only by leafhoppers and only from plant to plant. Both adult *G. nigrifrons* males and females transmit MCDV from infected to healthy plants, but females do not transmit the virus transovarially (Choudhury and Rosenkranz, 1983; Nault *et al.*, 1973). Nymphs transmit MCDV as well, but lose their inoculativity following a molt (Nault *et al.*, 1973). Both the lesser lawn leafhopper, *Graminella sonora* (= *Deltocephalus sonorus*) (Ball), and the gray lawn leafhopper, *Exitianus exitiosus* (Uhler), transmit MCDV experimentally, but much less efficiently than *G. nigrifrons* (Knoke *et al.*, 1983; Nault and Knoke, 1981). In one study, 41.4% of *G. nigrifrons* transmitted MCDV, but only 7.1% of *G. sonora* did (Nault and Knoke, 1981). In another, the relative rates of transmission by *G. nigrifrons*, *G. sonora*, and *E. exitiosus* were 1.0, 0.18, and 0.23, respectively, with about one-third of the *G. nigrifrons* transmitting (Knoke *et al.*, 1983). After a 48-hr acquisition-access period on infected maize, at least 86% of *G. nigrifrons* individuals eventually transmitted MCDV when inoculativity was followed for six weeks (Knoke *et al.*, 1983). Known nonvectors include *Baldulus tripsaci* (Kramer and Whitcomb), *Dalbulus elimatus* (Ball), *D. maidis* (Delong and Wolcott) (Nault and Bradfute, 1979), *Macrosteles fascifrons* (Stal), and *Stirellus bicolor* (Van Duzee) (Nault, personal communication).

G. *nigrifrons* can acquire MCDV from diseased plants and inoculate healthy plants, both in as little as 15 min (Choudhury and Rosenkranz,

1983) and with no intervening latent period in the vector (Choudhury and Rosenkranz, 1983; Nault et al., 1973). However, transmission is more efficient with longer acquisition-access periods (Nault, 1977). *G. nigrifrons* retains MCDV for two to four days after acquisition in what has been described as a semipersistent relationship (Nault et al., 1973). The retention time depends on temperature, and ranges from two days at 30°C to four days at 15, 20, and 25°C (Nault, 1977). Choudhury and Rosenkranz (1983) also reported retention times of two to four days at daily temperatures of 20–32°C. MCDV transmission was first reported to be persistent, in which inoculativity persisted for weeks (Rosenkranz, 1969), but this observation was later attributed to reacquisition of the virus by the leafhoppers during the week-long intervals that they were held on healthy test plants (Choudhury and Rosenkranz, 1983; Nault and Bradfute, 1979). Consistent with such reacquisition, *G. nigrifrons* can acquire MCDV from infected plants within three to five days after inoculation, which is two to four days before symptoms appear (Choudhury and Rosenkranz, 1983; Nault and Bradfute, 1979).

The term "semipersistent" to describe the two- to four-day retention of MCDV by *G. nigrifrons* was first used by Nault et al. (1973) and was based on Sylvester's (1969) definitions of aphid–virus relationships. Disagreeing with this terminology, Ling and Tiongco (1979) and Choudhury and Rosenkranz (1983) used the term "transitory," arguing that semipersistent implies retention times intermediate between nonpersistent and persistent, and is therefore inappropriate because there is no known nonpersistent transmission among leafhoppers. However, because neither term addresses the mechanism of transmission, both are less than ideal, and Nault and Bradfute (1979) suggest that new terminology (i.e., transitory) should be avoided until this relationship is better understood. Moreover, because semipersistent does have a precise technical meaning for aphid–virus relationships, it is reasonable to use it to describe similar retention times in leafhopper-virus systems, despite the absence of a nonpersistent relationship for leafhopper-borne viruses.

No transstadial passage, no latent period, and short retention times prompted Nault and Bradfute (1979) to speculate that MCDV may adsorb to a site or sites in the foreguts of vectors. Expanding on this idea, Harris (1979) proposed an ingestion–egestion hypothesis in which virus is ingested during feeding, attaches to the foregut, and is egested during later probes. This hypothesis is in accord with observations of MCDV-like particles attached to the foregut of insects that had fed on MCDV-infected maize, but not those that had fed on healthy controls (Harris, 1981; Harris and Childress, 1981). Unfortunately, these particles were not shown to be infective, transmissible, or even to be MCDV. Also, no mechanism of attachment and detachment of the virus was hypothesized.

RTSV is also semipersistently transmitted, predominantly by *Nephotettix virescens* (Distant) (= *N. impicticeps* Ishihara). The virus can be acquired in a 30-minute acquisition-access period, retained by the insect for about three days, and transmitted in a 10-min inoculation-access period (Hibino *et al.*, 1979). Other reported vectors of RTSV include *N. cincticeps* (Uhler), *N. nigropictus* (Stal), *N. malayanus* (Ishihara and Kawase), *N. parvus* (Ishihara and Kawase), and *Recilia dorsalis* (Mot-schulsky) (Hibino, 1983; Ling, 1972; 1973).

Rice waika virus (RWV) from Japan is transmitted mainly by *N. cincticeps* (Saito, 1977) and is apparently identical or very closely related to RTSV (Hibino, 1983; Saito, 1977). However, no RTBV-like particles have been found in RWV-infected plants and rice waika may be caused by a spherical particle alone (Hibino, 1983). RTSV can be transmitted alone, but RTBV is dependent on RTSV for its transmission (Hibino *et al.*, 1978). RWV can also assist in RTBV transmission (Hibino, 1983).

B. Epidemiology

Johnsongrass, a perennial weedgrass, appears to be the main and perhaps the only overwintering reservoir for MCDV (Nault *et al.*, 1976). Supporting this view, MCDV is found in field-collected johnsongrass (Reeves *et al.*, 1978; Nault and Bradfute, 1979; Pirone *et al.*, 1972) and MCD epiphytotics are limited to areas where johnsongrass is abundant (Gordon and Nault, 1977). In a study in southern Indiana, 4 of the 27 johnsongrass plants sampled were infected with MCDV (Reeves *et al.*, 1978). All other known plant hosts of MCDV are annuals (Nault *et al.*, 1976) and are unlikely to function as overwintering hosts.

Most of the natural spread of MCDV is probably due to *G. nigrifrons* (Nault *et al.*, 1973) because MCD incidence and spread are highly correlated with the numbers of *G. nigrifrons* (All *et al.*, 1977; Kuhn *et al.*, 1975). However, it is unclear whether *G. nigrifrons* overwinters in the northern areas of MCD occurrence or migrates into these regions each year (Gustin and Stoner, 1968; Stoner and Gustin, 1967). In either case, large numbers of *G. nigrifrons* build up in these areas each year on small grains and grasses, especially ryegrass (*Lolium perenne* L.), barnyardgrass (*Echinochloa crusgalli* (L.) Beauv.), crabgrass (*Digitaria sanguinalis* (L.) Scop.), and bermudagrass (*Cynadon dactylon* (L.) Pers.) (Knoke *et al.*, 1983). MCD is endemic throughout the region of overlap of the distributions of johnsongrass and *G. nigrifrons* (Gordon and Nault, 1977; Nault *et al.*, 1973).

The likelihood of MCDV infection appears to vary throughout the growing season. In Ohio, disease potential, as measured by the percentage of seedlings becoming infected after a week's exposure to field conditions, was first detected in early June, peaked in mid-July, and then sharply declined (Louie *et al.*, 1974). In other studies, variations in *G. nigrifrons* populations have been noted, but were not correlated with disease potential. Peak populations of *G. nigrifrons* were observed in late May and early June in Alabama (Stevens *et al.*, 1976) and Kentucky (Sedlacek and Freytag, 1983); in late July with lesser peaks in late May–early June and early September in one study in Ohio (Knoke and Louie, 1981); and in June and late July–early August in another (Knoke *et al.*, 1983).

G. *nigrifrons* prefers grasses and small grains to maize as feeding hosts (Knoke *et al.*, 1983), a trait that Stevens *et al.* (1976) felt might result in its being more mobile in maize fields and thus more efficient at spreading MCDV than if it were well-adapted to maize and moved about less. However, the rate of movement of G. *nigrifrons*, measured by following the movement of rubidium-tagged insects released from a point source, was quite slow, only 15 m in four days (Alverson *et al.*, 1980). J.K. Knoke (personal communication) also estimated slow movement, about 6 m per month, based on the rate of disease spread from a point source. G. *nigrifrons* appears to be attracted to maize in the early whorl (four-leaf) stage, especially in no-tillage fields that have a yellow background (Sedlacek and Freytag, 1982; 1983). The numbers of G. *nigrifrons* on maize drop dramatically after the early whorl stage (Sedlacek and Freytag, 1982; 1983).

C. Control

Control of MCDV is attempted most often through the use of tolerant maize varieties. In two tests, no immune lines were found among 104 dent corn inbreds and 158 commercial and experimental sweet corns, although some were tolerant (Guthrie *et al.*, 1982; Knoke *et al.*, 1981).

The use of systemic insecticides, usually carbofuran (2,3-dihydro-2,2-dimethyl-7-benzofuranyl methylcarbamate), can reduce MCD incidence (Bhirud and Pitre, 1972; Keaster and Fairchild, 1968; Pitre, 1968). In one study (Kuhn *et al.*, 1975), carbofuran applied in the furrow at the time of planting reduced MCD incidences about 75% and 50% and increased yields 37% and 125% for a moderately susceptible and a highly susceptible hybrid, respectively. The G. *nigrifrons* population was reduced 65–74% in these plots. The efficacy of carbofuran for reducing MCD incidence was later confirmed in the field (Stevens *et al.*, 1977) and greenhouse (Rains and Christensen, 1983).

N. *virescens* is very susceptible to insecticides, and the use of insecticides to control this insect also controls the tungro disease. In fact, insecticide treatment is the most common method of controlling both rice diseases caused by insect-borne viruses, and plant damage caused by the insects themselves (Heinricks, 1979). Planting rice varieties that are resistant to the vector may be of some value in controlling rice tungro (Pathak, 1970), but not if disease pressure is high (Heinricks, 1979).

Other recommendations for the control of MCD have included spraying with oils that interfere with leafhopper probing (Harris, 1981; Simons and Zitter, 1980), planting early so that plants are not exposed to high G. *nigrifrons* populations until they are older and more tolerant of infection (All, 1983; Ford and Milbrath, 1981; Keaster *et al.*, 1969; Pitre, 1968), and eradicating johnsongrass, the overwintering MCDV host, and other grasses that could support large numbers of leafhoppers (All, 1983; Kuhn et al., 1975). Several workers have suggested that a combined approach involving the use of resistant maize, johnsongrass eradication, and systemic insecticide may give the best control of MCD (All, 1983; All et al., 1976; 1977; Crawford et al., 1978). However, economic constraints, particularly the cost of insecticide, often preclude such comprehensive control efforts, and the single most effective control strategy remains the planting of resistant varieties (All et al., 1977; Crawford et al., 1978; Gordon, 1974a,b).

The genetics of MCDV resistance is poorly understood, in large part because MCDV is not mechanically transmissible. Thus, genetic studies must rely on insect transmission, which can lead to variable inoculations due to variable insect behavior. Consequently, it is difficult to control disease pressure from experiment to experiment and impossible to separate resistance to the virus from resistance to the vector. Other factors leading to uncertainties in the genetics of MCDV resistance have been: (1) the possibility of the confounding effect of pathogens other than MCDV in field studies, particularly maize dwarf mosaic virus; (2) masking or altering of symptom expression (and therefore disease rating) caused by hybrid vigor; and (3) variable symptom expression depending on the age of the plant at the time of infection (Findley et al., 1983). Moreover, because most current breeding efforts are aimed at tolerance, not immunity, even "resistant" plants are often infected and symptomatic, making genetic studies difficult to quantify and interpret (Findley et al., 1981). This probably explains the conflicting views of MCD resistance. For example, Naidu and Josephson (1976) reported as many as four genes for MCD resistance, but Scott and Rosenkranz (1977) found only three, on chromosomes 1, 3, and 4. In yet another study, Grogan and Rosenkranz (1968) found no evidence for dominant resistance genes and concluded that resistance resulted from an additive effect of several genes.

A possible source of MCDV resistance may be the perennial teosinte, Z. diploperennis (Iltis, Doebley, and Guzman), which is both immune to MCDV (Nault et al., 1982) and interfertile with maize (Iltis et al., 1979). Experiments to transfer the teosinte resistance gene(s) into maize at first suggested that the resistance to MCDV in Z. diploperennis is controlled by two dominant complementary genes with several minor genes also involved (Findley et al., 1983). However, difficulty has been encountered in stabilizing the alleged resistance genes in a homozygous state, prompting recent speculation that resistance results from an additive effect involving several genes (W. R. Findley and L. R. Nault, personal communication).

IV. PURIFICATION

MCDV and RTSV have been purified by extracting infected leaf tissue in phosphate or EDTA buffers ranging in pH from 6.0 to 7.0 and in concentration from 0.01 to 0.5 M, clarifying with chloroform, carbon tetrachloride, or heating, concentrating by centrifugation, polyethylene glycol precipitation, or precipitation at low ionic strength. Both viruses can be further purified by rate-zonal centrifugation in sucrose density gradients and isopycnic centrifugation in cesium chloride. Detailed procedures are given by Gingery (1976; 1978), Gordon and Nault (1977), and Louie et al. (1974) for MCDV, and by Galvez (1968) and Omura et al. (1983) for RTSV.

Attempts to demonstrate the infectivity of purified preparations of MCDV by reintroducing the virus into plants via membrane-fed or needle-injected leafhoppers have been unsuccessful (Gingery and Nault, unpublished work), and the evidence that the MCDV particle is a pathogen, although strong, remains circumstantial. By membrane feeding of *N. impicticeps*, Galvez (1968) reported infectivity associated with zones in sucrose density gradients containing spherical particles from rice tungro-diseased rice.

V. PARTICLE PROPERTIES

The properties of MCDV and RTSV are given in Table I. Rapid sedimentation rates and high buoyant densities distinguish MCDV and RTSV from other viruses of similar morphology.

Little is known about the viral proteins. Two or three bands ranging from 18,000 to 30,000 daltons have been detected after polyacrylamide gel electrophoresis of sodium dodecyl sulfate-treated purified virions (Gingery et al., 1981; Gingery, unpublished work).

TABLE I. Properties of Maize Chlorotic Dwarf and Rice Tungro Spherical (RTSV) Viruses

Property	MCDV	RTSV
Particle diameter	30 nm[a-c]	30–33 nm[d,e]
Buoyant density in CsCl	1.507 g/ml[f,g]	1.551 g/ml[e]
		1.49 g/ml[h]
Sedimentation coefficient	183 ± 6 S[i]	175 ± 5 S[d]
Absorbance at 260 nm (1 mg/ml, 1 cm light path)	5.9[f]	
A260/280	1.92 ± 0.05[f]	1.75[e]
		1.41 ± 0.08[d]
Ultraviolet absorbance	258 nm (max),[f]	259 nm (max),[d]
	236 nm (min)	238 nm (min)
Virion molecular weight (daltons)	8.8×10^6[f]	
Percent RNA in virion	36[f]	
RNA molecular weight	3.2×10^6[f]	
Molar percentage of nucleotides	G24, A30, C17, U29[f]	

[a] Bradfute et al. (1972a). [b]Bradfute et al. (1972b). [c]Pirone et al. (1972). [d]Galvez (1968). [e]Omura et al. (1983). [f]Gingery (1976). [g]Gordon and Gingery (1974). [h]Gingery (unpublished work). [i]Gordon (personal communication).

VI. SEROLOGY AND IMMUNOCHEMISTRY

Antisera to MCDV have been prepared by Gordon and Nault (1977) and Reeves *et al.* (1978) and to RTSV by Omura *et al.* (1983) and Saito (1977). MCDV can be serologically detected by agar gel double diffusion (Gordon, 1977), immune density gradient centrifugation (Gordon and Nault, 1977), immunofluorescence (Gingery, 1978), enzyme-linked immunosorbent assay (ELISA) (Reeves *et al.*, 1978), and immunospecific electron microscopy (Derrick and Brlansky, 1976). RTSV can be detected by precipitin ring interface and agar gel double-diffusion tests (Omura *et al.*, 1983). MCDV and RTSV are unrelated serologically (Gordon, Gingery, and Ling, unpublished work), but RTSV and RWV are very closely related serologically (Saito, 1977).

VII. GENOME PROPERTIES AND REPLICATION

Some properties of the RNAs of MCDV and RTSV are listed in Table I. There is no information on their translational strategies or replication.

VIII. RELATIONS WITH CELLS AND TISSUES

Two unique inclusions are found in vascular and occasionally mesophyll cells of MCDV-infected leaves. One is a dense granular inclusion that contains embedded 31-nm MCDV-like particles; the other is a striated sheet inclusion that in some ways resembles the pinwheel inclusions associated with Potyviruses (Bradfute *et al.*, 1972a,b; Bradfute and Robertson, 1981; Francki *et al.*, 1985; Pirone *et al.*, 1972). Isometric viruslike particles are also found individually and aggregated in the cytoplasm and central vacuole of phloem cells (Bradfute *et al.*, 1972b; Harris and Childress, 1983). Harris and Childress (1983) also report laminate inclusions, numerous vesicles, and fibrillar material of unknown composition in MCDV-infected cells.

REFERENCES

All, J. N., 1983, Integrating techniques of vector and weed-host suppression into control programs for maize virus diseases, in: *Proceedings of the International Maize Virus Disease Colloquium and Workshop* (D. T. Gordon, J. K. Knoke, L. R. Nault, and R. M. Ritter, eds.), pp. 243–247, The Ohio State University, Ohio Agricultural Research and Development Center, Wooster.

All, J. N., Kuhn, C. W., and Jellum, M. D., 1976, The changing status of corn virus diseases: Potential value of a systematic insecticide, *Georgia Agricultural Research* 17(4):4.

All, J. N., Kuhn, C. W., Gallaher, R. N., Jellum, M. D., and Hussey, R. S., 1977, Influence of no-tillage-cropping, carbofuran, and hybrid resistance on dynamics of maize chlorotic dwarf and maize dwarf mosaic diseases of corn, *J. Econ. Entomol.* 70:221.

Alverson, D. R., All, J. N., and Kuhn, C. W., 1980, Simulated intrafield dispersal of maize chlorotic dwarf virus by *Graminella nigrifrons* with a rubidium marker, *Phytopathology* **70:**734.

Ayers, J. E., Boyle, J. S., and Gordon, D. T., 1978, The occurrence of maize chlorotic dwarf and maize dwarf mosaic viruses in Pennsylvania in 1977, *Plant Dis. Rep.* **62:**820.

Bhirud, K. M., and Pitre, H. N., 1972, Bioactivity of systemic insecticides in corn: Relationship to leafhopper vector control and corn stunt disease incidence, *J. Econ. Entomol.* **65:**1134.

Bradfute, O. E., and Robertson, D. C., 1981, Electron microscopy of viruses and virus-infected cells of maize, in: *Virus and Viruslike Diseases of Maize in the United States* (D. T. Gordon, J. K. Knoke, and G. E. Scott, eds.), pp. 25–32, Southern Cooperative Series Bulletin 247, Ohio Agricultural Research and Development Center, Wooster.

Bradfute, O. E., Gingery, R. E., Gordon, D. T., and Nault, L. R., 1972a, Tissue ultrastructure, sedimentation and leafhopper transmission of a virus associated with maize dwarfing disease, *J. Cell Biol.* **55:**25a.

Bradfute, O. E., Louie, R., and Knoke, J. K., 1972b, Isometric viruslike particles in maize with stunt symptoms, *Phytopathology* **62:**748.

Choudhury, M. M., and Rosenkranz, E., 1973, Differential transmission of Mississippi and Ohio corn stunt agents by *Graminella nigrifrons, Phytopathology* **63:**127.

Choudhury, M. M., and Rosenkranz, E., 1983, Vector relationship of *Graminella nigrifrons* to maize chlorotic dwarf virus, *Phytopathology* **73:**685.

Crawford, J. L., Kuhn, C. W., and Jellum, M. D., 1978, Corn virus disease control, *University of Georgia College of Agriculture Cooperative Extension Service Circular* No. 635 (revised).

Damsteegt, V. D., 1976, A naturally occurring corn virus epiphytotic, *Plant Dis. Rep.* **60:**858.

Derrick, K. S., and Brlansky, R. H., 1976, Assay for viruses and mycoplasmas using serologically specific electron microscopy, *Phytopathology* **66:**815.

Findley, W. R., Josephson, L. M., and Dollinger, E. J., 1981, Breeding for disease resistance in corn, in: *Virus and Viruslike Diseases of Maize in the United States* (D. T. Gordon, J. K. Knoke, and G. E. Scott, eds.), pp. 137–140, Southern Cooperative Series Bulletin 247, Ohio Agricultural Research and Development Center, Wooster.

Findley, W. R., Nault, L. R., Styer, W. E., and Gordon, D. T., 1983, *Zea diploperennis* as a source of maize chlorotic dwarf virus resistance: A progress report, in: *Proceedings of the International Maize Virus Disease Colloquium and Workshop* (D. T. Gordon, J. K. Knoke, L. R. Nault, and R. M. Ritter, eds.), pp. 255–257, The Ohio State University, Ohio Agricultural Research and Development Center, Wooster.

Ford, R. E., and Milbrath, G. M., 1981, Environmental factors influencing disease development: Corn diseases caused by viruses and spiroplasma, in: *Virus and Viruslike Diseases of Maize in the United States* (D. T. Gordon, J. K. Knoke, and G. E. Scott, eds.), pp. 88–91, Southern Cooperative Series Bulletin 247, Ohio Agricultural Research and Development Center, Wooster.

Francki, R. I. B., Milne, R. G., and Hatta, T., 1985, Maize chlorotic dwarf virus group, in: *Atlas of Plant Viruses*, Volume I, (Francki, R. I. B., Milne, R. G., and Hatta, T.), CRC Press, Boca Raton, Fla.

Galvez, G. E., 1968, Purification and characterization of rice tungro virus by analytical density-gradient centrifugation, *Virology* **35:**418.

Gingery, R. E., 1976, Properties of maize chlorotic dwarf virus and its ribonucleic acid, *Virology* **79:**311.

Gingery, R. E., 1978, An immunofluorescence test for maize chlorotic dwarf virus, *Phytopathology* **68:**1526.

Gingery, R. E., Gordon, D. T., Nault, L. R., and Bradfute, O. E., 1981, Maize chlorotic dwarf virus, in: *Handbook of Plant Virus Infections and Comparative Diagnosis* (E. Kurstak, ed.), pp. 19–32, Elsevier/North Holland Biomedical Press, Amsterdam.

Gordon, D. T., 1977, Routine serological assays for diagnosis of maize virus diseases, in: *Proceedings of the International Maize Virus Disease Colloquium and Workshop*

(L. E. Williams, D. T. Gordon, and L. R. Nault, eds.), pp. 99–102, Ohio Agricultural Research and Development Center, Wooster.

Gordon, D. T., and Nault, L. R., 1977, Involvement of maize chlorotic dwarf virus and other agents in stunting diseases of *Zea mays* in the United States, *Phytopathology* **67**:27.

Gordon, D. T., Findley, W. R., Knoke, J. K., Louie, R., Nault, L. R., Bradfute, O. E., Dollinger, E. J., and Gingery, R. E., 1974a, Distinguishing symptoms and latest research findings on corn virus diseases in the United States, in: *Proceedings of the 29th Annual Corn and Sorghum Research Conference*, pp. 153–173, American Seed Trade Association, Washington, D. C.

Gordon, D. T., Findley, W. R., Knoke, J. K., Louie, R., Nault, L. R., Bradfute, O. E., Dollinger, E. J., and Gingery, R. E., 1974b, Maize dwarf mosaic and maize chlorotic dwarf diseases in the United States, in: *Proceedings of the Corn Disease Conference* (Feb. 11–12, 1974), pp. 52–86, Purdue University, Lafayette, Indiana.

Gordon, D. T., Bradfute, O. E., Gingery, R. E., Knoke, J. K., Louie, R., Nault, L. R., and Scott, G. E., 1981, Introduction: History, geographical distribution, pathogen characteristics, and economic importance, in: *Virus and Viruslike Diseases of Maize in the United States* (D. T. Gordon, J. K. Knoke, and G. E. Scott, eds.), pp. 137–140, Southern Cooperative Series Bulletin 247, Ohio Agricultural Research and Development Center, Wooster.

Grogan, C. E., and Rosenkranz, E. E., 1968, Genetics of host reaction to corn stunt virus, *Crop. Sci.* **8**:251.

Gustin, R. D., and Stoner, W. N., 1968, Biology of *Deltocephalus sonorus* (Homoptera: Cicadellidae), *Ann. Entomol. Soc. Am.* **61**:77.

Guthrie, W. D., Tseng, C. T., Knoke, J. K., and Jarvis, J. L., 1982, European corn borer and maize chlorotic dwarf virus resistance—susceptibility in inbred lines of dent maize, *Maydica* **27**:221.

Harris, K. F., 1979, Leafhopper and aphids as biological vectors: Vector–virus relationships, in: *Leafhopper Vectors and Plant Disease Agents* (K. Maramorosch and K. F. Harris, eds.), pp. 217–308, Academic Press, New York.

Harris, K. F., 1981, Arthropod and nematode vectors of plant viruses, *Ann. Rev. Phytopathol.* **19**:391.

Harris, K. F., and Childress, S. A., 1981, Mechanism of maize chlorotic dwarf virus (MCDV) transmission by its leafhopper vector, *Graminella nigrifrons*, in: *81st Annual Meeting of the American Society of Microbiology* (1–6 March, 1981), p. 251, Dallas, Texas.

Harris, K. F., and Childress, S. A., 1983, Cytology of maize chlorotic dwarf virus infection in corn, *Int. J. Trop. Plant Dis.* **1**:135.

Heinricks, E. A., 1979, Control of leafhopper and planthopper vectors of rice viruses, in: *Leafhopper Vectors and Plant Disease Agents* (K. Maramorosch and K. F. Harris, eds.), pp. 529–560, Academic Press, New York.

Hibino, H., 1983, Transmission of two rice tungro-associated viruses and rice waika virus from doubly and singly infected source plants by leafhopper vectors, *Plant Dis.* **67**:774.

Hibino, H., Roechan, M., and Sudarisman, S., 1978, Association of two types of virus particles with penyakit habang (tungro disease) of rice in Indonesia, *Phytopathology* **68**:1412.

Hibino, H., Saleh, N., and Roechan, M., 1979, Transmission of two kinds of rice tungro-associated viruses by insect vectors, *Phytopathology* **69**:1266.

Iltis, H. H., Doebley, J. F., Guzman, H. R., and Pazy, B., 1979, *Zea diploperennis* (Gramineae): A new teosinte from Mexico, *Science* **203**:186.

Keaster, A. J., and Fairchild, M. L., 1968, Reduction of corn virus disease incidence and control of southwestern corn borer with systemic insecticides, *J. Econ. Entomol.* **61**:367.

Keaster, A. J., Zuber, M. S., Fairchild, M. L., and Loesh Jr., P. J., 1969, Effect of planting dates on the incidence and severity of corn virus diseases, *Agron. J.* **61**:363.

Knoke, J. K., and Louie, R., 1981, Epiphytology of maize virus diseases, in: *Virus and Viruslike Diseases of Maize in the United States* (D. T. Gordon, J. K. Knoke, and G. E. Scott, eds.), pp. 92–102, Southern Cooperative Series Bulletin 247, Ohio Agricultural Research and Development Center, Wooster.

Knoke, J. K., Anderson, R. J., Findley, W. R., Louie, R., Abt, J. J., and Gordon, D. T., 1981, The reaction of sweet corn hybrids to maize dwarf mosaic strains and maize chlorotic dwarf virus, *Ohio Agricultural Research and Development Center Research Bulletin* No. 1135.

Knoke, J. K., Anderson, R. J., Louie, R., Madden, L. V., and Findley, W. R., 1983, Insect vectors of maize dwarf mosaic virus and maize chlorotic dwarf virus, in: *Proceedings of the International Maize Virus Disease Colloquium and Workshop* (D. T. Gordon, J. K. Knoke, L. R. Nault, and R. M. Ritter, eds.), pp. 130–138, The Ohio State University, Ohio Agricultural Research and Development Center, Wooster.

Kramer, J. P., 1967, A taxonomic study of *Graminella nigrifrons*, a vector of corn stunt disease, and its cogeners in the United States (Homoptera:Cicadellidae:Deltocephalinae), *Ann. Entomol. Soc. Am.* **60**:604.

Kuhn, C. W., Jellum, M. D., and All, J. N., 1975, Effect of carbofuran treatment on corn yield, maize chlorotic dwarf and maize dwarf mosaic virus diseases, and leafhopper populations, *Phytopathology* **65**:1017.

Ling, K. C., 1966, Nonpersistence of the tungro virus of rice in its leafhopper vector, *Nephotettix impicticeps, Phytopathology* **56**:1252.

Ling, K. C., 1972, *Rice Virus Diseases,* International Rice Research Institute, Los Banos, Philippines.

Ling, K. C., 1973, *Synonymies of Insect Vectors of Rice Viruses,* International Rice Research Institute, Los Banos, Philippines.

Ling, K. C., and Tiongco, E. R., 1979, Transmission of rice tungro virus at various temperatures: A transitory virus-vector interaction, in: *Leafhopper Vectors and Plant Disease Agents* (K. Maramorosch and K. F. Harris, eds.), pp. 349–366, Academic Press, New York.

Louie, R., Knoke, J. K., and Gordon, D. T., 1974, Epiphytotics of maize dwarf mosaic and maize chlorotic dwarf diseases in Ohio, *Phytopathology* **64**:1455.

Main, C. E., Nusser, S. M., and Bragg, A. W., 1984, Crop losses in North Carolina due to plant diseases and nematodes, *North Carolina State University Department of Plant Pathology Special Publication,* No. 3.

Matthews, R. E. F., 1982, Classification and nomenclature of viruses: Fourth report of the International Committee on Taxonomy of Viruses, *Intervirology* **17**:1.

Naidu, B., and Josephson, L. M., 1976, Genetic analysis of resistance to the corn virus disease complex, *Crop Sci.* **16**:167.

Nault, L. R., 1977, Vectors of maize viruses, in: *Proceedings of the International Maize Virus Disease Colloquium and Workshop* (L. E. Williams, D. T. Gordon, and L. R. Nault, eds.), pp. 111–115, Ohio Agricultural Research and Development Center, Wooster.

Nault, L. R., and Bradfute, O. E., 1979, Corn stunt: Involvement of a complex of leafhopper-borne pathogens, in: *Leafhopper Vectors and Plant Disease Agents* (K. Maramorosch and K. F. Harris, eds.), pp. 561–586, Academic Press, New York.

Nault, L. R., and Knoke, J. K., 1981, Maize vectors, in: *Virus and Viruslike Diseases of Maize in the United States* (D. T. Gordon, J. K. Knoke, and G. E. Scott, eds.), pp. 77–84, Southern Cooperative Series Bulletin 247, Ohio Agricultural Research and Development Center, Wooster.

Nault, L. R., Styer, W. E., Knoke, J. K., and Pitre, H. N., 1973, Semipersistent transmission of leafhopper-borne maize chlorotic dwarf virus, *J. Econ. Entomol.* **66**:1271.

Nault, L. R., Gordon, D. T., Robertson, D. C., and Bradfute, O. E., 1976, Host range of maize chlorotic dwarf virus, *Plant Dis. Rep.* **60**:374.

Nault, L. R., Gordon, D. T., Damsteegt, V. D., and Iltis, H. H., 1982, Response of annual and perennial teosintes *(Zea)* to six maize viruses, *Plant Dis.* **66**:61.

Omura, T., Saito, Y., Usugi, T., and Hibino, H., 1983, Purification and serology of rice tungro spherical and rice tungro bacilliform viruses, *Ann. Phytopath. Soc. Japan* **49**:73.

Pathak, M. D., 1970, Genetics of plants in pest management, in: *Concepts of Pest Management* (R. L. Rabb and F. E. Guthrie, eds.), pp. 138–157, North Carolina State University, Raleigh, 1970.

Pirone, T. P., Bradfute, O. E., Freytag, P. H., Lung, M. C. Y., and Poneleit, C. G., 1972, Virus-like particles associated with a leafhopper transmitted disease of corn in Kentucky, *Plant Dis. Rep.* **56**:652.

Pitre, H. N., 1968, A preliminary study of corn stunt vector populations in relation to corn planting dates in Mississippi. Notes on disease incidence and severity, *J. Econ. Entomol.* **61**:847.

Rains, B. D., and Christensen, C. M., 1983, Effect of soil-applied carbofuran on transmission of maize chlorotic dwarf virus and maize dwarf mosaic virus to susceptible field corn hybrid, *J. Econ. Entomol.* **76**:290.

Reeves, J. T., Jackson, A. O., Paschke, J. D., and Lister, R. M., 1978, Use of enzyme-linked immunosorbent assay (ELISA) for serodiagnosis of two maize viruses, *Plant Dis. Rep.* **62**:667.

Rosenkranz, E., 1969, A new leafhopper-transmissible corn stunt disease agent in Ohio, *Phytopathology* **59**:1344.

Saito, Y., 1977, Interrelationship among waika disease, tungro disease and other similar diseases of rice in Asia, in: *Symposium on Virus Diseases of Tropical Crops* (Tropical Agriculture Research Series No. 10), pp. 129–135, Tropical Agricultural Research Center, Tsukuba, Ibaraki, Japan.

Saito, Y., Hibino, H., Omura, T., and Usugi, T., 1981, Transmission of rice tungro bacilliform virus and rice tungro spherical virus by leafhopper vectors, *Proceedings of the 5th International Congress for Virology* No. 213.

Scott, G. E., and Rosenkranz, E. E., 1977, Location of genes conditioning resistance to the corn stunting disease complex in maize, *Crop Sci.* **17**:923.

Scott, G. E., Rosenkranz, E. E., and Nelson, L. R., 1977, Yield loss of corn due to corn stunt disease complex, *Agron. J.* **69**:92.

Sedlacek, J. D., and Freytag, P. H., 1982, Seasonal occurrence of *Graminella nigrifrons* (Forbes) in corn and pasture ecosystems in Kentucky, in: *Abstracts of the 37th Annual Meeting of the Northcentral Branch of the Entomological Society of America* No. 30 (23–25 March, 1982), Entomological Society of America, Sioux Falls, South Dakota.

Sedlacek, J. K., and Freytag, P. H., 1983, Field biology of *Graminella nigrifrons* (Forbes): Vector of maize chlorotic dwarf virus in Kentucky, in: *Abstracts of the 38th Annual Meeting of the North Central Branch of the Entomological Society of America*, No. 98 (15–17 March, 1983) Entomological Society of America, St. Louis, Missouri.

Shikata, E., 1979, Rice viruses and MLO's, and leafhopper vectors, in: *Leafhopper Vectors and Plant Disease Agents* (K. Maramorosch and K. F. Harris, eds.), pp. 515–527, Academic Press, New York.

Simons, J. N., and Zitter, T. A., 1980, Use of oils to control aphid-borne viruses, *Plant Dis.* **64**:542.

Stevens, C., Gudauskas, R. T., and Karr, G. W., Jr., 1976, Seasonal incidence of two viral diseases of corn, *J. Ala. Acad. Sci.* **47**:130.

Stevens, C., Gudauskas, R. T., Karr, G. W., and Estes, P. M., 1977, Effects of carbofuran on incidence of maize chlorotic dwarf and maize dwarf mosaic in corn, *J. Ala. Acad. Sci.* **48**:57.

Stoner, W. N., and Gustin, R. D., 1967, Biology of *Graminella nigrifrons* (Homoptera: Cicadellidae), a vector of corn (maize) stunt virus, *Ann. Entomol. Soc. Am.* **60**:496.

Sylvester, E. S., 1969, Virus transmission by aphids-viewpoint, in: *Viruses, Vectors, and Vegetation* (K. Maramorosch, ed.), pp. 159–173, Academic Press, New York.

CHAPTER 10

Parsnip Yellow Fleck Virus, Type Member of a Proposed New Plant Virus Group, and a Possible Second Member, Dandelion Yellow Mosaic Virus

A. F. MURANT

I. INTRODUCTION

Parsnip yellow fleck virus (PYFV) (Murant and Goold, 1968; Murant, 1974) has isometric particles approximately 30 nm in diameter, which contain a single species of ssRNA of unusually high molecular weight, approximately 3.3×10^6 (Murant et al., 1981). PYFV is transmitted by aphids in a semipersistent manner, but only in association with a helper virus, anthriscus yellows (AYV) (Murant and Goold, 1968). The two viruses resemble each other in particle composition (Hemida and Murant, 1985, 1986) but, whereas PYFV infects cells throughout the leaf and is transmissible experimentally by mechanical inoculation, AYV is restricted to phloem cells and is not mechanically transmissible. AYV is therefore unlikely to be classified

A. F. MURANT ● Scottish Crop Research Institute, Invergowrie, Dundee DD2 5DA, Scotland, United Kingdom.

with PYFV, and will not be considered in detail in this chapter. PYFV is proposed here as the type member of a new plant virus group.

The only other virus known to have properties similar to those of PYFV is the dandelion yellow mosaic virus (DYMV) of Bos *et al.* (1983). DYMV has isometric particles approximately 30 nm in diameter, and is transmissible mechanically, and by aphids in what may be a semipersistent manner. Although more information is needed on its particle properties and vector relations, it is treated here as a tentative member of the proposed PYFV group.

II. HOSTS, DISEASES, AND GEOGRAPHICAL DISTRIBUTION

A. Parsnip Yellow Fleck Virus

1. Host Range

PYFV has been reported only in a few European countries and has a restricted natural host range, being found only in species of Umbelliferae (Table I). Experimentally, it can be transmitted to a restricted range of hosts in a few other families.

PYFV induced systemic infection in 11 out of 19 species of Umbelliferae but in only five out of 25 species in four out of eight other families (Murant and Goold, 1968). The PYFV isolates so far discovered belong to one or the other of two major serotypes. One serotype (the parsnip strain) occurs in parsnip (*Pastinaca sativa*), celery (*Apium graveolens*), and hogweed (*Heracleum sphondylium*), but not carrot (*Daucus carota*) or cow parsley (*Anthriscus sylvestris*). The other serotype (the *Anthriscus* strain) behaves in exactly the opposite way. Both serotypes infect some other species of Umbelliferae, such as chervil (*Anthriscus cerefolium*) and coriander (*Coriandrum sativum*). There are also differences between the serotypes in ability to infect herbaceous test plants. For example, isolates of the *Anthriscus* serotype do not infect *Nicotiana glutinosa* or *Petunia hybrida*, and infect *N. clevelandii* rarely and usually only in inoculated leaves, whereas isolates of the parsnip serotype infect all these species readily (Murant and Goold, 1968; Hemida and Murant, 1986, and unpublished data). Within each serotype, individual isolates display minor variations in virulence or symptom expression.

2. Symptoms

Symptoms induced by PYFV in naturally infected crop plants are as follows:

Parsnip plants infected with isolates of the parsnip serotype may at first show a prominent systemic vein yellowing, but more characteristically exhibit small yellow flecks or a more definite yellow and green

TABLE I. Hosts and Geographical Distribution of PYFV and DYMV

Virus	Main natural hosts and diseases caused	Geographical distribution	Families with systemic hosts	Local lesion assay hosts	References
PYFV, parsnip strain	Celery ("yellow net") *Heracleum sphondylium* (symptomless) Parsnip ("yellow fleck")	Europe (U.K., German Democratic Republic, Netherlands)	Amaranthaceae Chenopodiaceae Portulacaceae Solanaceae	*Chenopodium amaranticolor, C. quinoa*	Murant and Goold (1968) Singh (1980) Van Dijk and Bos (1985)
PYFV, *Anthriscus* strain	*Anthriscus sylvestris* (symptomless) Carrot ("viral dieback")				
DYMV	Dandelion ("yellow mosaic") Lettuce ("necrosis")	Europe (U.K., German Federal Republic, Czechoslavakia, Netherlands, Scandinavia)	Amaranthaceae Chenopodiaceae Compositae Solanaceae	*C. amaranticolor, C. quinoa, Gomphrena globosa*	Bos *et al.* (1983) Brčák (1979) Brčák and Polák (1966) Čech and Branišová (1973) Hein (1963) Kassanis (1947)

mosaic; the plants are stunted but not killed (Murant and Goold, 1968). Isolates from celery too are of the parsnip serotype, and most of them induce yellow spotting, netting, and leaf distortion symptoms in celery (Singh, 1980; Lennon, 1984), similar to the "celery yellow net" symptoms described by Hollings (1964); some isolates induce severe stunting, necrosis, and chlorotic flecking (Pemberton and Frost, 1986; Lennon, 1984). In carrot, symptoms induced by isolates of the *Anthriscus* serotype range from mild mottle or yellow mosaic, sometimes accompanied by reddening, to rapid necrosis, "dieback," and death (Van Dijk and Bos, 1985).

The following plants are useful diagnostic hosts:

1. *Anthriscus cerefolium* (chervil), *Coriandrum sativum* (coriander). Seedlings inoculated when young develop blackening and shrivelling of the first systemically infected leaves, the necrosis spreading back along the petioles into the crown, causing death of the plant.
2. *Chenopodium amaranticolor*. Tiny necrotic local lesions; some isolates sometimes induce systemic necrosis and leaf distortion.
3. *C. quinoa*. Necrotic local lesions 2–3 mm in diameter. Some isolates induce systemic necrotic flecks and leaf distortion at some times of year.
4. *Nicotiana benthamiana*. Large chlorotic local lesions, often becoming necrotic; systemic vein clearing, followed by progressive necrosis and death within three to four weeks.
5. *N. clevelandii*. The parsnip strain induces local necrotic spots and rings and systemic veinal necrosis; later the plants partially recover, developing a light and dark green mottle. The *Anthriscus* strain induces no symptoms and usually no infection.
6. *Spinacia oleracea* (spinach). Local chlorotic spots; systemic yellow flecks, mottle, or yellowing of the entire leaf.

PYFV is best propagated in *Spinacia oleracea*, *Nicotiana benthamiana*, or (parsnip strain only) *N. clevelandii*. *Chenopodium quinoa* is the best local-lesion assay host, although *C. amaranticolor* may also be used. Chervil and coriander are useful indicator plants for tests of aphid transmission, and unlike most other cultivated species of Umbelliferae, are also hosts of the helper virus, AYV.

For descriptions of the symptoms induced by PYFV in other natural and experimental hosts see Murant and Goold (1968) and Van Dijk and Bos (1985).

B. Dandelion Yellow Mosaic Virus

Like PYFV, DYMV has a restricted natural host range (Table I), being reported only from two species of Compositae, dandelion (*Taraxacum officinale*), and lettuce (*Lactuca sativa*). Yellow mosaic disease of dandelion was first described in England by Kassanis (1944, 1947). Affected plants showed a bright yellow mosaic with rings and oak-leaf patterns. A virus could be transmitted from these plants by aphids, and with difficulty by manual inoculation, to lettuce, which developed a variety of foliar symptoms, including veinal chlorosis and necrosis, interveinal spotting and bronzing, followed by thickening, narrowing, and curling of the leaves, severe stunting of the plant, and complete prevention of heading. What was apparently the same virus was also found occurring naturally in lettuce. The virus was therefore named dandelion yellow mosaic virus. Similar diseases of dandelion and lettuce have since been reported from a few other European countries (Table I).

A virus isolated from affected lettuce in the Netherlands was purified and partially characterized by Bos et al. (1983) and shown to be serologically related to two other Dutch isolates from lettuce and a Czechoslovakian isolate from dandelion. Isolates from both lettuce and dandelion induced symptoms typical of DYMV in lettuce (although they differed somewhat in virulence) and were considered to be the same as the DYMV of Kassanis (1947). However, only one isolate (from lettuce) was returned to dandelion, in which it induced no symptoms. Moreover, this isolate was not one that had been identified serologically as DYMV. Kassanis (1947) too had difficulty in returning virus isolates from lettuce to dandelion, and although he achieved this on three occasions there is no evidence that the yellow mosaic symptoms observed were caused by the same virus that caused the necrosis in lettuce. It is therefore not clear what role, if any, DYMV plays in the etiology of dandelion yellow mosaic disease. In recent observations of dandelion plants from England (A.F. Murant and G. H. Duncan, unpublished data), two plants with typical yellow mosaic symptoms did not contain DYMV (as shown by manual inoculation and immunosorbent electron microscopy tests), whereas the virus was easily detected by these means in a third plant showing only mild mottling symptoms.

Experimentally, DYMV induced systemic infection in five out of nine species of Compositae but in only nine out of 23 species in three out of seven other families (Bos et al., 1983). The only plants suitable as diagnostic hosts, besides lettuce, are Chenopodium amaranticolor and C. quinoa, which develop small chlorotic local lesions and systemic leaf stippling (Bos et al., 1983). These Chenopodium species are also useful for local lesion assay and for propagating the virus for purification.

III. TRANSMISSION, EPIDEMIOLOGY, AND CONTROL

A. Parsnip Yellow Fleck Virus

1. Transmission

Experimentally, PYFV is transmissible by mechanical inoculation but there are no reports of seed transmission.

In nature, PYFV is transmitted by the aphids *Cavariella aegopodii* and *C. pastinacae*, but only from plants that also contain the helper virus AYV (Murant and Goold, 1968). Neither virus is transmitted by *C. theobaldi* or *Myzus persicae*. Experimentally, *C. aegopodii* allowed to feed first on a source of AYV alone and then on a source of PYFV alone can transmit both viruses to test plants, but aphids allowed to feed on the sources in the reverse order transmit only AYV (Elnagar and Murant, 1976a). *C. aegopodii* already carrying AYV (but not virus-free aphids) can also acquire PYFV by feeding through membranes on crude extracts of PYFV-infected plants or on purified preparations of PYFV particles (Elnagar and Murant, 1976b). Aphid transmission of PYFV was not assisted by carrot red leaf Luteovirus, by the Potyviruses celery mosaic, parsnip mosaic or potato virus Y, or by the ungrouped virus parsnip mottle (Elnagar and Murant, 1976b). Aphids injected with purified preparations of PYFV did not transmit, nor did aphids injected with extracts of other aphids carrying AYV plus PYFV.

2. Vector Relations

The vector relations of both PYFV and AYV are of the semipersistent type and were studied in detail by Elnagar and Murant (1976a). Apterous nymphs and apterous adults of *C. aegopodii* were equally efficient as vectors, but alates were less efficient. *C. aegopodii* transmitted the viruses after a minimum acquisition access time (AAT) on doubly infected plants of 10–15 min and a minimum inoculation access time (IAT) of 2 min (with no latent period), the frequency of transmission of each virus increasing with increasing AAT or IAT up to 24 hr. Starving the aphids for 3 hr before the acquisition feed did not affect the frequency of transmission of either virus. Infective adult aphids retained the ability to transmit the viruses for up to four days (more usually only one day),but infective nymphs ceased to transmit after moulting. Aphids already carrying AYV acquired PYFV from singly infected plants in a minimum AAT of only 2 min, i.e., less than the time they needed to acquire PYFV plus AYV from doubly infected plants; this is in line with other evidence that AYV is located in phloem tissue (Murant and Roberts, 1977), whereas PYFV is distributed throughout the leaf (Murant *et al.*, 1975).

In electron microscope studies, Murant *et al.* (1976) found viruslike particles at a site in the foregut of *C. aegopodii* carrying AYV plus PYFV, but not in nonviruliferous aphids. However, it is not clear whether any of these particles were those of PYFV because similar particles were found in aphids carrying AYV alone.

3. Epidemiology

PYFV can be separated from AYV (and thus rendered no longer transmissible by aphids) not only by mechanical inoculation, but also by transmission to plants that are immune to AYV, such as carrot, celery, or parsnip (Murant and Goold, 1968). Therefore, although one or another serotype of PYFV causes recognized diseases in these three species, the virus does not spread from plant to plant within crops—all infected plants observed are primary infections (Murant and Goold, 1968; Waterhouse and Murant, 1981; Van Dijk and Bos, 1985; Lennon, 1984). Nevertheless, PYFV was the most common virus found in a survey of parsnip crops in southern England (Murant and Goold, 1968). Lennon (1984) found that PYFV infection of celery crops in England was not extensive and was usually confined to the outermost rows in the fields. PYFV infection of ware crops of carrot is rare (Waterhouse and Murant, 1981; Van Dijk and Bos, 1985) but, in the Netherlands at least, infection of seed crops is apparently common: symptoms of early-season dieback, now attributed to infection with the *Anthriscus* serotype of PYFV, were reported (Anon., 1955) to recur annually and occasionally to lead to severe losses in seed production.

Van Dijk and Bos (1985) recorded dieback symptoms attributable to infection with PYFV in a range of wild and cultivated species of Umbelliferae, including five that were observed to develop symptoms in natural populations. Such infections are presumably self-eliminating and are of little importance in the epidemiology of the virus. However, *Anthriscus sylvestris* is commonly symptomlessly infected with the *Anthriscus* serotype, and *Heracleum sphondylium* with the parsnip serotype (Murant and Goold, 1968; Tomlinson and Carter, 1970; Elnagar and Murant, 1974; Bem and Murant, 1979; Van Dijk and Bos, 1985); and *Cavariella* spp. found on both species readily transmit PYFV from them. Although *A. sylvestris* is commonly infected with AYV, attempts to detect this virus in, or to transmit it to, *H. sphondylium* have so far failed; the helper virus in this species may therefore differ from AYV, at least in host range (Murant, 1984).

The picture that emerges is that *Heracleum sphondylium* is the main natural source of isolates of PYFV that occur in celery and parsnip, whereas *Anthriscus sylvestris* is the main overwintering host of PYFV isolates that occur in carrot. *Cavariella aegopodii* occurs on both wild hosts; *C. pastinacae* occurs on *H. sphondylium*, but not on *A. sylvestris*. A third *Cavariella* sp. common on *H. sphondylium*, *C. theobaldi*, does not transmit AYV or PYFV. *C. aegopodii*, which feeds on carrot, celery, parsnip, and several other umbelliferous species, is therefore probably the only important vector of the *Anthriscus* serotype of PYFV and the most important vector of the parsnip serotype in cultivated plants.

In autumn, *C. aegopodii* migrates to willow, on which it produces a sexual generation that lays eggs to survive the winter. The insect that emerges from the egg gives rise eventually to winged forms that migrate in very large numbers in late May or early June and may feed on wild plants of *A. sylvestris* or *H. sphondylium* before arriving in crops of carrot or other species of Umbelliferae (Dunn, 1965; Dunn and Kirkley, 1966). In mild winters, apterae may survive on perennating umbelliferous plants.

4. Control

The diseases caused by PYFV in celery, parsnip, and in ware crops of carrot are not sufficiently serious or prevalent to justify specific control measures. Normal insecticidal treatments, even those directed primarily against other insects (e.g., carrot fly, *Psila rosae*), will effect good control of *Cavariella* spp. and hence of the viruses carried by them. Of these viruses, PYFV is less important in ware crops of carrot than is the complex of carrot red leaf and carrot mottle viruses, which cause the disease known as motley dwarf (Watson *et al.*, 1964). PYFV infection seems a more serious problem in carrot seed crops than in ware crops, at least in the Netherlands, and might justify specific control measures; however, none seem to be available except perhaps to grow the crops in areas where the population of vector aphids is low. *Anthriscus sylvestris* and *Heracleum sphondylium* are ubiquitous in hedgerows and wasteland in Britain, the Netherlands, and other western European countries, so that control by the removal of sources of infection would be impractical. Attempts to prevent infection of carrot seed crops by the use of a systemic insecticide met with only limited success (Van Dijk and Bos, 1985), no doubt because of the short minimum inoculation access times needed (only 2 min), and because all infections in carrot are primary ones. Breeding for resistance to PYFV seems unlikely to be worthwhile, especially because no natural sources of resistance are known in carrot, celery, or parsnip.

B. Dandelion Yellow Mosaic Virus

DYMV is difficult to transmit by mechanical inoculation. Using sap from infected dandelion or lettuce leaves as inoculum, Kassanis (1947) was unable to obtain infection of dandelion but was able to infect lettuce, provided that the inoculum contained an abrasive, was prepared from leaves having severe symptoms, and was not diluted more than 1/50. Bos *et al.* (1983) found that preparation of inoculum in 0.03 M phosphate, pH 7, containing 0.25% sodium metabisulfite, 0.5% sodium DIECA, and 7% Norit SX-1 (an activated charcoal) enabled the virus to be transmitted to most plants of a susceptible species.

Kassanis (1947) obtained no evidence for seed transmission of DYMV in dandelion or lettuce, but found that it was transmitted by the aphids *Aulacorthum solani, Myzus ascalonicus,* and *M. ornatus* from lettuce to lettuce and from dandelion to lettuce; transmission from lettuce to dandelion was achieved by *M. ornatus* but only in three out of 100 attempts. *M. persicae* was not a vector of DYMV in Kassanis's tests but transmitted the virus studied by Bos *et al.* (1983).

The efficiency of transmission of DYMV by *M. ornatus* was not improved by starving the aphids before the acquisition feed and was greatest when they were allowed to feed for one day on the source (Kassanis, 1947). These limited data suggest that DYMV may have a semipersistent mode of transmission. Little is known about the epidemiology or importance of DYMV, but Moore (1946) listed several instances of disease in lettuce in southern and eastern England for which it may have been responsible. Kassanis (1947) reported that DYMV spreads more rapidly in lettuce than in dandelion, although most outbreaks occurred in the vicinity of infected dandelion, which he considered important as an overwintering host. In small field plots of lettuce, spread from lettuce source plants was rapid but localized, and Kassanis suggested that control of the disease in field-grown lettuce would readily be achieved by strict control of weeds, especially dandelion. Control by the use of resistant varieties seems unlikely to be successful because all cultivars of lettuce tested by Bos *et al.* (1983) were susceptible to the virus.

IV. PURIFICATION

A. Parsnip Yellow Fleck Virus

Infectivity of PYFV in sap of *Spinacia oleracea*, assayed on *C. quinoa*, had a dilution end-point of 10^{-4} to 10^{-5} (the number of lesions being closely proportional to the dilution), a thermal inactivation point (10 min) of 55–65°C, and a longevity *in vitro* of four to seven days at 18°C, or >15 months at −15°C (Murant and Goold, 1968; Abu Salih, 1968). Infectivity in lyophilized sap survived more than six years (A. F. Murant, unpublished data).

PYFV reaches only low concentration in sap of plants grown under warm summer glasshouse conditions. Detailed studies by Abu Salih (1968) showed that in *Spinacia oleracea* the virus has a low temperature optimum and multiplied best in growth chambers held at 15°C and illuminated for 8 hr per day at 10,000 lux from Philips Warm White fluorescent tubes. Systemically infected leaves developed their highest virus concentration about 19 days after the plants were inoculated.

Murant and Goold (1968) used a mixture of *n*-butanol and chloroform to purify isolates of both serotypes, but subsequent work has shown that this method may be too drastic for some isolates and/or under some conditions. A more reliable procedure (S.K. Hemida and A.F. Murant, unpublished data) is as follows. Harvest *Spinacia oleracea* leaves from plants grown in the conditions specified above. Extract each 100 g material in 300 ml 0.06 M phosphate buffer (pH 7.0) containing 0.01 M disodium ethylene diamine tetra-acetate and 0.1% thioglycerol. Squeeze the extract through cheesecloth and clarify it by adding either (1) *n*-butanol to 8.5% (v/v) or (2) an equal volume of diethyl ether. Only the latter method may be successful with sensitive isolates, especially those of the *Anthriscus* serotype. Yields of nucleoprotein particles (bottom component) are in the range 5–20 mg/kg leaf material. In addition there may be as much or more top component (protein shells).

B. Dandelion Yellow Mosaic Virus

In sap of lettuce assayed on *C. amaranticolor* or *C. quinoa*, the infectivity of a Dutch isolate of DYMV studied by Bos *et al.* (1983) had a dilution end-point of 10^{-4} to 10^{-5}, a thermal inactivation point between 60 and 65°C, and a longevity *in vitro* of barely 24 hr; infectivity survived in samples stored over $CaCl_2$ for up to 6.5 years.

Bos *et al.* (1983) obtained "very low" yields of virus particles by using the following purification procedure. Harvest systemically infected *C. quinoa* leaves 13–18 days after inoculation of the plants. Extract each 1 kg tissue in 2 liters 0.5 M potassium phosphate buffer (pH 7) containing 0.1% thioglycollic acid, strain through cheesecloth, and store the filtrate at −20°C. Thaw, and clarify by adding *n*-butanol to 8% (v/v). Separate and concentrate the virus particles by differential centrifugation, resuspending the high-speed pellets in 0.01 M tris-HCl, pH 9.

V. PARTICLE PROPERTIES

A. Parsnip Yellow Fleck Virus

Particle preparations of both serotypes of PYFV contain isometric particles approximately 30 nm in diameter. There are usually two sedimenting components: "empty" protein shells (top component) and nucleoprotein particles (bottom component) (Hemida and Murant, 1985). The top component particles are fragile and are easily destroyed by some purification procedures, e.g., the *n*-butanol/chloroform procedure used by Murant and Goold (1968). The components may be separated by sedimentation in sucrose density gradients or by isopycnic banding in CsCl or Cs_2SO_4 solutions. The top and bottom components of an isolate of the parsnip serotype had sedimentation coefficients $(s^0_{20,w})$ of 60 and 152 Svedberg units, buoyant densities in CsCl of 1.29 and 1.49 g/cm³, and A_{260}/A_{280} ratios of 0.8 and 1.7, respectively (S.K. Hemida and A.F. Murant, unpublished data).

Bottom component particles of PYFV contain a single species of single-stranded RNA. Murant *et al.* (1981) determined the molecular weight of the RNA of an isolate of the parsnip serotype to be 3.3×10^6 (9650 nucleotides) by electrophoresis of glyoxylated RNA in agarose gels. More recent experiments, in which the same technique was used but the molecular weights of the marker RNA species were slightly revised in the light of new sequence data, gave values of 3.5×10^6 for isolates of the parsnip serotype and 3.3×10^6 for isolates of the *Anthriscus* serotype (S.K. Hemida and A.F. Murant, unpublished data). This RNA molecule, which is exceptionally large for a virus with small isometric particles only 30 nm in diameter, presumably accounts for the high sedimentation coefficient of bottom component particles; it comprises about 43% of the weight of the particles, estimated by Reichmann's (1965) method.

Particles of top and bottom component contain three major polypeptides, of molecular weight (x10^{-3}) 31, 26, and 22.5 (parsnip serotype) or 31, 26, and 24 (*Anthriscus* serotype) (Hemida and Murant, 1985). Data from chemical cleavage and immunoblotting experiments (S.K. Hemida, M.A. Mayo, and A.F. Murant, unpublished data) indicate that within each serotype the three proteins are distinct.

B. Dandelion Yellow Mosaic Virus

Particle preparations of a Dutch isolate of DYMV contained isometric particles approximately 30 nm in diameter, all sedimenting as nucleoprotein particles; they had a sedimentation coefficient ($s^0_{20,w}$) of 159 svedbergs, an A_{260}/A_{280} ratio of 1.67, and a buoyant density in Cs_2SO_4 of 1.42 g/cm^3 (Bos *et al.*, 1983). No component representing "empty" particles was detected in sucrose density gradients; particles penetrated by negative stain were seen by electron microscopy but these were thought to be "damaged" nucleoprotein particles.

In preliminary experiments (S.K. Hemida and A.F. Murant, unpublished data), particles of DYMV have been found to contain a single RNA species which comigrates electrophoretically with that of the *Anthriscus* serotype of PYFV and therefore its approximate molecular weight is 3.3×10^6.

DYMV particles have three polypeptide species. Preliminary estimates of their molecular weight (x10^{-3}) are 32, 29.5, and 27 (H.J. Vetten, unpublished data) and 29, 25, and 24 (S.K. Hemida and A.F. Murant, unpublished data).

VI. SEROLOGY

A. Parsnip Yellow Fleck Virus

PYFV is moderately to highly immunogenic, antiserum titers being in the range 1/256 to 1/4096 in gel diffusion tests. The two serotypes have a serological differentiation index (SDI) of 4 to 5.

The virus content in sap of field-infected plants and of glasshouse-grown test plants is usually too low to enable reactions to be obtained in gel diffusion tests, but *Spinacia oleracea*, grown in the conditions mentioned in Section IV and picked at the optimum time, yields sap that gives obvious lines of precipitate. The passive hemagglutination test (Abu Salih *et al.*, 1968a) was shown to be only about ten times less sensitive than infectivity assay for detecting the virus and was used as an assay method in a detailed study of its multiplication (Abu Salih, 1968). Latex flocculation was eight times less sensitive than the passive hemagglutination test but 80 times more sensitive than tube precipitin tests for detecting PYFV (Abu Salih *et al.*, 1968b). ELISA was used by Lennon

(1984) to detect PYFV in *Nicotiana clevelandii* test plants but has not been used so far to detect the virus in field-infected plants. ISEM has been used to detect PYFV in plant extracts and to look for serological relationships (Murant *et al.*, 1986; Murant and Duncan, 1985).

B. Dandelion Yellow Mosaic Virus

An antiserum to DYMV, obtained from a rabbit after a prolonged series of intravenous, subcutaneous, and intramuscular injections (Bos *et al.*, 1983), had a titer of 1/256 in agar double-diffusion tests. Gel diffusion tests were useful with purified virus preparations but did not detect the virus in sap from infected lettuce. However, this could be done with ISEM and with ELISA, especially following the modified procedure of Flegg and Clark (1979). ISEM has been used too to detect DYMV in dandelion (A.F. Murant and G.H. Duncan, unpublished data).

VII. GENOME PROPERTIES

A. Parsnip Yellow Fleck Virus

Infective PYFV RNA molecules bind to oligo(dT)-cellulose under high salt conditions, and therefore probably have a polyadenylate sequence. The infectivity of RNA preparations is greatly decreased by treatment with proteinase K, which suggests that the molecules possess a genome-linked protein needed for infectivity (Hemida and Murant, 1985, 1986).

In cDNA hybridization tests, about 12% homology was found between the genomes of the parsnip and *Anthriscus* serotypes (Murant *et al.*, 1986).

B. Dandelion Yellow Mosaic Virus

Preliminary studies (S.K. Hemida and A.F. Murant, unpublished data) indicate that infective DYMV RNA molecules probably have a polyadenylate sequence, because they bind to oligo(dT)-cellulose under high salt conditions.

VIII. RELATIONS WITH CELLS AND TISSUES

A. Parsnip Yellow Fleck Virus

PYFV seems to occur in most types of cell throughout the leaf. In leaf tissue cells of chervil, *Nicotiana clevelandii*, and spinach, PYFV induces the formation of characteristic inclusion bodies, usually adjacent to the cell nucleus (Murant *et al.*, 1975). In young inclusion bodies, mitochondria and Golgi bodies are arranged around the periphery and the inner part consists of ribosome-studded vesicles (probably derived from the endoplasmic reticulum), other smaller vesicular structures, and groups of randomly oriented straight tubules about 30 nm in diameter and up to 1 μm long. Older inclusions also have mitochondria around the periphery but are much larger, lack the ribosome-studded vesicles, and have few of the smaller vesicles; they are made up almost entirely of the 30-nm tubules together with whorls of membranes.

Another characteristic feature of PYFV-infected cells is the production of cell wall outgrowths up to 15 μm long ensheathing cytoplasmic tubules about 45 nm in diameter; these in turn contain usually a single row of viruslike particles 25–30 nm in diameter. Tubules containing viruslike particles may also be found free in the cytoplasm, and in the sieve tubes (Murant *et al.*, 1975).

B. Dandelion Yellow Mosaic Virus

No ultrastructural studies are reported with DYMV, but membranous tubules containing single rows of virus particles were seen in partially purified preparations (Bos *et al.*, 1983) and probably occur too in infected cells.

IX. POSSIBLE AFFINITIES OF PARSNIP YELLOW FLECK AND DANDELION YELLOW MOSAIC VIRUSES

The data presented above show that PYFV and DYMV have many properties in common. Like PYFV, DYMV has a restricted natural and experimental host range, occurs in low concentrations in plants, and has isometric particles of approximately 30 nm diameter some of which are penetrated by negative stain. It is transmitted by aphids in what may be a semipersistent manner. DYMV particles contain three polypeptide species and a single RNA species, all of which are of similar molecular weight to those of PYFV, and DYMV RNA, like that of PYFV, contains a polyadenylate sequence. Despite these resemblances, antiserum to DYMV did not react with PYFV particles in gel diffusion or immunoelectron microscopy tests (Murant *et al.*, 1986).

There are similarities in particle composition between these two viruses and AYV. In recent studies, Hemida and Murant (1986) found that AYV particles contain a single RNA species of molecular weight approximately 3.7×10^6, possessing a polyadenylate sequence and at least three protein species of molecular weight ($\times 10^{-3}$) approximately 33.5, 25.5, and 23.5. However, AYV is serologically unrelated to PYFV (Murant and Duncan, 1985) and differs from it in other important respects. It is confined to the phloem, is not manually transmissible, and induces the formation of "currant bun" inclusions consisting of a presumably proteinaceous matrix in which virus particles are embedded (Murant and Roberts, 1977). AYV is therefore unlikely to be classified in the same virus group as PYFV.

The AYV/PYFV complex somewhat resembles the complex of rice tungro spherical virus (RTSV) and rice tungro bacilliform virus (RTBV), which are transmitted in a semipersistent manner by leafhoppers (*Nephotettix* spp.) (Hibino *et al.*, 1978, 1979). RTSV acts as a helper for RTBV and, like AYV, induces the formation of "currant bun" inclusions in phloem tissue. RTSV is tentatively placed in the maize chlorotic dwarf virus (MCDV) group (Matthews, 1982; Chapter 9, this volume). Despite these resemblances, no serological reaction was detected in ISEM between AYV or PYFV and antisera to RTSV, RTBV, or MCDV (Murant and Duncan, 1985). Antiserum to radish yellow edge virus (Natsuaki, 1985), a "cryptic" virus which has a double-stranded RNA genome but, like AYV, is phloem-limited and induces "currant bun" inclusions, was also included in these tests but with negative results.

In attempts to detect possible relationships to other plant viruses, Murant and Goold (1968) obtained no reactions in gel diffusion tests between PYFV and antisera to arabis mosaic, carnation ringspot, cucumber mosaic, pea enation mosaic, raspberry ringspot, strawberry latent ringspot, tobacco necrosis, tomato black ring, or tomato bushy stunt viruses.

Thus, with the exception of DYMV, no other plant virus is known that resembles PYFV in the combination of its biological properties and the physicochemical properties of its particles. There therefore seem good grounds for creating a new plant virus group, of which PYFV would be the type and, at present, the only definitive member. DYMV seems best regarded as a tentative member until more is known of its vector relations and particle properties. Pending further information on the properties of both viruses, it seems best to use the term "parsnip yellow fleck virus group" rather than to coin a special group name.

In considering the possible wider affinities of PYFV and DYMV, and of AYV, it is noteworthy that the particle properties of all three viruses are much like those of the Picornaviruses of vertebrates and insects (Matthews, 1982). These have genomes composed of a single piece of single-stranded RNA of molecular weight of approximately 2.5×10^6, with a genome-linked protein at the 5' terminus and a polyadenylate sequence at the 3' terminus, and have three or four polypeptide species in the protein coat. PYFV, DYMV, and AYV may therefore be the first examples of plant-infecting Picornaviruses. However, no serological reactions were obtained between PYFV (either serotype) and antisera to the insect-infecting Picornaviruses, cricket paralysis virus, and drosophila virus C, or to the picornalike virus described by D'Arcy *et al.* (1981) from the aphid *Rhopalosiphum padi* (Murant *et al.*, 1986, and unpublished data).

REFERENCES

Abu Salih, H. S., 1968, Sensitive serological tests and their use in studies on the accumulation of parsnip yellow fleck virus in *Spinacia oleracea*, Ph.D. Thesis, University of St. Andrews.

Abu Salih, H. S., Murant, A. F., and Daft, M. J., 1968a, Comparison of the passive haemagglutination and bentonite flocculation tests for serological work with plant viruses, *J. Gen. Virol.* **2**:155.

Abu Salih, H. S., Murant, A. F., and Daft, M. J., 1968b, The use of antibody-sensitized latex particles to detect plant viruses, *J. Gen. Virol.* **3**:299.

Anon., 1955, Ziekten en plagen in 1954: Het "zwart," *Jaarversl. Plantenziekt. Dienst,* **1954**:123.

Bem, F., and Murant, A. F., 1979, Transmission and differentiation of six viruses infecting hogweed (*Heracleum sphondylium*) in Scotland, *Ann. Appl. Biol.* **92**:237.

Bos, L., Huijberts, N., Huttinga, H., and Maat, D. Z., 1983, Further characterization of dandelion yellow mosaic virus from lettuce and dandelion, *Neth. J. Plant Pathol.* **89**:207.

Brčák, J., 1979, Czech and Scandinavian isolates resembling dandelion yellow mosaic virus, *Biol. Plantarum* **21**:298.

Brčák, J., and Polák, Z., 1966, Importance of wild hosts of plant viruses, *Meded. Rijs-fac. Landbw. Gent* **31**:967.

Čech, M., and Branišová, H., 1973, Some problems with the isolation of the *Taraxacum* mosaic virus, in: *Plant Virology, Proc. 7th Conf. Czechosl. Plant Virol., High Tatras, 1971,* pp. 263–266, Slovak Academy of Sciences, Bratislava.

D'Arcy, C. J., Burnett, P. A., Hewings, A. D., and Goodman, R. M., 1981, Purification and characterization of a virus from the aphid *Rhopalosiphum padi*, *Virology* **112**:346.

Dunn, J. A., 1965, Studies on the aphid, *Cavariella aegopodii* Scop. I. On willow and carrot, *Ann. Appl Biol.* **56**:429.

Dunn, J. A., and Kirkley, J., 1966, Studies on the aphid, *Cavariella aegopodii* Scop. II. On secondary hosts other than carrot, *Ann. Appl. Biol.* **58**:213.

Elnagar, S., and Murant, A. F., 1974, Viruses from umbelliferous plants, *Rep. Scott. Hortic. Res. Inst.,* **1973**:66.

Elnagar, S., and Murant, A. F., 1976a, Relations of the semi-persistent viruses, parsnip yellow fleck and anthriscus yellows, with their vector, *Cavariella aegopodii, Ann. Appl Biol.* **84**:153.

Elnagar, S., and Murant, A. F., 1976b, The role of the helper virus, anthriscus yellows, in the transmission of parsnip yellow fleck virus by the aphid *Cavariella aegopodii, Ann. Appl. Biol.* **84**:169.

Flegg, C. L., and Clark, M. F., 1979, The detection of apple chlorotic leafspot virus by a modified procedure of enzyme-linked immunosorbent assay (ELISA), *Ann. Appl. Biol.* **91**:61.

Hein, A., 1963, Über ein Vorkommen des Salatnekrosevirus in Westdeutschland, *Nachrichtenbl. Deutsch. Pflanzenschutzdienstes, Braunschweig* **51**:17.

Hemida, S. K., and Murant, A. F., 1985, Characterisation of parsnip yellow fleck virus (PYFV), *Rep. Scott. Crop Res. Inst.,* **1984**:185.

Hemida, S. K., and Murant, A. F., 1986, Aphid-transmitted viruses resembling picornaviruses, *Rep. Scott. Crop Res. Inst.,* **1985**:148.

Hibino, H., Roechan, M., and Sudarisman, S., 1978, Association of two types of virus particles with penyakit habang (tungro disease) of rice in Indonesia, *Phytophathology* **68**:1412.

Hibino, H., Saleh, N., and Roechan, M., 1979, Transmission of two kinds of rice tungro-associated viruses by insect vectors, *Phytopathology* **69**:1266.

Hollings, M., 1964, Some properties of five viruses of celery (*Apium graveolens* L.) in Britain, *J. Hortic. Sci.* **39**:130.

Kassanis, B., 1944, A virus attacking lettuce and dandelion, *Nature (London)* **154**:16.

Kassanis, B., 1947, Studies on dandelion yellow mosaic and other virus diseases of lettuce, *Ann. Appl. Biol.* **34**:412.

Lennon, A. M., 1984, The role of umbelliferous weeds as hosts of celery viruses, Ph.D. Thesis, University of Manchester.

Matthews, R. E. F., 1982, Classification and nomenclature of viruses. Fourth Report of the International Committee on Taxonomy of Viruses, *Intervirology* **17**:1.

Moore, W. C., 1946, A virus disease of cos lettuce (dandelion yellow mosaic virus), *Trans. Brit. Mycol. Soc.* **29**:252.

Murant, A. F., 1974, Parsnip yellow fleck virus, *Commonw. Mycol. Inst./Assoc. Appl. Biol. Descr. Plant Viruses* No. 129.

Murant, A. F., 1984, A new helper for parsnip yellow fleck virus (PYFV) in hogweed (*Heracleum sphondylium*), *Rep. Scott. Crop Res. Inst.*, **1983**:189.

Murant, A. F., and Duncan, G. H., 1985, Characterization of parsnip yellow fleck virus (PYFV), *Rep. Scott. Crop Res. Inst.*, **1984**:186.

Murant, A. F., and Goold, R. A., 1968, Purification, properties and transmission of parsnip yellow fleck, a semi-persistent, aphid-borne virus, *Ann. Appl. Biol.* **62**:123.

Murant, A. F., and Roberts, I. M., 1977, Virus-like particles in phloem tissue of chervil (*Anthriscus cerefolium*) infected with anthriscus yellows virus, *Ann. Appl. Biol.* **85**:403.

Murant, A. F., Roberts, I. M., and Hutcheson, A. M., 1975, Effects of parsnip yellow fleck virus on plant cells, *J. Gen. Virol.* **26**:277.

Murant, A. F., Roberts, I. M., and Elnagar, S., 1976, Association of virus-like particles with the foregut of the aphid *Cavariella aegopodii* transmitting the semi-persistent viruses anthriscus yellows and parsnip yellow fleck, *J. Gen. Virol.*, **31**:47.

Murant, A. F., Taylor, M., Duncan, G. H., and Raschké, J. H., 1981, Improved estimates of molecular weight of plant virus RNA by agarose gel electrophoresis and electron microscopy after denaturation with glyoxal, *J. Gen Virol.* **53**:321.

Murant, A. F., Hemida, S. K., Robinson, D. J., and Roberts, I. M., 1986, Aphid-transmitted viruses resembling picornaviruses, *Rep. Scott. Crop Res. Inst.* **1985**:150.

Natsuaki, T., 1985, Radish yellow edge virus, *Assoc. Appl. Biol. Descr. Plant Viruses* No. 298.

Pemberton, A. W., and Frost, R. R., 1986, Virus diseases of celery in England, *Ann. Appl. Biol.* **108**:319.

Reichmann, M. E., 1965, Determination of ribonucleic acid content of spherical viruses from sedimentation coefficients of full and empty particles, *Virology* **25**:166.

Singh, H., 1980, Studies on virus diseases of celery, Ph.D. Thesis, University of Manchester.

Tomlinson, J. A., and Carter, A. L., 1970, Virus diseases of umbelliferous plants. Hogweed, *Rep. Nat. Veg. Res. Sta.* **1969**:110.

Van Dijk, P., and Bos, L., 1985, Viral dieback of carrot and other Umbelliferae caused by the *Anthriscus* strain of parsnip yellow fleck virus, and its distinction from carrot motley dwarf, *Neth. J. Plant Pathol.* **91**:169.

Waterhouse, P. M., and Murant, A. F., 1981, New virus isolates from umbelliferous plants, *Rep. Scott. Crop Res. Inst.* **1980**:102.

Watson, M., Serjeant, E. P., and Lennon, E. A., 1964, Carrot motley dwarf and parsnip mottle viruses, *Ann. Appl. Biol.* **54**:153.

Index